T0221545

# Michael Polanyi and His Generation

# Michael Polanyi and His Generation

## Origins of the Social Construction of Science

MARY JO NYE

The University of Chicago Press
Chicago and London

The University of Chicago Press, Chicago 60637
The University of Chicago Press, Ltd., London
© 2011 by The University of Chicago
All rights reserved. Published 2011.
Paperback edition 2013
Printed in the United States of America

22  21  20  19  18  17  16  15  14  13      3  4  5  6  7

ISBN-13: 978-0-226-61063-4 (cloth)
ISBN-13: 978-0-226-10317-4 (paperback)
ISBN-13: 978-0-226-61065-8 (e-book)
DOI: 10.7208/chicago/978022610658.001.0001

Library of Congress Cataloging-in-Publication Data

Nye, Mary Jo.
    Michael Polanyi and his generation : origins of the social construction of
science / Mary Jo Nye.
        p. cm.
    Includes bibliographical references and index.
    ISBN-13: 978-0-226-61063-4 (cloth : alk. paper)
    ISBN-10: 0-226-61063-2 (cloth : alk. paper) 1. Polanyi, Michael,
1891–1976—Influence. 2. Polanyi, Michael, 1891–1976—Friends and
Associates. 3. Jewish scientists—Hungary—Intellectual life. 4. Jewish
Scientists—Germany—Intellectual life. 5. Science—Philosophy—History—
20th century. 6. Science-Social aspects. I. Title.
    BS945.P584N94 2011
    192—dc22
                                                                      2011003356

*For my dear friends Yocheved and Sabetai Unguru*

# CONTENTS

ACKNOWLEDGMENTS

It is a pleasure to be able to thank those who generously have helped me in my research and revisions on this project as it has taken shape over the years. Some of the initial research was supported by the National Science Foundation (grant no. SBR-9321305), and later work was supported by the Thomas Hart and Mary Jones Horning Endowment in the Humanities at Oregon State University. It was a real benefit for my work and a wonderful experience to have residence as a Visiting Scholar in Ursula Klein's research group at the Max Planck Institute for the History of Science in Berlin for brief periods during 1999, 2002, and 2004. I am very grateful to Ursula and to her colleagues, as I am to my colleagues in the History Department at Oregon State University, especially Paul Farber, for their interest and support.

I thank Robert A. Nye for reading and advising me on every aspect of the conception and execution of this project and for the insightful criticism and invaluable enthusiasm that he brought to all drafts and to the final manuscript. I have many debts to him and to my daughter Lesley N. Nye. I am grateful to Alan J. Rocke and Karen M. Darling for encouraging me to write this manuscript and for offering suggestions about it from the beginning. At the University of Chicago Press, I thank my editor Karen M. Darling, editorial associate Abby Collier, and manuscript editor Mary Gehl for their guidance and expert work in helping me complete this book.

I am indebted to Alan J. Rocke for his detailed comments on the entire manuscript. I also thank Martin X. Moleski, SJ, who critiqued the manuscript in its entirety and provided me with materials that I had not previously seen. Father Moleski completed the splendid biography of Michael Polanyi that was begun by the late William T. Scott. It has been essential to the final version of this project. Comments by an anonymous reviewer for the University of Chicago Press have especially helped improve the introduction and some of the arguments about Polanyi and Popper.

I am very grateful to Gábor Palló for his careful reading and expert and insightful comments on the first chapter. I value highly our discussions over the years, particularly during my visit to Budapest in September 2003 when Gábor took the time to show me his city. I thank Endre J. Nagy for discussion about Polanyi and Budapest. I am indebted to Dieter Hoffmann for his detailed review and suggestions on chapter 2 and to Wolfgang Lefevre and Margaret Schabas for their advice on chapter 5. I thank Gerald Holton for the time that he took in discussing chapter 7 with me and John L. Heilbron and Ronald N. Giere for additional comments on that chapter. I am grateful to John C. Polanyi for reading some of my work and for giving permission to publish direct quotations from his father's papers.

I have benefitted from comments and suggestions on various parts of this project at different stages, including earlier related published articles. I would like to acknowledge Diana Kormos Buchwald, Moritz Epple, the late Marjorie Grene, Arne Hessenbruch, the late Frederic L. Holmes, William B. Husband, Derry W. Jones, Trevor Levere, David Luft, Lewis Pyenson, Ljubisa Radovic, Jeffry Ramsey, Lisa T. Sarasohn, Sason Shaik, Ruth L. Sime, Mary Singleton, Jeffrey Sklansky, Gunther Stent, Donald G. Truhlar, and Andrea Woody. Thanks, too, to Keay Davidson for helpful insights; to Mary Ellen Bowden and Pnina Abir-Am for information; to Karl Hufbauer, Jordi Cat, Joshua Nall, and Alan Richardson for copies of unpublished papers; and to Alan Richardson, Ronald N. Giere, Friedrich Stadler, Karoly Kokai, Elisabeth Nemeth, Daniel Kuby, Donata Romizi, Catherine Herfeld, Beatrice Collina, and other members of the July 2009 Vienna International Summer University for many helpful discussions. Erica Jensen aided with finding book reviews. Kristin Johnson and Linda Richards provided some useful materials. Yasu Furukawa and Annette Vogt provided helpful information about Polanyi files at the Max Planck Society archives. Susanne Kovacs and Gabor Kovacs provided me with a rough translation of a paper in Hungarian by Gábor Palló. Thanks to Derry W. Jones and Brian Gowenlock for accounts of their experiences of Polanyi's teaching at Manchester.

I am grateful for authorization to consult archives and to publish quotations from papers at the Michael Polanyi Papers at the Special Collections Research Center of the University of Chicago Library (MPP), the Max Planck Society Archives in Berlin-Dahlem (MPG), the Rockefeller Archive Center at Pocantico Hills in Sleepy Hollow, New York (RFA), the John Rylands University Library Archives at the University of Manchester (UMA), the Nobel Archive of the Royal Swedish Academy of Sciences in Stockholm

(PKVA), the Sources for the History of Quantum Physics at the American Institute of Physics (SHQP), the Papers of Patrick Maynard Stewart Blackett, OM FRS, Baron Blackett of Chelsea, at the Royal Society Library (BP), and the Linus and Ava Helen Pauling Papers in Special Collections at Oregon State University's Valley Library (PP).

I thank Eckart Henning, Marion Kazemi, and Susanne Uebele, who helped me at the MPG; Erwin Levold and Thomas Rosenbaum at the RFA; James Peters at the UMA; Alice Schreyer at the MPP; Joe Anderson at the AIP; and Clifford Mead at the PP. I am grateful to Maria Asp, Tore Frängsmyr, Karl Grandin, and Anne Wiktorsson for hospitality and aid at the PKVA. I thank Giovanna Blackett Bloor for giving permission for access to her father's papers at the Royal Society Library and for permission to quote from them. Sandra Cumming and Mary Sampson were most helpful at the Royal Society during my visits in London.

I have drawn upon some of my earlier published essays and articles in writing this book, most closely in parts of chapter 4, where there is some duplication with "Working Tools for Theoretical Chemistry: Polanyi, Eyring and Debates over the 'Semi-Empirical Method,'" *Journal of Computational Chemistry* 28 (2007): 98–108. There is overlap with other of my publications: in chapter two with "Historical Sources of Science-as-Social Practice: Michael Polanyi's Berlin," *Historical Studies in the Physical and Biological Sciences* 37, no. 2 (2007): 411–36; in chapter 3 with "Laboratory Practice and the Physical Chemistry of Michael Polanyi," in *Instruments and Experimentation in the History of Chemistry*, edited by F. L. Holmes and Trevor Levere (Cambridge, MA: MIT Press, 2000), 367–400; as well as "At the Boundaries: Michael Polanyi's Work on Surfaces and the Solid State," in *Chemical Sciences in the Twentieth Century*, edited by Carsten Reinhardt (Berlin: Wiley-VCH, 2001), 246–57; "Michael Polanyi's Theory of Surface Adsorption: How Premature?" in *Prematurity and Scientific Discovery*, edited by Ernest. B. Hook (Berkeley: University of California Press, 2002), 151–63; in chapter 6 with "Re-reading Bernal: History of Science at the Crossroads in 20th-Century Britain," in *Aurora Torealis: Studies in the History of Science and Ideas in Honor of Tore Frängsmyr*, edited by Marco Beretta, Karl Grandin, and Svante Lindqvist (Sagamore Beach, MA: Science History Publications), 237–60; and in chapter 7 with "Science and Politics in the Philosophy of Science: Popper, Kuhn, and Polanyi," in *Science as Cultural Practice*, volume 1: *Cultures and Politics of Research from the Early Modern Period to the Age of Extremes*, edited by Moritz Epple and Claus Zittel (Berlin: Akademie Verlag, 2010), 201–16.

As a member of the1960s graduate-school generation in the history of science, I was in a cohort of scholars who began to pay attention to the social situation of science as well as to its intellectual history. Many of us were led into concern with the social aspects of science by the political events of the period. We also were influenced by Thomas Kuhn's 1962 *Structure of Scientific Revolutions*. It was Kuhn's brief reference to Michael Polanyi that first introduced me, like many others, to Polanyi's name. At the time I did not realize that Polanyi the philosopher of science was a physical scientist, like Kuhn, but a much more distinguished one.

Kuhn's *Structure* became a focus for heated debates inside and outside the history and philosophy of science. Philosophical disputes ranged over the strengths and weaknesses of logical positivism, probabilistic verificationism, and Karl Popper's falsificationism as alternative accounts to Kuhn's of how science works. Popper's misnamed *Logic of Scientific Discovery* (1959) offered a stark contrast to Kuhn's *Structure*, since Popper argued that philosophy of science has nothing to say about discovery, because discovery is inaccessible to rational analysis. For Popper, the main business of science is testing theories for anomalous results rather than seeking data for further verification. Popper's critical philosophy was thoroughly logical in its prescriptions for science, while Kuhn's philosophical history was disconcertingly psychological and social in its descriptions of science.

Kuhn's *Structure* presaged revolutionary developments still to come. One of the earliest was the 1967 manifesto by Kuhn's student Paul Forman at Berkeley. As a graduate student in Madison, I heard rumors of Forman's PhD dissertation almost immediately, but it was not readily available. Its main argument appeared in 1971 in Russell McCormmach's new journal *Historical Studies in the Physical Sciences*, where Forman laid out his case that Werner Heisenberg's quantum theory of indeterminacy had its origin in physicists' capitulation to popular philosophical resistance in Weimar Germany to scientific determinism.[1]

"Forman's Thesis," as it came to be called, did not reach a broad public audience, in part because it never appeared as a book. The same was true of another radical document published in 1975 by Sabetai Unguru, who had studied the history of ancient and medieval mathematics under David Lindberg at the University of Wisconsin. Unguru enraged many senior historians with his article "On the Need to Rewrite the History of Greek Mathematics." They objected to his denial that the history of mathematics is a history of equivalent, universal, and transcendent mathematical truths and to his claim that ancient Greek geometry was situationally determined in its own unique framework, which constituted a world apart from the algebra of later centuries.[2]

These manifestos exemplified a new mood in the history and philosophy of science and a move toward the social history and sociology of science. One of the first academic programs with this orientation in the United States was the University of Pennsylvania's Department of the History and Sociology of Science, founded in 1967. It was preceded in 1966 by the Science Studies Unit at the University of Edinburgh, where David Edge collaborated with Roy MacLeod at the University of Sussex in 1971 to start *Science Studies*, which later became *Social Studies of Science*. By the end of the 1970s, the term "social epistemology of science" had been coined for the general problem of understanding how "the history of science would be correctly conceived as essentially social history" using the insight that "knowledge is constitutively social," as Barry Barnes and Steven Shapin wrote in 1977.[3] The study of the nature and origin of scientific knowledge would be based in social relations, in social studies of science, and in what David Bloor christened the "strong program in the sociology of science."[4] The title of Shapin's 2010 collection of his earlier essays captures the mood at that time: *Never Pure: Historical Studies of Science as if It Was Produced by People with Bodies, Situated in Time, Space, Culture, and Society, and Struggling for Credibility and Authority.*[5]

Some of Michael Polanyi's work was incorporated by the Edinburgh group into the new field of the social construction of science, particularly themes from Polanyi's book *Personal Knowledge* (1958) and the collection of essays *The Tacit Dimension* (1966). Kuhn had brought attention to Polanyi by a single reference that led many scholars in social studies of science to *Personal Knowledge*. Kuhn wrote that Polanyi had "brilliantly developed" an idea similar to Kuhn's notion of a scientific paradigm. This, said Kuhn, was to be found in Polanyi's argument "that much of the scientist's success depends upon 'tacit knowledge,' i.e., upon knowledge that is acquired through practice and that cannot be articulated explicitly."[6] The themes of scientific practice and tacit knowledge, in contrast to empirical logic and explicit rules, became major sites for inquiry in the new social epistemology of science in the late 1960s and the 1970s.[7]

The argument of my book is that the origins of this new social conception of science lie in a historical period considerably earlier than that of the 1960s. The roots are to be found in the scientific culture and political events of Europe in the 1930s, when scientific intellectuals struggled to defend the universal status of scientific knowledge and to justify public support for science in an era of economic catastrophe, the rise of Stalinism and Fascism, and increasing demands from governments for applications of science to industry and social welfare. Scientists disagreed among themselves upon the best means to meet these challenges and to defend the integrity of their scientific work. Polanyi was among the leading protagonists in these cultural debates, which arose in the 1930s in Great Britain and continued in their original framework into the 1950s.

Unlike Popper or Kuhn, Polanyi was first and foremost a scientist. The Royal Society of Chemistry in Great Britain still awards the Polanyi Medal in Gas Kinetics every two years. Born in Budapest in 1891, Polanyi completed a medical degree in 1913 and a PhD in physical chemistry in 1917 at the University of Budapest. He studied physical chemistry in Karlsruhe in Germany during the summer term of 1912 and the academic year 1913–1914. He was a medical doctor in the Austro-Hungarian Army during the war, although he often was in Budapest on sick leave. After working briefly in Budapest with Georg von Hevesy, who would later be a Nobel laureate in Chemistry, Polanyi returned to Karlsruhe, and then in 1920 he moved to Berlin to a position at the Kaiser Wilhelm Institute for Fiber Chemistry.

In 1923 he became director of the chemical-kinetics research group in Fritz Haber's Institute for Physical Chemistry and Electrochemistry, where the Fiber Institute was housed. When offered alternative research or teaching positions in industry or universities, including an offer at Manchester in late 1932, Polanyi turned them down because of what he regarded as the superior research and intellectual environment of Berlin. He was fortunate in receiving a second offer to head the physical chemistry laboratory at Manchester after he and other scientists of Jewish origin realized that they had to leave Germany following the rise of the National Socialist Party to power in early 1933.

After the Second World War, when Polanyi's interests decisively shifted from chemical research to economics, politics, and philosophy, he received the opportunity in 1948 to exchange his chemistry professorship for a chair of "social studies" at Manchester. He accepted a chair in social philosophy at the University of Chicago in 1951, but he was unable to take up the position because of the refusal of the McCarthy-era U.S. State Department to issue him an immigrant visa. He retired from Manchester in 1959 and moved to Merton College at Oxford as a senior research fellow. He died in 1976 at the age of eighty-four.

Two major biographical studies have been written about Polanyi. The first is a biographical memoir coauthored by the physicist Eugene Wigner and Robin Hodgkin, a Quaker educational theorist. A second, much more extensive biography was begun by the late William T. Scott, who was a physical chemist and friend of Polanyi's, and completed by Martin X. Moleski, SJ, a philosopher and professor of religious studies at Canisius College.[8] In contrast to these biographies, my book on the historical origins of the social epistemology of science and the social construction of science is not intended as a biography. While my book follows a fundamentally chronological narrative beginning with Polanyi's childhood and education in Budapest, there is some fast-forwarding and backtracking, and the chapters are organized thematically rather than in strictly sequential narrative.

The leitmotif of the book is Polanyi's scientific life and political culture in Hungary, Germany, and Great Britain from the First World War to the cold war. The problem is an explanation of how he and other natural scientists and social scientists of his generation—including J. D. Bernal, Ludwik Fleck, Karl Mannheim, and Robert K. Merton—and the next—notably Thomas Kuhn—arrived at a strong new conception of science as a socially based enterprise that does not rely on empiricism and reason alone, but on social communities, behavioral norms, and personal commitments that ultimately strengthen rather than weaken the growth of scientific knowledge.

As discussed in chapter 1, a crucial part of Polanyi's story is that he grew up in the Hapsburg dual monarchy as a Hungarian intellectual of nonobservant Jewish origins with a commitment to secular science, internationalism, and political liberalism. He was part of the cosmopolitan elite or "Hungarian phenomenon" of intellectuals that included among his close friends the Hungarian physicists Eugene Wigner and Leo Szilard, the sociologist Karl Mannheim, and the writer Arthur Koestler, all of whom, like Polanyi, left Hungary after the First World War and then had to leave Germany and Austria in the 1930s. Not only did they seem to have special gifts of creativity, but some of them, including Wigner, Szilard, and Polanyi, as well as John von Neumann and Edward Teller, came to have considerable influence in science policy in their adopted homes, particularly in the United States. All of them had to grapple with foreign status and with the social processes of acculturation against the odds of ethnic prejudice—experiences that contributed to their wide-ranging originality and creativity.

Chapters 2, 3, and 4 follow the thread of Polanyi's scientific career and his chemical researches—including both disappointments and triumphs— from the 1920s to the early 1940s. Chapter 2 focuses on Weimar Germany and particularly on Berlin as a scientific capital in which Polanyi mastered the craft of doing world-class science and running a world-class chemical laboratory with support and collaboration from Hungarian, German, and Austrian colleagues, as well as from visiting young scientists from Great Britain and the United States. What he learned and did in Berlin constituted the experiential basis for his later arguments about the social nature of science, the ideal structure of the scientific community, and the freedom of scientists to run their own affairs.

As seen in chapter 3, Polanyi's disappointment in colleagues' reactions to some of his early chemical work infused his later writings on dogmatic traditions in science. As discussed in chapter 4, the reception of his semi-empirical theory of reaction dynamics led to philosophical reflections on the insufficiency of first principles and precise mathematics by themselves to guarantee scientific progress. Chapter 4 also sets the stage for Polanyi's defection from his chemical career as he struggled to reproduce at Manchester the research facilities that he had enjoyed in Berlin and as he tried to adjust to the political and intellectual culture of a new country while observing from afar the descent of Europe into the abyss of Nazism and Fascism.

After more than thirty years of a high-profile research and administrative career in physical chemistry, Polanyi turned away from chemical research in the late 1930s to his studies in economics, politics, and the philosophy of science. As discussed in chapter 5, Polanyi devoted himself to an analysis of the causes of the Great Depression, making a film and writing articles and books on Keynesian free-market economics. He used economic analogies to suggest a model for the scientific community as an internally spontaneous and free-market community, arguing against Socialist planning schemes and against government control of science. One of the Socialists whom he had in mind was the economic historian Karl Polanyi, who was Michael's elder brother. In opposition to Michael Polanyi's classically liberal views, Karl Polanyi argued that the free-market system was outmoded and that it emerged in the nineteenth century only because of local conditions of time and place, not as the result of natural or historical law. Karl's stress on the role of institutions in economic systems eventually found an echo in Michael's work on the social dimensions of science.

Within the British scientific community, as analyzed in chapter 6, J. D. Bernal's Marxist-oriented promotion of the social relations of science movement in Great Britain in the 1930s became a prime target of Polanyi's essays and lectures. Bernal's 1939 book on the "social function" of science and on the integral relationship between pure and applied science fueled Polanyi's writings against science planning in Britain, leading him to help found a "freedom in science" countermovement aimed against the British scientific Left. Polanyi and Bernal had unbridgeable differences over Marxism and the Soviet Union that perhaps not surprisingly prevented Polanyi from seeing the striking similarity of their views on the operation of social norms of behavior within the scientific community and the inadequacy of a history of science told as the march of disembodied ideas. Polanyi's political writings in this period included arguments for the social nature of science, the roles of scientific apprenticeship and craftsmanship, the insight that science is traditional and dogmatic rather than revolutionary and skeptical, and the notion of a logical gap between competing interpretive frameworks that can be breached only by a process of psychological conversion.

In contrast to his political differences with Bernal, there was agreement between Polanyi and the Viennese-born Popper on broad political matters, but not on the nature of science or on the role of skepticism in science and politics. Popper, like Polanyi, expressed some of his most influential philosophical ideas in explicitly political writings during the 1930s and 1940s. Chapter 7 focuses on the ideas and relationships among Popper, Polanyi, and the younger Kuhn, with some attention to the explicit or implicit political agendas in their philosophical writings. By the late 1950s, Kuhn had joined Polanyi in combating Popper's rationalist view that everyday scientific thinking is naturally skeptical and revolutionary, rather than conservative and committed to a framework. Polanyi and Kuhn each argued that Popper failed to understand what it is that scientists actually do.

Chapter 8 begins with a discussion of Polanyi's dense and difficult effort at a philosophy of science in *Personal Knowledge: Towards a Post-Critical Philosophy*, a book that baffled most philosophers of science, pleased many religious thinkers, and provided important insights to historians and sociologists of science. Polanyi incorporated themes from his earlier lectures and essays into *Personal Knowledge*, including the necessity for "pure" science and scientific autonomy, the craft nature of scientific practice, the roles of "schools of research" and "apprenticeship," the tension in science between innovation and tradition, the authority structure of science, and the existence of specifically scientific norms and values. He had become familiar with Gestalt psychology, leading him to develop a terminology of "subsidiary awareness" and "focal awareness" to distinguish the observer's subsidiary awareness of the particulars of a thing from his focus on the wholeness of a thing.

As noted by Kuhn in *Structure*, "tacit" knowledge was an important pre-occupation in *Personal Knowledge*, with the claim that there are two kinds of knowledge: explicit, articulated, and formal knowledge on the one hand and tacit, unarticulated, and nonformalized knowledge on the other hand. Polanyi argued that the first cannot be achieved without the second, just as "focal" knowing relies on "subsidiary" awareness. The rule-bound knowing of empiricism and logic is linked to objectivity, and the tacit knowing of know-how, intuition, and passion is linked to subjectivity. For Polanyi, personal knowledge is the unification of the objective and subjective aspects of all knowing, including, prominently, scientific knowing.[9]

There was a larger program in *Personal Knowledge*, however, than just the philosophy of science. Polanyi aimed to undermine what he believed to be the false ideal of "objectivity" in post-Enlightenment scientific and "critical" thinking—a rationalist outlook, he believed, that prevents unification of the biological and physical sciences and, more broadly, the natural sciences and social sciences.[10] He identified the ideal of objectivity with the elimination of realism from the philosophy of science, and he reiterated forcefully and at length in *Personal Knowledge* his earlier statements of faith in scientific discovery as an effort to make contact with reality in something like a "prayerful search for God."[11] Calling the reader's attention to the atrocities committed by totalitarian regimes during the 1930s and 1940s, he condemned "objectivity" for its alleged moral blindness and indifference to human freedom. In so writing, Polanyi distanced his philosophy from mainstream philosophy of science, despite the value of many of his insights and observations.[12]

Chapter 8 analyzes the largely negative reaction to *Personal Knowledge* on the part of its intended audience in the philosophy of science, as well as its more favorable reception among interested scientists, theologians, religiously concerned intellectuals, and proponents of intelligent design. Ironically, given the incorporation of some of Polanyi's themes into the sociology of science, one of his principal targets in *Personal Knowledge* was sociology in general and Mannheim's sociology of knowledge in particular. Mannheim's sociology and the early work of Merton which Polanyi ad-

mired are discussed in chapter 8. The chapter concludes with a description
of the rise by the late 1960s of a movement in the United States and Great
Britain toward broadening scientific education and public discussions of
science to include the relationship between science and society. Within this
framework, Jerome Ravetz, John Ziman, and David Edge were among those
members of Kuhn's generation who began shifting historical and contemporary studies of science in Great Britain away from networks of ideas and
logic toward social and sociological studies of scientific practices and scientific communities.

Kuhn's *Structure of Scientific Revolutions* came to be seen at the end of the
twentieth century as the "harbinger" of the "strong program" in the sociology of scientific knowledge (SSK) and of the "constructivist movement."[13]
In playing this role, as I argue in the epilogue, Kuhn was a transitional figure in three generations of pioneers in the social epistemology of science.
The first generation included Polanyi, Fleck, Mannheim, Bernal, and Merton, all born during the period 1890 through 1910. The second transitional
generation included Kuhn, Ziman, Ravetz, and Edge among its members.
A third generation—the "1960s generation" born in the 1940s—enrolled
Harry Collins, Steven Shapin, and Bruno Latour. Members of the second
two generations all acknowledge some debt or inspiration to the first generation, and they routinely refer to the importance of Polanyi's notions of
tacit knowledge, apprenticeship, and the social nature of science.

What distinguishes the first generation from their intellectual children
and grandchildren is that Polanyi's generation—Mannheim included—felt
a deep reverence for natural science and mathematics. These intellectuals
shared a conviction of the transcendence and universalism of scientific
thinking—a conviction found equally in the tenets of logical empiricism
and in the Vienna Circle, just as it is found in Popper's philosophy of critical rationalism. This view of science was rooted in this first generation's
common culture of the 1930s. In contrast, Kuhn's generation in social studies of science was not firmly committed to scientific truth with a capital "T"
or to the cumulative and universal character of scientific knowledge. Unlike
the next generation, however, Kuhn fought hard against any epistemological relativism that would devalue the scientific enterprise as a whole. For
the first two generations who adopted a social epistemology of science, the
problem to be explained about scientific knowledge was the almost sacred
mystery of its overall stability and reliability.

The 1960s generation in science studies pursued a program of examining and unmasking the mechanisms by which scientists gain credibility within and outside their disciplinary communities in local times and

places. Influenced by political events and movements of the 1960s, this group was largely suspicious of the claims of the scientific elite for the privileged status and civic virtue of its scientific knowledge. Within social studies of science, the programs of sociology of scientific knowledge as practiced by Shapin and Collins, for example, and of scientific constructivism, as pressed by Bruno Latour, challenged traditional history and philosophy of science by transforming the subject from the study of scientific knowledge into the study of scientific belief.

The historical narrative of this book demonstrates how the major problems and solutions in twentieth-century epistemology of science were embedded in historical events and political cultures. Although the period of the 1930s was the first era in which reflections on the social nature of science began to appear, these statements were marked by strong ideological and political differences among their proponents, as was true, too, among the 1960s generation that self-consciously created the field now known as science studies (as briefly discussed in the epilogue). Michael Polanyi was unusual among twentieth-century philosophers of science in his career experiences as a major scientist, and he was unusual among scientists in his definitive departure from his original life's vocation into the fields of economics, politics, and philosophy. My own first encounter with Polanyi by way of a footnote in Kuhn's *Structure of Scientific Revolutions* is symptomatic of the ways in which Polanyi's views often have been absorbed by readers through descriptive snippets in paragraphs or footnotes that demonstrate little appreciation of the scientific life and political culture that created his strongly principled world view and inspired the intellectual battles that he fought. This book aims to provide a fuller picture of these 1930s origins of the social construction of science.

# ONE

# Scientific Culture in Europe and the Refugee Generation

Among scientists and philosophers who first clearly recognized and explained science as communities of social practices, rather than as a system of empirically verified ideas, was the physical chemist Michael Polanyi.* His experience of living in the international scientific center of Weimar Berlin and having to leave it in 1933 was central not only to his personal life but to his philosophy of science and to the development of a social epistemology of science. Polanyi's principal career lay in his scientific work, which he expanded to economic and political analysis and finally to philosophical writings. His life was idiosyncratic, like all individual lives, but many of his most significant experiences were shared in particular with a generation of European refugee intellectuals who were arguably unique in the history of science.

Polanyi was a member of the 1930s central European refugee generation of Jewish origins who left their home countries because of the politics of anti-Communism or anti-Semitism. Like others of that generation, two of the central events of Polanyi's youth were the Great War of 1914–1918, concluded by the punitive Versailles Treaty, and the Bolshevik Revolution. Like so many of his scientific colleagues who were members of his generation, Polanyi had embarked upon a successful scientific career by the 1930s, when he had to make his way into a new social culture as he moved in 1933 from Germany to England. Among the 1930s refugee scientists, Polanyi was a member of a subgroup that is identified by historians as the twice-exiled "Hungarian phenomenon." These scientists left Hungary in

---

* In this chapter alone, Hungarian accent marks are used in the spelling of Hungarian names, with the exceptions of the names of Michael Polanyi (Mihály Polányi), Karl Polanyi (Károly Polányi), and their immediate family members. In later chapters, Anglicized spellings are used for the most part for Hungarians who settled elsewhere.

late 1919 or early 1920 for Germany, Austria, and elsewhere under pressure from the attitudes and policies of a new postrevolutionary conservative regime that adopted an explicitly Christian and nationalist ideology in a Hungarian nation that previously had been a multiethnic state. The Hungarian scientists' first exile is reflected in the description of the "socially unattached intelligentsia" that becomes capable of a synthetic perspective in the Hungarian sociologist Karl Mannheim's *Ideology and Utopia* (1929).[1]

Michael Polanyi was one of the older members of the twice-exiled refugees. Many of them had attended the same schools, lived in the same neighborhoods, or known each other in Budapest. They were friends and colleagues in Germany before moving on to Great Britain, the United States, South America, and other countries. Polanyi worked closely in Budapest with György (George) von Hevesy, who became an émigré at the same time as Polanyi. Polanyi's circle of friends in Berlin included the émigrés Jenő (Eugene) Wigner, Leó Szilárd, János Neumann (John von Neumann), and Dénes (Dennis) Gábor. Thódor Kármán (Theodore von Karman), who was ten years older than Polanyi and had studied in Budapest, arrived in Germany earlier and advised the younger generation on opportunities in Germany before he was lured in 1930 from the Aeronautical Institute at Aachen to Pasadena, California, where he became director of the Guggenheim Aeronautical Laboratory. Ede (Edward) Teller, almost a decade younger than Polanyi, left Budapest in 1926 for Germany. These were the most prominent of dozens of Hungarians who worked in physical sciences and mathematics in German scientific centers before the anti-Semitic policies of the Third Reich forced a second Hungarian emigration to other scientific centers.

While later chapters in this book follow Polanyi's scientific career and his shift to economics, politics, and philosophy, this chapter focuses on Polanyi's early life in Budapest; the circumstances of the double exiles of Polanyi, his family, and scientific colleagues; his later experiences of discrimination in the United States; and his reflections on assimilation and Jewish identity. In Budapest, the young Polanyi absorbed a cosmopolitan, liberal, and transnational European point of view that characterized many assimilated or baptized Jews of early-twentieth-century Budapest. His family upbringing was secular, educated, literary, and scientific, with a strong dose of left-wing politics, which he came eventually to reject, from his mother, oldest sister, and brothers. In his biography of the young Viennese philosopher Karl Popper, the historian Malachi Cohen argues that the vision of the multiethnic society promulgated among liberal politicians and thinkers of the Hapsburg monarchy influenced Popper toward his vision of the "Open Society."[2] Similarly, the roots of Polanyi's "republic of science" lay in his Hapsburg heritage, which is the subject of this chapter. His career experiences in scientific Berlin, his free-market economics, and his political anti-Communism are subjects for later chapters.

The Budapest upbringing was crucial in other respects as well. Nowhere in Europe were Jews more assimilated, secularized, and "Westernized" than in Budapest before the Great War. The confrontation with a rise in anti-Semitism in Budapest in 1919, linked to the politics of the new nationalist regime, which was magnified into legal anti-Semitism in Germany in the mid-1930s, forced Polanyi into rethinking the meaning of Jewish identity and his relationship to religion and to Christianity. His religious reflections strongly influenced his later philosophy of science. More significantly, his commitment to Jewish assimilation was part of his commitment to modern science, to the old Hapsburg Empire and internationalism, and to his political criticism of the kind of radical and revolutionary politics that Central European anti-Communists came to identify as Jewish. As with many émigrés, Polanyi's experiences led him to acknowledge his potentially marginalized status as cultural outsider and encouraged in him the wide-ranging intellectual exploration and combativeness that helped constitute his originality. In this vein, Arthur Koestler, one of Polanyi's family friends from Budapest who wound up in England, ruminated on the fates of Hungarians, "the only people in Europe without racial and linguistic relatives in Europe; therefore they are the loneliest on the continent. This . . . perhaps explains the peculiar intensity of their existence . . . Hopeless solitude feeds their creativity, their desire for achieving."[3]

## The Polanyi Family and Imperial Budapest

Michael (Mihály or "Misi" Polányi) Polanyi was born on March 11, 1891, in Budapest. At the time, his family lived in an elegant apartment that occupied an entire spacious floor of a large building at number 12 on Andrássy ut, a large boulevard modeled on the Champs-Elysées in Paris and running for more than a mile between the verdant City Park (Városliget) to the northwest and the Elizabeth (Erzsébet) Square to the southeast. The boulevard is lined by some of Budapest's finest architecture: the State Opera House is near the building where the Polanyis lived, as is the Basilica of Saint Stephen, which had been under construction for decades and was completed in 1905. Across from the Opera House was the palatial Dreschler Café, one of five famous coffeehouses along a four-block stretch of Andrássy Avenue, including the Japan, the Hall of Arts, the Opera, and the Abbazia. The tracks for Europe's first underground electric subway lay under the avenue.[4]

Polanyi's older brothers and sisters had been born in Vienna. Laura, called "Mausi" by everyone, and the eldest, was born in 1882; Adolf in 1883; Karl (Károly or "Karli") in 1886; and Sophie (Zsófia) in 1888. Paul (Pál or "Pali") was born around 1893 and institutionalized due to severe mental disability. He died in late adolescence.[5] In Budapest, the children had English and French tutors, a German governess, Hungarian servants, and a fencing instructor. They owned a stable where a groom lived and took care of their horses, and their father joined them in riding on Sundays. The family spoke German with their mother, Cecile (Cecil in Hungarian), who understood but did not fluently speak Hungarian and had grown up in Vilna in then Russian Lithuania. Their father, Mihály, spoke fluent French. The fine apartment's public rooms were open to artists, writers, and other intellectuals of the Budapest avant-garde who came to Cecile's weekly salons.[6]

The family was in many ways typical of wealthy and assimilated Jews of Budapest. The Compromise (*Ausgleich*) of 1867 had established the dual monarchy of Austria-Hungary, in which the Cisleithanian and Transleithanian (Hungarian) regions of the state were governed by separate parliaments and prime ministers. Austria-Hungary shared a common emperor who also was Hungary's king, a military force, and specified ministries. Budapest, which was unified in 1872 from the three towns of Óbuda and Buda (on the hilly western side of the Danube River) and Pest (on the flat eastern side), became the second capital of the monarchy.[7] The population of Hungary included many non-Hungarians, and the Hungarian-speaking Magyars encouraged Magyarization of the Jews, most of whom were "upper-land" urban Jews who had migrated from Moravia, Austria, and Germany. In November 1867, the Hungarian National Assembly passed a law making Jews equal in status to Christians in civil and political rights, a law unique in central Europe at the time.[8]

Michael Polanyi's great-grandfather, Mihály Pollacsek, and his wife were one of the first Jewish families in Hungary to obtain permission around 1800 to lease forest lands in Moravia from the crown on which they set up lumber mills. The name Pollacsek had been the family name for at least one generation since the Holy Roman Emperor Joseph II had issued an edict in 1782 requiring Jews to have surnames.[9] Mihály Pollacsek's son Adolf, who was born in Slovakia, was a construction engineer, and his wife, Sophie Schlesinger, came from a family with a large textile mill. They ran the family lumber mills, a flour mill, and a distillery and brewery. Among their six children was Michael Polanyi's father Mihály Pollacsek, who was educated at a Catholic boarding school in Vienna. He studied at the Eidgenossische Technische Hochschule (ETH) in Zurich and in Stuttgart and Dresden before becoming an engineer, first for river control in Pest, and then, beginning in 1873, for the Swiss National Railway. By 1885, he was working for the City Railway Company of Vienna, where he met Cecile Wohl, the daughter of a scholarly Jewish rabbi and graduate of the Titov Gymnasium in Vilna. Her father had sent her to Vienna to get her away from Socialist and anarchist student movements in Vilna.[10]

Around 1890, Mihály and Cecile moved to Budapest with their four children. The population of Budapest was expanding and reached 733,000 in 1900, making Budapest the sixth largest city in Europe. It was an overwhelmingly Catholic city with Protestants numbering about 25 percent of the Hungarian population. While the population of Hungary as a whole was about 5 percent Jewish in 1900, the Jewish population of Budapest grew from 16 percent in 1872 to 21.5 percent in 1900 with increasing migrations from the "lowland" countries of Poland, Lithuania, and Russia. There was no Jewish quarter, but there was a visible Jewish community in the area of Orczy House, on Király Street, where the newest Jewish arrivals found synagogues, apartments, baths, restaurants, cafes, and shops. The twin-towered synagogue on Dohány Street, which was completed in 1859, is now the largest active synagogue in Europe.[11] The large Jewish population in Budapest inspired the ugly epithet "Judapest" employed by Vienna's anti-Semitic mayor Karl Lueger by way of contrasting Budapest with his own city.[12]

The financial aristocracy of Budapest was largely Jewish in their origins, as were many of the city's architects, artists, and entrepreneurs. Half of Budapest's lawyers and doctors and a third of its engineers were Jewish. In the early 1900s, there were sixteen Jewish members of Parliament and two dozen Jewish professors in the universities. In 1910, Samu Hazai became minister of war; in 1912, János Teleszky became minister of finance; and in 1913, Ferenc Heltai, the nephew of the founder of Zionism, Tódor (Theodor) Herzl, became mayor of the city. All were of Jewish origins. Jews were one among a number of non-Magyar groups in the modernizing class of bankers, industrialists, entrepreneurs, and modern professionals who constituted a new middle class in a country previously divided between the large Magyar nobility and an even larger peasantry. As Gábor Palló writes, the traditionally suppressed Hungarians—most prominently the Germans, Greeks, Slovaks, and Jews—got their chance at assimilation and influence in this period.[13]

Mihály Pollacsek quickly felt local pressure to Magyarize his family name and in 1904 changed his children's names, although not his and Cecile's, to Polányi.[14] The Polanyis were firmly secular and nonobservant Jews. Cecile's father was liberal in his views, interested in Christianity, and critical of Jewish ritual. He regularly sent Christmas greetings to the Polanyi family, who set up and lit a tree during the Christmas holidays. The only religious instruction given to the children was in mandatory classes at the *gymnasium*, which included two hours of Jewish religious instruction each week for "Israelites."[15]

As a civil engineer and entrepreneur in the railroad industry, Michael Polanyi's father was working in a rapidly expanding business in the late 1800s. Budapest was the central hub of the national railroad network, and it lay on the path of the Orient Express from Paris to Istanbul. The Hungarian national railroad network expanded from 7,025 miles in 1890 to 10,632 in 1900, and the rail density of Hungary in length of track per 100,000 people was only slightly less than in France and greater than in Austria and imperial Germany.[16] When Michael Polanyi was 9 years old, the railroad line that his father had been building from the Danube Valley into Slovakia and Poland was washed out by 3 months of steady rain. The government refused to invoke the "Act of God" clause in the construction contract. Determined to pay some 2,000 workers their wages and to repay his creditors, Mihály declared bankruptcy in 1900. The family moved into a smaller, fourth-floor apartment at number 9 on Ferencziek Square, southeast of their old flat and near the Erzsébet Bridge in the inner city. Mihály found a job traveling for the Frankfurt Trade Fair company and regaled his family with stories of science and industry in German cities.[17]

The children were all tutored at home until they were twelve years old and ready to enter gymnasium schools. The eldest, Mausi, attended the Lutheran Boys' Gymnasium as a special student (for a higher fee) and then entered the National Women's Educational Association Gymnasium when it opened in 1896. It was the first gymnasium in Budapest for girls. She entered the University of Budapest in 1900 and finished her doctorate in history in 1909 after taking off three years to work and marrying Sándor (Alexander) Striker, who owned a textile mill in Austrian Bohemia, in 1904. She was able to move back to Andrássy Avenue (number 83) with her new husband, whose financial means helped members of the family. Adolf was studying Japanese at the Oriental Trading Academy and law at the University of Budapest. Karl was a student at the Minta Gymnasium, where Michael matriculated in 1900 with assistance from a Brill Armin Foundation scholarship for poor Jewish students.[18]

The Minta (Model) Gymnasium was one of Budapest's leading humanistic secondary schools, located in the city's center on Trefort Street just behind the university; it was established as a teaching school for educators at the university. It became a school attended by members of the Budapest elite. The Minta Gymnasium was established in 1872 by Mór Kármán, the father of Tódor Kármán, who himself attended Minta, as did Teller and the low-temperature physicist Miklós Mór Kürti (Nicholas Kurti). As a demonstration and modern school, the Minta stressed creativity rather than piety and patriotism. Michael studied Hungarian and German literature, Latin and Greek, religion, history, art, geography, natural history, descriptive geometry, mathematics, and physics. Since English was not taught formally in the school, he continued reading English books on his own. He wrote his first scientific paper, in German, in his last year at the gymnasium, on the subject of the specific heats of gases using kinetic theory. He later characterized the paper as "nonsensical," but it was a start on a subject at the heart of his later scientific career.[19]

The bankruptcy of Polanyi's father had come at the start of his gymnasium years. Just more than halfway through Polanyi's years at Minta, a worse tragedy struck in January 1905 when his father suddenly died from pneumonia following a hot-and-cold cycle at a Budapest sauna. At the time, Karl Polanyi was nineteen years old. Laura was married, Adolf was working in Japan, and Sophie and Michael were living at home with their mother, who had become depressed and neurotic but invited friends occasionally for discussions or met with them in cafés on Andrássy Avenue. Michael was fourteen at the time of his father's death.

In the next years, as he finished Minta and entered the University, Michael Polanyi found himself observing and thinking about the many political activities of his elder siblings, particularly Karl, who was studying law at the University of Budapest. Karl was expelled in 1907 for his defense of Professor Gyula Pikler against students and officials who objected to Pikler's relativist and sociological approach and who charged that he denied the validity of Christianity. Karl finished his law degree in 1909 at the University of Koloszvár, about midway between Budapest and Bucharest, with Bódog Somló, who, like Pikler, was of Jewish origin and an enthusiast for positivism and Spencerian ideas on the evolution of social institutions and living species. Somló attracted many progressive students at the time, such as Karl Polanyi and György Lukács (George Lukacs).[20]

Somló and Pikler had been founding members in 1900 of the Society of Social Science, whose second president, in 1901, was Count Gyula Andrássy, a patriot, monarchist, and aristocratic liberal. He resigned in 1906 when there was a split in the group along lines of conservative and liberal politics. The society began publishing the periodical *Huszadik Század* (The twentieth century), the first scholarly review of sociology and political science in Hungary, in 1900. Its editor was Oszkár (Oscar) Jászi, whose article "Scientific Journalism" in the first issue warned that "every explicit or implicit, courageous or cowardly expression of a reactionary world view will be proscribed . . . reactionary political theory cannot, by definition, be political theory."[21] Weekly meetings of the editorial board convened at the Modern Kávéház (Café Modern), where Jászi pushed for a scientific outlook for twentieth-century Hungary that would be cosmopolitan, rational, and critical. Karl Polanyi signed on as a writer. Articles in *Huszadik Század* tackled questions of suffrage, land reform, education, and competing nationalities in a greater Hungary where huge estates still were held by the church and the aristocracy, and landless peasants lived in conditions just short of serfdom.[22] By 1906, some of the leaders of the group were hoping to establish a kind of free university for workers, with Jászi writing, "We must create a new ethic, a new morality, in the place of the collapsed morality of religion or metaphysics."[23] This was too much for Count Andrássy.

Another leading magazine for liberal thought in Budapest was *Szabadgondolat* (Free thought), a magazine established in 1911 by the Galileo Circle (Galilei Kör).[24] This group originated in 1908 as an organization founded by Karl Polanyi, the later mathematician György (George) Pólya and like-minded friends, some of them former members of the Marxist-oriented Socialist Students Club, founded in 1902 by Adolf Polanyi with Ernö Pór and Egon Szécsi (whose sister Adolf married in 1907). A notice of a club meeting led to a brawl with young men from an organization called A Thousand Christian Youth and to police interference. Adolf left shortly afterward for Japan.[25]

The Galileo Circle was a student organization and subsidiary of the Hungarian Association of Free Thinkers. Its aim was "the defense and propagation of unbiased science." Pikler was the group's advisor, and Karl became its first president and a member of *Szabadgondolat's* editorial board in 1913. The Galileo Circle numbered several hundred members, although meetings usually had only fifteen to twenty in attendance. The circle's leaders sponsored lectures and set up reading rooms. Karl wrote articles for the periodical but also penned witty aphorisms, many of them anticlerical in tone, which served as fillers between articles in the magazine.[26] Michael Polanyi, just entering the university, participated in the Galileo Circle, joining its Committee on Natural Science and giving occasional talks on physics and chemistry.[27]

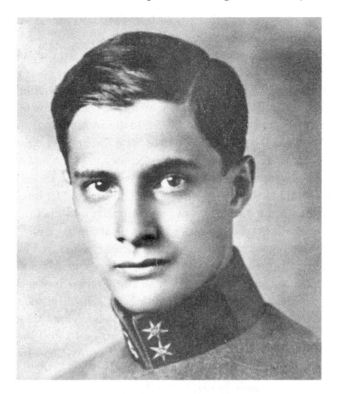

Figure 1.1. Michael Polanyi in military uniform, ca. 1915.
(Courtesy of the Special Collections Research Center,
University of Chicago Library, and John C. Polanyi.)

Michael started medical studies at the University of Budapest in 1908, along with one-quarter of his Minta classmates. One of his good friends was Pólya, whose family was originally Jewish but baptized their children into the Roman Catholic Church. Pólya and Polanyi together heard physics lectures given by the much revered Baron Loránd (Roland) Eötvös, whom Albert Einstein called the "prince of physics" for Eötvös's experimental work on the equivalence of gravitational and inertial mass.[28] Polanyi's first scientific publications were drafted while he was a medical student in Budapest, when he worked as a laboratory assistant in Ferenc Tangl's Institute of Pathology and Physiological Chemistry. The nineteen-year-old student published a research paper in 1910 on the chemistry of hydrocephalic liquid and a second paper in 1911 on blood serum. The first paper appeared in a Hungarian journal but also in German in the *Biochemische Zeitschrift* alongside the blood-serum paper. It was while studying colloidal gels that resemble tissues in the body that Polanyi first encountered the phenomenon of adsorption that would later become a theme in his work in physical chemistry.[29]

Polanyi received his medical degree in April 1913 and spent the next academic year in Karlsruhe in Germany, only to be caught up in the declaration of war on Serbia by Austria-Hungary on July 28, 1914, following the assassination in Sarajevo of the Austrian archduke Franz Ferdinand and his wife by a Serbian student. Michael enlisted in the army and was assigned to a regimental hospital in Zombor, north of Novi Sad and not quite halfway between Budapest and Sarajevo. By mid-October 1914, he had developed a mild case of diphtheria but returned to Zombor in early 1914, becoming sick again with kidney inflammation and a chronic bladder infection. For the rest of the war he was sometimes on furlough and sometimes serving in light military duties, taking his furloughs in Budapest. While in hospitals or recuperating, he was able to write scientific papers and complete a doctoral thesis in physical chemistry.[30] Karl, too, was in and out of Budapest during the war years. He served as a cavalry officer in Galicia, which now is the southern part of Poland, and left the army near the war's end with injuries and a serious depressive illness.[31] In a letter to his Aunt Irma in Vienna, Karl expressed the sense of powerlessness to stop the carnage: "Humanity is a Golem which stares with horror on its own frozen mask, the tortured soul at the terrible machine."[32]

In late 1915, Michael joined a Sunday afternoon group that began meeting at the home of the poet and dramatist Béla Balázs. The Sunday circle included Károly (Karl) Mannheim and Lukács, who was one of Karl's closest friends. Lukács led discussions that began at 3 p.m. and continued well into the night—sometimes until 3 a.m. The subjects for discussion often were ethical problems and often ones inspired by the writings of Dostoyevsky, Tolstoy, and Kierkegaard.[33] Decades later, in 1944, Michael Polanyi wrote Mannheim a letter in which he reflected upon these youthful discussions:

> My religious interests were awakened by reading *The Brothers Karamazov* in 1913. I was 22. For the following ten years I was continuously striving for religious understanding and for a time, particularly from 1915 to 1920, I was a converted Christian on the lines of Tolstoy's confession of faith. Towards the middle twenties my religious convictions began to weaken and it was only in the last 10 years that I have returned to them with any degree of conviction. My faith in God never failed me entirely since 1913, but my faith in the divinity of Christ (for example) has been with me only for rare moments.[34]

Polanyi became closer during the war years to Karl's old friend Oszkár Jászi, who joined with like-minded progressives to establish the Radical Party in 1914, which founding members characterized as the party of the "under-forties." The aim was to appeal to reformist-minded citizens who were discontent with the more conservative political programs of the Christian Socialists by offering a new party composed of middle-class radicals (craftsmen, tradesmen, and entrepreneurs), non-Marxist Socialists, and Marxist Socialists who did not belong to the Social Democratic Party. Karl Polanyi became the secretary for the party, which had its headquarters on the fashionable Andrássy Avenue.[35]

During the course of the war, 3.4 million men were conscripted into the Hungarian defense force from Hungary and Croatia, of whom 530,000 were killed, 1.4 million wounded, and 833,000 taken as prisoners of war from greater Hungary. With worsening economic conditions at home, shortages of food supplies, and a mounting toll in war casualties, hunger protests and demonstrations increased. News of the Russian Revolution in February 1917 and the abdication of the czar was followed by the Bolshevik Revolution of October 1917 and Russia's withdrawal from the war effort in March 1918, a year after the entry of the United States into the war. After the Central Powers began collapsing toward the end of the summer of 1918 and Bulgaria sued for peace in late September 1918, it became clear, as Count Mihály Károlyi said on October 16, 1918 in the Budapest Parliament, "We have lost the war; now it is important not to lose the peace." The next day Hungary's prime minister, István Tisza, acknowledged to Parliament the truth of Károlyi's statement: "We have lost the war."[36] On October 25, Károlyi's New Independence Party formed a coalition with the Social Democrats and the Radicals and established a Hungarian National Council. It promulgated a twelve-point program that demanded immediate conclusion of a separate peace, independence from Austria, reconciliation with the non-Magyar nationalities without endangering Hungary's territorial integrity, land reform, and universal suffrage in a country where only 6 percent of the population was permitted to vote.[37]

With the Romanians, Czechs, Croats, Slovaks, and Ukrainians all seceding from the dual monarchy, a bloodless "Chrysanthemum Revolution" took place on the night of October 30, 1918, with military units loyal to the National Council occupying strategic sites. The next day, Tisza was murdered in his villa by several armed soldiers who reputedly blamed him for Hungary's involvement in the long and tragic war. Károlyi sent a very large wreath to the Tisza family the next day and they promptly threw it on the garbage heap. An armistice was signed with Italy on November 3, 1918, marking the end of the war for the Austro-Hungarian Empire, eight days before the armistice of November 11, 1918 signed by Germany with the Entente allies. On November 11, Emperor Karl I (King Karl IV of Hungary), who had succeeded the eighty-six-year-old Franz Joseph upon his death in 1916, abdicated and renounced "taking part in the affairs of government." A Republic of Austria was declared the next day. The Károlyi government proclaimed a Republic of Hungary on November 16, 1918, and the National Council elected Károlyi President on January 11, 1919.[38]

Jászi served in the early weeks of the republic as the minister of nationalities, hoping to transform the old Hapsburg Empire into a federation along the lines of Switzerland. Michael Polanyi shared this goal and he wrote of his vision of a larger European confederation in a pamphlet and in an article published toward the end of the war in Jászi's journal *Huszadik Század*. Calling for international peace, Polanyi warned against the "materialist prejudice" that convinces people that war is in their interest, and he criticized the "tacit presuppositions" and nationalisms that subvert the goal of a peaceful Europe. For peace to be insured, he argued, national sovereignties must he eliminated, and a European state established that is united by a system of law and order and by a European army.[39]

Both Jászi's goals and Polanyi's long-term vision were hopeless, however, as nationalism swept through the various national and ethnic groups of what had been greater Hungary. Laura (Mausi), Karl, and Michael all wanted to participate in the new government. Karl was general secretary of Jászi's Radical Party, and Mausi decided to become a candidate for the parliamentary elections, planned for April 1919, on the basis of a universal suffrage that would include women. Michael became secretary to Dr. Max H. Goldzieher in the Ministry of Health and received the task of drafting plans for demobilization of the soldiers of the Hungarian Army.[40] Another scientist who served in the government was Thódor Kármán who had returned to Hungary to serve in the army, taking leave from his professorship in aerodynamics and mechanics at the University of Aachen. Kármán served in Károlyi's Ministry of Education.[41]

The Communist Party had been outlawed under the old regime but was now legalized. Karl had debated Leninists and Marxists in the dome hall of the university, and the October 1917 Bolshevik Revolution had inspired him to devote the entire December issue of *Szabadgondolat* to the problem of Bolshevism, asking three leftist intellectuals to write the leading articles: Jászi, Lukács, and Eugene Varga. Lukács and Varga both joined the Hungarian Communist Party, but Jászi took a point of view similar to Karl's that dictatorship cannot be regarded as a legitimate transitional condition. Jászi argued for "civic radicalism" as a continuation of the best part of classical liberalism and rejected the strategy of class war in favor of what he called "the organs of intellectual and moral forces, by the development of cultural ideals, and on the basis of the consensual and fair co-operation of every social group."[42]

The journalist and insurance clerk Béla Kun returned to Budapest from Russian captivity on November 16, 1918. The Károlyi government's arrest of Kun and other Communists for provoking disorder in February 1919 brought an outcry from Karl Polanyi and other Galileo Circle members, despite their opposition to Communism. The situation worsened in Budapest when Károlyi received a document on March 20, 1919, from the Entente representative in Budapest, French Colonel Ferdinand Vyx, informing Károlyi that Romanian, Czechoslovak, and Yugoslav troops would be allowed to advance to demarcation lines deep within Hungary. Feeling betrayed by the allies of the old Entente, Károlyi resigned and transferred power to the Social Democrats without knowing that they had concluded a pact with the Communists whom he had just imprisoned. Although nominally only in charge of foreign affairs, Kun immediately took control of the government, created a Red Army to take back territory lost to the Czechs and Romanians, nationalized Hungary's estates and all businesses with more than twenty employees, eliminated political opposition in the government, and took extreme measures to counter hoarding and black-market operations.

Michael Polanyi resigned immediately from his government post and accepted an academic position teaching courses as an assistant to Georg von Hevesy at the University of Budapest. In contrast to Polanyi's decision, his former Sunday Circle colleague Lukács became a commissioner for education and culture under Kun's government, and Eugene Varga, who was an economist, became commissioner of finance and chairman of the Supreme Council of the National Economy.[43] Kármán stayed on in the new government in his post at the Ministry of Education.[44] On May 2, 1919, Karl Polanyi wrote Lukács from the hospital that he, too, would join the Communist Party, but he did not have the opportunity to act on his decision since he was taken to Vienna for recuperation from an operation.[45]

The Hungarian Soviet Republic lasted just over four months. At the end of July 1919, a Romanian Army began to move toward Budapest despite the fact that the terms of an Entente ultimatum to end fighting in Hungary required the Romanians to remain in place. Hungarian Red Army officers had resigned in protest of Kun's withdrawal of the Red Army in response to the Entente's decree. Kun promised that the Russians would come to the aid of Hungary, but no help arrived. When the Romanians were within fifty miles of Budapest, Kun and most of the government's leaders fled to Vienna. The Romanian troops walked into the city's streets on August 4, 1919. By then a Hungarian counterrevolutionary group had rallied in Szeged and formed a new transitional Hungarian government with Admiral Miklós Horthy de Nagybánya, the last commander of the Austro-Hungarian Navy, as minister of war. As the newly reconstituted Hungarian National Army advanced toward Budapest, the Romanians decamped, and the National Army marched into the capital on November 16, 1919, with Horthy at the head on a white horse. In his first public speech he called Budapest a "guilty" or "sinful" city for its revolutionary past, launching the theme of a new Hungary that would be "Christian" and "national."[46] The dream of a cosmopolitan and European greater Hungary in which Jews and other non-Magyar ethnic populations had a welcome public role was dead. Many of the liberal and reform-minded leaders of the original Republican revolution felt both responsibility and despair. Of these last years in Budapest, Karl later wrote Jászi that his activities in Hungary had been ethically a success but politically a failure:

> It was due to the Galilei Circle's failing that there was not available in 1918 a generation, welded in one with the peasantry and with the national minorities, in long-standing, stern battles . . . whose responsibility? Mine. I had been leading the Circle in an anti-political direction. Neither with the working-class, nor with the peasantry, nor with the national minorities did I try to achieve, or even seek some unity based on action . . . I have never been a politician; I had no talent that way, no interest even.[47]

## Two Waves of Migration: 1920 and 1933

Horthy became the regent of Hungary and the head of state in a Kingdom of Hungary with an absent king. Horthy remained regent, refusing to support restoration of the monarchy and its king Karl IV, who died in Madeira in 1922. In October 1944, after Horthy tried to withdraw Hungary from the Axis, he was arrested by the Germans and freed by the Americans in the spring of 1945. While Hungary was unique in central and Eastern Europe during the 1930s and in the early 1940s for the relative safety in which 800,000 Jews and baptized Jews lived, especially in Budapest, deportation and mass murders were accomplished very quickly after German occupation began in March 1944. Hungarians assisted Germans in deporting 437,000 Jews to Auschwitz during the late spring and summer. A larger number of Hungarian Jews, especially Jews living outside Budapest, survived the war than Jews in other central European countries. In Budapest, over 100,000 died, while some 25,000 members of the Jewish population were hidden by friends or survived by other means. Horthy's anti-Semitism was said not to have been natively brutal, as in so many other countries, but it was part of his political platform when he rode into Budapest in 1919.[48]

Horthy's ascension to political power in 1919 was based in a double outrage. Politically, Kun's Hungarian Soviet Republic had disturbed all classes: the landowners, the commercial bourgeoisie, and laborers who had expected a land redistribution policy for them rather than nationalization of properties. Militarily, after initial victories, Kun's Red Army failed to defend Hungarians against the troops of invading national groups in the chaos following the First World War. The second outrage was the Versailles Treaty, or the Trianon Treaty as it was called in Hungary, which turned over two-thirds of the larger Hungary of the dual monarchy to the states of Ukraine, Romania, Czechoslovakia, and Yugoslavia, with a small slice of Hungary's land and people given, too, to Austria. Before the war, Hungary was the sixth largest country in land and the seventh largest in population in Europe. After the Versailles Treaty it became the fifteenth largest in land and the eleventh largest in population.[49] As Jászi later argued in his book *The Dissolution of the Habsburg Monarchy*, a smaller Magyar Hungary no longer needed the Magyarized Jews in a struggle against the potential political power within Hungary of non-Magyar national, ethnic, and linguistic minorities. After the Great War, both elite and popular sentiment blamed the Kun regime for the terms of the Versailles Treaty and blamed as well the secular and modernizing reformers of the Galileo Circle and periodicals like *Huszadik Század*. Thirty-two of the forty-five commissars of the Hungarian Soviet Republic, or "Commune," had been Jewish, including Kun. The "Jewish Republic," it was said, had stabbed the Christian Hungarian nation in the back by establishing the Commune and by providing an anti-Bolshevik rationale for dismantling larger Hungary in the final terms of the Versailles Treaty.[50] As Teller's father told his son, fearing the worst, anti-Semitism was inevitable because "too many of the communist leaders are Jews."[51]

In fact, many Jews had equally suffered with Christians under Kun's short-lived Commune when many industrialists, landowners, and merchants were stripped of their assets or arrested. Communists requisitioned the law office rooms of Teller's family apartment and forbade Ede Teller's father from practicing law because of his wealth.[52] Wigner's father lost his position as director and co-owner of the Mauthner Brothers Tannery when the regime nationalized the tannery and appointed a commissar to run it.[53] When a couple of Communist officials arrived at the elegant home of Laura Polanyi and her husband Sándor Striker, the Strikers escaped arrest only because one of the commissars recognized them as fellow members in earlier years of the Galileo Circle.[54]

Estimates vary wildly for the numbers of people executed and arrested during the regimes of 1919. While it is generally agreed that some three hundred to six hundred people were killed under Kun's "Red Terror," estimates of totals for the early 1920s under Horthy's rule range from six hundred to five thousand. Many more were interned or arrested as political opponents for shorter or longer periods of time during the "White Terror."[55] When Communists, Socialists, and some Jews began fleeing Budapest, the Strikers decided to stay in their home in Buda, which they had bought after returning at war's end from living in Vienna. Karl Polanyi decided to remain in Vienna, where he had gone to a rest home in May 1919. He lived in Vienna until 1933. Sophie Polanyi and her husband Egon Szécsi, who was a lawyer and a former member of Adolf Polanyi's Marxist-oriented Socialist group, fled to Italy with their children, stopping in Vienna on the way. Michael was living with the Strikers at their home in Buda, and he departed for Karlsruhe in December 1919 after he was removed from his post at the University of Budapest.[56]

Prospects in Hungary for Jewish and leftist scientists and educated professionals further dimmed with the passage in 1920 of the first Numerus Clausus law in Europe that restricted admissions to the university, and therefore to the professions, partly on the basis of race. Identification cards required indication of religious affiliation. The overwhelming majority of Jews before 1920 belonged to the Israelite congregation as their religious denomination and recorded "Israelite" on their cards. With the aim of undercutting the power of non-Magyar populations, the Numerus Clausus law of July 22, 1920, sought both to curtail the total number of students admitted to higher education and to restrict admissions to students of "impeccable moral bearing," Hungarian loyalty, and appropriate educational background. In addition, the law stipulated that the student bodies must be approximately representative of the various nationalities and "races" (népfaj) living in Hungary. Since the percentage of Jewish or "Israelite" population in post-Versailles Hungary was approximately 6 percent, the new proportional restriction set a target for future university admissions that would substantially reduce the numbers of Jews who previously had made up 21 percent of law schools, 52 percent of medical schools, 21 percent of liberal arts faculties, 34 percent of pharmacy schools, and 37 percent of the Technical University during the 1917–1918 academic year. (In 1913–1914, the percentages were 18, 47, 15, 31, and 33, respectively.)[57]

In the fall of 1919, before enactment of the Numerus Clausus law, the percentage of Jewish students enrolled in higher education had fallen from 36 percent in the previous year to 6 percent. The postwar precipitous decline indicates that many Jewish students did not find the Hungarian political atmosphere safe in the wake of Kun's Commune and the takeover by Horthy. By 1920–1921 the percentage of Jews in higher education increased from 6 percent to 12 percent, and enrollments in the early 1920s ranged from approximately one-fifth to one-half of post-1900 maximum percentages for Jews, with only the Technical University in fact approaching a percentage as low as the 6 percent target of the 1920 Numerus Clausus law. By the late 1920s, as a consequence of greater stability in Hungary and of voices raised against discrimination and anti-Semitism, a 1928 amendment to the original law rewrote the "racial" and nationalities section and made the applicant's loyalty to the nation and excellent educational record the primary requirements for university admission.[58]

By the early 1920s, some students of Jewish ancestry no longer fell under the restrictions of the original 1920 law, since, in immediate response to the anti-Semitism of the early Horthy regime and to the original law of 1920, thousands of Jews joined the Roman Catholic, Calvinist, or Lutheran churches, enabling them no longer to have to record "Israelite" as their religious denomination.[59] Wigner's family converted when his father decided that the whole family should convert to Christianity because Jews had so strongly become associated with Communism. Antal Wigner chose the Lutheran denomination because both father and son had been educated at the Lutheran Gymnasium.[60] Leó Szilárd, who experienced physical abuse from anti-Semitic students at the Technical University in Budapest, converted to Calvinism and left for Berlin at the end of 1919. This was around the same time that Michael Polanyi was baptized into the Roman Catholic Church, in October 1919, just before he left Budapest for Karlsruhe.[61] Teller, whose family did not convert to Christianity, left Budapest for Karlsruhe in 1926 at the age of eighteen and then moved to Munich to study physics with Arnold Sommerfeld. Teller's experiences of prejudice at the gymnasium in the early 1920s included memories of his mathematics teacher's sneering remark: "So you are a genius Teller? Well, I don't like geniuses."[62]

As discussed in detail in the next chapter, life in Weimar Germany was one of economic and political instability that was especially hard in the early 1920s and the early 1930s. Polanyi and other Hungarian refugees settled there and built outstanding scientific careers. Among Polanyi's clos-est friends in Berlin were Szilárd, Wigner, and Neumann. After leaving Bu-dapest in December 1919, Szilárd studied physics and engineering first at the Technische Hochschule in Berlin and then at the University of Berlin, where he worked with Max von Laue, Max Planck, and Albert Einstein. Szi-lárd completed his doctoral dissertation under Laue on statistical fluctua-tions and thermodynamics in 1922 and his *Habilitationsschrift* in 1925 on the relationship between entropy and information. From 1922 to 1924, Szilárd worked at the Kaiser Wilhelm Institute for Fiber Chemistry in the Berlin suburb of Dahlem, where Polanyi acquired his first research posi-tion in 1920 after spending six months in Karlsrule. By the late 1920s, Szi-lárd was a *Privatdozent* at the University of Berlin, where he offered some courses jointly with Neumann, Erwin Schrödinger, Harmut Kallmann, Fritz London, and Lise Meitner.[63]

Wigner arrived in 1921 to study at the Berlin Technische Hochschule. He contacted Polanyi with an introduction from their mutual Hungarian friend, the chemical engineer Paul Beer, and Wigner began working eigh-teen hours a week at the Fiber Chemistry Institute in 1923. One evening Polanyi invited Wigner to join him and the Viennese chemist Hermann Mark at Polanyi's residence, where Wigner was awed by the breadth of scientific discussion.[64] Mark was Jewish on his Hungarian father's side, al-though his family had converted to Lutheranism.[65] Wigner worked with Mark on the theory of crystal symmetries and he collaborated with Polanyi on rates of chemical reactions, spending a few hours in the Dahlem labora-tory and many more hours doing calculations at home. After Wigner's fa-ther asked him to return to Budapest in 1925 to work in his father's leather factory, Polanyi visited the family while in Budapest and persuaded them to allow Wigner to accept a research post with Karl Weissenberg at the Uni-versity of Berlin to assist Weissenberg with calculations applying symmetry principles and the mathematics of group theory to atoms in X-ray crystal-lography. Wigner published an article jointly with Polanyi in 1925 on the formation and dissociation of two-atom molecules, using some of Niels Bohr's theory on quantized energy levels, and this work went into Wigner's doctoral thesis which he formally completed under Polanyi in 1928. They also coauthored a paper in 1928 on unimolecular reactions at high pres-sures, but Wigner was disappointed that not only did coworkers in Po-lanyi's laboratory not share Wigner's preoccupation with quantum theory but that Polanyi himself felt the theory was overly mathematical for him.[66]

In addition to his collaboration with Polanyi, Wigner worked with Richard Becker in theoretical physics at the University of Berlin and collaborated with Neumann, who was a Privatdozent at the university. Some of Wigner's papers of 1926 and 1927 laid foundations, as Hermann Weyl was doing in Göttingen, for the use of symmetry principles and group theory in quantum theory, and Wigner and Neumann coauthored an important series of five papers in this field in 1928 and 1929. When Princeton University recruited Neumann in 1930, he negotiated an arrangement for Wigner to share a post with him, and they began splitting the academic year between Princeton and Berlin.

While Szilárd's and Wigner's fathers were engineers and businessmen like Polanyi's father, Neumann's father was a banker, and "Johnny" was a child prodigy who could divide eight-digit numbers in his head when he was six years old. After his father's death, Neumann converted to the Roman Catholic Church in 1929. Wigner and Neumann, who was one year younger than Wigner, went to the same Lutheran high school in Budapest, minutes away from the Ring Road and Andrássy Boulevard, and they occasionally walked home together. At school they both were inspired by the legendary mathematics teacher Laszlo Rátz.[67] Neumann received a diploma in chemical engineering at the ETH in Zurich in 1926, the same year that he received the PhD in mathematics from the University of Budapest with a thesis on set theory. From 1926 to 1929, he taught as Privatdozent at the University of Berlin and moved to Hamburg one year before he received the appointment at Princeton.[68]

Szilárd, Wigner, and Neumann, along with Polanyi, were the core of the Hungarian expat community of physicists and chemists in Berlin from the early 1920s on. Another of their Hungarian friends was Imre Bródy, who was working in Göttingen and occasionally came to Berlin before deciding to return to Hungary in 1923 to work at Tungsram. He invented the krypton-filled fluorescent light, doing some of his work in collaboration with Polanyi, who remained a consultant at General Electric Lighting Tungsram after leaving Budapest.[69] Another Budapest refugee was Kürti, who had studied at Minta and left Budapest to study in Paris and traveled to Berlin, where he took his degree in low-temperature physics in 1931 with Franz Simon at the Technische Hochschule.[70] Yet another was Gábor, who finished a diploma in 1924 and a doctoral degree in 1927 in electrical engineering at the Berlin Technische Hochschule, after which he went to work for Siemens Halske where one of his first inventions was a high-pressure quartz mercury lamp. Gábor, like Szilárd, Wigner, Neumann, and Polanyi, was a regular attendee of the weekly afternoon physics colloquia run by Max von Laue at the University of Berlin.[71] The émigrés often met at the Romanische Café on the Kurfürstendamm not far from rooms rented by

Szilárd. Wigner and Szilárd could be seen nearby walking together through the green paths of the Tiergarten near Brandenburg Gate, just as they had strolled in earlier years under the trees of Budapest's City Park.[72]

By the mid-1930s, the émigré Hungarians were nearly all dispersed from Berlin in their second migration. The general story of the dismissal and exile of Jewish scientists under National Socialism is well known. Flight began in earnest after passage of the Law for the Restoration of the German Civil Service of April 7, 1933. Technically it applied only to German citizens, exempting Hungarian and Austrian citizens from its legal provisions. The law demanded dismissal of political enemies and non-Aryans who were employees of the state. Non-Aryans were defined as all persons with at least one Jewish grandparent, irrespective of religion. Employment forms required entries under the categories of both "religion" and "race." Exemptions from dismissal for non-Aryans were made for World War I Jewish frontline soldiers in the April 1933 law but rescinded two years later. The National Socialist regime promulgated a Numerus Clausus law, allowing admission of Jews "to universities and to the professions of attorney and physician" only in proportion to their numbers in the German population, that is, less than 1 percent.[73] For those who initially held out hope of moderation of these policies or the demise of the regime, their hope dimmed with the passage in 1935 of the Nuremberg laws, which deprived Jews of citizenship and prohibited marriage between Jews and other Germans. Despair followed in 1938 with the annexation of Austria in March and Kristallnacht, the "night of broken glass" in November, when Jewish shops, department stores, cemeteries, and synagogues were destroyed in Germany and Austria; firearms were confiscated from Jews; and some thirty thousand Jewish were arrested and placed in concentration camps in Dachau, Buchenwald, and Sachsenhausen.

A list put together in the autumn of 1934 by the Reich Ministry for Education and Science listed 614 university teachers who had been dismissed or forced to retired from higher education institutions. Of these, 80 were dismissed for ostensibly political reasons and 107 for "simplification of administration," while the majority (384) were dismissed as non-Aryans. The universities of Berlin, Frankfurt, and Breslau accounted for 40 percent of the total. By 1938, perhaps as many as 39 percent of all German higher educational faculty had been dismissed or forced to retire on racial or political grounds, with the percentage among economists and social scientists at around 47 percent. When non-university research scientists, such as those at the Kaiser Wilhelm Institutes, are taken into account in the figures, the number reaches 2,000 individuals.[74] These dismissals destroyed some of the most important research groups in physical sciences in Germany, such as the Göttingen institutes of physics and mathematics and the Kaiser Wilhelm Gesellschaft Institute of Physical Chemistry, as well as the division of social sciences and economics at the University of Frankfurt, an institution established shortly before the outbreak of World War I with funds contributed by Jewish citizens.[75]

The vast majority of émigrés, whether they left in the early 1930s or the mid-1930s, ended up in Great Britain or the United States. They included the Hungarian scientists Polanyi, Szilárd, Wigner, Teller, Kürti, and Gábor. Others went to Turkey, South America, and Palestine. Some scientific administrators saw great opportunity for their own institutions in hiring Hungarian, Austrian, German, and other émigrés, for example, Frederick Lindemann in his recruitment of Kürti, Kurt Mendelssohn, Heinz London and his brother Fritz London, Franz Simon, and the non-Jewish Erwin Schrödinger to Oxford.[76] The modernization program of Kemal Ataturk in Turkey provided attractive offers to some refugees. In Copenhagen, Niels Bohr worked closely with the Rockefeller Foundation and other agencies not only to select and find places for important émigré scientists but, as Finn Aaserud argues, to set specific scientific research agendas that speeded up the development of molecular biology and nuclear physics.[77]

In the social sciences, the Rockefeller Foundation aided the New School for Social Research in New York City in hiring refugee social scientists in a graduate school established in 1933 and called the University in Exile. Its dean was Emil Lederer, a refugee economist from Berlin.[78] Help in placing refugee scientists and other professionals came not just by word of mouth but through organizations that were set up fairly quickly to document displaced persons and help coordinate at least temporary employment for them. The Hungarian-born Philip Schwarz, who had held a professorship in pathology and anatomy at Frankfurt before fleeing to Zurich in March of 1933, founded the Notgemeinschaft Deutscher Wissenschaftler im Ausland (Emergency Society of German Scholars Abroad). With assistance from the Rockefeller Foundation, he published a list in 1936 of 1,500 displaced scholars with assistance from the Rockefeller Foundation. Scientists in Great Britain established the Academic Assistance Council with Ernest Rutherford as president and Sir William Beveridge as one of the secretaries. Beveridge, who was director of the London School of Economics, had been on holiday in Vienna in March 1933 when he happened to see an evening paper with a list of names of professors already dismissed in Germany, some of whom he knew. Szilárd, who had just fled Berlin for Vienna, talked with Beveridge at that time about creating a university for refugees in Switzerland, but Beveridge countered with a plan for a British organization which was set up in the Royal Society premises at Burlington House. Beveridge invited Szilárd to serve as an advisor and contact person, and Szilárd worked with the Council's coordinator Tess Simpson until he left for the United States in 1937.[79] By the end of the Second World War and relying mostly on private funding, the renamed Society for the Protection of Science and Learning registered 2,541 refugee scholars, most of them Germans and Austrians.

In the United States, the Emergency Committee in Aid of Displaced German (later, Foreign) Scholars was founded with private funding. Its origin lay partly in the early efforts of the biologist Leslie Dunn and his colleagues at Columbia University to set up a fund at Columbia to aid displaced scholars. Stephen Duggan, director of the Institute of International Education, became director and Edward R. Murrow assistant director, before he began his famous broadcasting career. The committee ultimately received some 6,000 applications for assistance, of which only a small proportion, 335 in number, were granted.[80] Wigner and Rudolf Ladenburg, a physicist who left Berlin for the University of Minnesota in 1932, sent a letter in December 1933 to 27 mostly German émigré physicists at American institutions, asking them to contribute funds and aid for placing 28 scientists whose names were attached to the letter. Of those scientist refugees who left countries controlled by Hitler's regime during the 1930s, 28 were Nobel laureates between 1915 and 1971, among them the German scientists Richard Willstätter, Fritz Haber, Albert Einstein, James Franck, Max Born, and Hans Bethe, the Austrian physicist Schrödinger, and the Hungarian scientists Hevesy, Gábor, and György (Georg) Békésy.[81]

Great Britain and the United States became countries in which Michael Polanyi and his family sought refuge from Germany and Austria. Cecile had continued to make her primary home in Budapest while making long visits to see her children. She died in Budapest in 1939. Adolf left his Hungarian ex-wife and four children in Italy and arrived in England with his second wife, who was Viennese, before embarking in June 1940 for Brazil. All his children survived the war, living in Italy, Hungary, the United States, and elsewhere. Karl left Vienna for England in 1933 and, after teaching for fifteen years in adult education in England, accepted a position at Columbia University in 1947 but made his permanent home in Toronto. Laura, her now-estranged husband Sándor, and their three children were each able to reach the United States during the period from 1938 to 1941. She settled in a small apartment on New York's Upper West Side and later became engaged in a historical project with Bennington College historian Bradford Smith that led her to original work on the subject of Captain John Smith (of Pocahontas fame) and his early-seventeenth-century travels in Transylvania and Hungary. Neither Laura nor her brothers could get Sophie and her husband out of Vienna in time. Egon was taken in March 1938 to Dachau, where he died in April 1941, and Sophie and her son died in the Warsaw Ghetto sometime after July 1942, where they had been sent from Vienna. Her daughter Edit survived in London.[82]

## Émigrés as Suspects and Polanyi's Thwarted Move to America

The Second World War began with the German invasion of Poland on September 1, 1939, following the late August Soviet-German Non-Aggression Pact. Michael Polanyi had just become a naturalized British citizen in late August 1939. Following the German invasion of Denmark and Norway in April 1940, Michael and Magda Polanyi decided to send eleven-year-old John with a group of Manchester children to Canada, where he would live with the family of a Toronto physician. Their older son, George, entered Magdalen College at Oxford in the fall. By the autumn of 1942, George was in the Royal Air Force, where he became a captain in counterintelligence in Europe. John returned from Toronto in January 1944 and developed a bacterial inflammation in the bone of his right thigh. The release of penicillin for civilian use saved his life.[83] At the outbreak of the war, Michael immediately began calling upon some of his friends, such as the Manchester physicist Patrick Blackett, to help him get war work, but was not successful, he assumed, because he was an émigré who still had relatives living in central Europe.[84]

The outbreak of the war created new difficulties for émigrés and refugees in England, especially following the rapid advances of German forces in May 1940 and rising anxieties about an invasion of Great Britain. As noted by Mitchell Ash and Alfons Söllner, many émigrés were judged by the British to be insufficiently suited to English life and advised in the 1930s to try their chances in the United States.[85] On the other hand, coordinators who were trying to place refugees in American universities were fearful of a backlash of anti-Semitism if young Jewish émigrés were perceived to be competing with American-born scholars for jobs or to be exceeding informal numerus clausus practices in American academic institutions. A prime example of the most informal of these practices was Robert Millikan's decision not to hire a second Jewish physicist at Caltech because he thought it best not to have "more than about one Jew anyway."[86]

In 1940, incidents of popular anti-Semitism increased in England with fears that some émigrés might be spies. Internment procedures were instituted for nonnaturalized émigrés, with sites at Liverpool and the Isle of Man, as well as in Canada, and many émigrés were detained for months as they underwent evaluation before their release. Suspicions existed in university towns and elsewhere. At Oxford, a registrar wrote to the Home Office in May 1940, "We are anxious about the aliens. Because we are a University town we have probably a much bigger percentage of them to the rest of the population than most other towns . . . We are taking certain steps ourselves to protect them, but they [our steps] cannot, in the nature of things, be half as effective as clearing out any suspicious characters there may be."[87]

Perhaps because of the anxiety about invasion and his inability to contribute to the war effort in Britain, and with his younger son scheduled to move to Canada, Polanyi cabled his friend Hugh S. Taylor in Princeton in late May 1940. Taylor was a prominent physical chemist and a friend and colleague of Wigner. Taylor immediately telephoned Warren Weaver, the natural sciences director at the Rockefeller Foundation, that Polanyi would like to come to the United States on however modest an appointment. Taylor assumed that Polanyi had not yet completed the process of British naturalization and was in danger of internment. Weaver, who had known Polanyi for some time, immediately promised that the Rockefeller Foundation would provide support of $2,000 a year for two years at Princeton University.

Unfortunately for Polanyi, Princeton in May 1940 was a town not unlike Oxford, even though there was no danger of invasion and the United States was not at war. Taylor wrote Weaver four days after their telephone conversation that because of the stress of the international situation and "because of the excitement that the presence of aliens of doubtful sympathies is actually creating in a small village such as Princeton at the present time, the University authorities have reached an informal opinion that it is undesirable at the present moment to do anything which might further complicate the situation in respect to such aliens at the present time . . . Specifically, therefore, the University authorities are unwilling to commit themselves to a program with respect to Prof. M. Polanyi such as we discussed by telephone on Friday afternoon."[88] Ironically, six years later Polanyi would receive an honorary degree at Princeton University, with a

Figure 1.2. Eugene Wigner and Leo Szilard in the 1930s.
(From E. P. Wigner, "On the Future of Physics," *Fizikai Szemle* 5
(1988): 174. (Courtesy of *Fizikai Szemle* and Gábor Palló.)

citation that described him as "a physical chemist who has devised new tools to determine how fast atoms react; a veteran campaigner against those who would take from science the freedom she requires for the pursuit of the truth."[89]

As is well known and well documented in the history of the Manhattan Project and the development of the atomic bomb in North America, many central European refugee scientists contributed in fundamental and essential ways to the Allied war effort, among them Polanyi's fellow Hungarian émigrés Wigner, Neumann, Szilárd, and Teller. The first announcement of uranium fission was at Princeton University's Monday Physics Department Journal Club on January 16, 1939. It came from Niels Bohr's colleague Léon Rosenfeld, who had arrived with Bohr that very day in New York City from Copenhagen. Bohr had been told by Austrian refugee Otto Frisch just before embarking for the United States of calculations that Frisch had done with his refugee aunt Lise Meitner of the nuclear energy that might be released by the uranium fission confirmed only a few weeks previously by her former Berlin colleagues Otto Hahn and Fritz Strassmann. Bohr discussed these developments on January 26 at a conference at George Washington University. By July, Szilárd and Wigner together approached Einstein to ask him to write Queen Elizabeth of Belgium, who was his friend, to safeguard uranium supplies in the then Belgian Congo. Consulting with Teller and trying to think of ways for Einstein to reach the office of President Franklin Delano Roosevelt, Szilárd got a suggestion from the Berlin émigré economist Gustav Stolper, whom he had met through Michael Polanyi's economics discussion group in Berlin in 1929. On Stolper's recommendation, Szilárd contacted the economist and New Deal advisor Alexander Sachs, who eventually delivered a letter signed by Einstein to President Roosevelt about the dangers of German development of an atomic bomb. The Hungarian émigrés were at the heart of the Manhattan Project from its very beginning.[90]

Fears of émigré central European scientists as potential spies persisted, however, in England and elsewhere, although some of the most effective spies turned out to be native-born English sons, such as the so-called Cambridge Four: Kim Philby, Donald Maclean, Guy Burgess, and Anthony Blunt. More disturbing for the scientific community was the arrest of British-born physicist Alan Nunn May, a collaborator of James Chadwick at Cambridge, who had worked with the British atomic energy team in Montreal and spent some time at the Manhattan Project Metallurgical Laboratory in Chicago. May, who had been a member of the British Communist Party, was found guilty in 1946 for passing information to the Russians.

Klaus Fuchs was an émigré spy, on the other hand. He was the son of a Lutheran pastor and member of the German Communist Party who fled Germany in 1933. Fuchs took a PhD with Nevil Mott in 1937 and then studied with the refugee Max Born in Edinburgh except for a brief period of internment. While working with Born in the summer of 1941, Fuchs was invited by the German refugee Rudolf Peierls to work with him in Birmingham on calculations for a uranium bomb. In March 1940, Peierls and Frisch had calculated that the amount of Uranium-235 needed for a bomb was considerably smaller than previously thought. Fuchs, who had accompanied Peierls to Columbia University to work on the Manhattan Project and collaborated at Los Alamos with dozens of other refugees physicists, was arrested as a spy at Harwell, the British nuclear energy facility, in 1950, where he was head of the theoretical division. He was passing on information to the Russians during the war, however, and not to the Germans. That same year the arrests in the United States of Harry Glass, David Greenglass, and Ethel and Julius Rosenberg unfolded, followed by the trial and execution of the Rosenbergs in 1951.[91]

The hunt for spies was a pervasive aspect of the anti-Communism that pervaded much of American and British thinking in debates at the United Nations about control of nuclear energy and in reaction to postwar Russian occupation and establishment of communist regimes in central and Eastern European countries after the war. Ironically, Michael Polanyi again was to find himself an outsider who was excluded from professional opportunity, this time on the grounds of suspicions about his political attitudes in the widening anti-Communism measures that followed President Harry Truman's executive order of 1947. This order prohibited federal employees from belonging to or sympathizing with groups considered by the attorney general to be "fascist, totalitarian, communist, or subversive."[92] By 1950, loyalty oaths were implemented in many state institutions and agencies, prominent among them the University of California system, where every employee from janitorial staff to graduate student teaching assistants and tenured faculty members had to swear not to be a member of the Communist Party. In September 1950, Congress passed the Internal Security Act, or, "McCarran Act," over the veto of President Truman. The act created a Subversive Activities Control Board to identify and register Communist and Communist-front organizations, prohibited passports for American citizens who were associated with these organizations, and authorized the State Department to deny visas to foreign nationals on similar grounds or, more vaguely, for past activities "prejudicial to the public interest." In June 1952 a new Immigration Act required applicants for U.S. visas to list all past and current memberships in organizations and societies.[93]

By 1958, under the watchful surveillance of the head of its passport division, Ruth B. Shipley, the State Department had refused passports on political grounds to approximately six hundred persons and delayed hundreds of others in permissions to travel abroad. In only a one-year period following May 1951, Shipley's office prevented three hundred Americans from traveling abroad, sometimes on the flimsiest of excuses. By late 1952, a Federation of American Scientists committee estimated that at least half of all foreign scientists seeking to enter the United States experienced some sort of difficulty in obtaining a visa. The rationale was as strongly based in punishing critics of American foreign policy as it was in preventing communist subversion.[94]

These measures for repression of political dissidence and political enemies in the United States affected Michael Polanyi in the early 1950s just as similar political limitations had in earlier decades in Hungary and Germany. In the spring of 1950, Polanyi gave a series of lectures at the University of Chicago called the "Logic of Liberty" and a speech titled "Freedom in Science." He saw old friends on this visit, including the sociologist Edward Shils and the British Austrian émigré economist Friedrich von Hayek. Early in 1951, Polanyi received the University of Chicago's offer of a chair in social philosophy to begin in the fall term of 1951, and he decided to accept it. Polanyi resigned from the professorship in social studies, which he had held at Manchester since March 1948.[95]

Things did not go well with Mr. M. B. Lundgren at the American consulate office in Liverpool, where Polanyi applied in January 1951 for a non-quota immigrant visa to the United States. After Polanyi still had heard nothing by August, even though he was slated to take up his new position in October, he talked with Shils in London. In September, Shils wrote the social sciences director at the Rockefeller Foundation, Joseph Willits, that Polanyi was coming into apparent difficulty because he was a visa applicant whose national origins were in a Communist, Iron Curtain country despite Polanyi's well-known opposition to Marxism and Soviet Communism. Shils also wrote the University of Chicago's dean of students, Robert Strozier, about the absurdity of the situation, attributing to Polanyi the waning in Great Britain of Marxist and Socialist arguments for the central planning of scientific research on a Soviet model.[96]

Shils hoped that the Rockefeller Foundation would intervene with the State Department, but this kind of action was contrary to the foundation's policies and it was especially inopportune at a moment, when Representative E. E. Cox of Georgia had singled out the Rockefeller Foundation in a speech in Congress on August 1, 1951, as an example of American philanthropic foundations that Cox alleged were promoting activities in conflict with national security. Within a year, the foundation finished a document that showed that of over fourteen thousand grants made since 1935, only three organizations and twenty-three persons had been explicitly criticized or listed by congressional committees investigating subversion including the House Un-American Activities Committee.[97] Lawrence A. Klimpton, president of the University of Chicago, contacted the State Department and other officials. His letter in early January 1952 to Shipley included a letter signed by three prominent University of Chicago faculty members—Samuel Allison, John Nef, and Cyril Stanley Smith, protesting the adverse effects of the McCarran Act on the University. This letter also was sent to some U.S. senators and to Secretary of State Dean Acheson.[98]

There were hopes at Chicago that Polanyi could delay taking up his official teaching duties until the spring quarter of 1952, while technically beginning his appointment in October so that he could receive a salary. This arrangement proved unfeasible, however, for both parties. Polanyi resigned from Chicago and, again, as in 1933, Manchester rehired him. In January 1952, Lundgren interviewed Polanyi in Liverpool, questioning him about two specific organizations with which the State Department was associating Polanyi's name. In early March 1952, Polanyi wrote a letter to the editor of the *Manchester Guardian* reporting his visa experience. The *Guardian* titled Polanyi's letter "American Political Tests" and followed up with an March 8,

1952, article titled "A New Inquisition." In a special October 1952 issue of the American journal *Bulletin of the Atomic Scientists*, Polanyi was one of the prominent cases reported as examples of the atrocities of the McCarran Act. Other horror stories about their experiences with the State Department came from Linus Pauling, Rudolf Peierls, Mark Oliphant, Jacques Monod, and Jacques Hadamard. The issue of the *Bulletin* was prefaced with an editorial written by Shils.[99]

Rumors had reached Polanyi over the summer that the problem might be his association with student societies in Budapest before the First World War. In fact, it turned out that he was being barred from settling in the United States on the grounds that his name had been listed as a sponsor or patron in publications of the Institute for Free German Culture in 1942 and the Society for Cultural Relations with the USSR from 1946 to 1947. In the first case, he had given an invited lecture in London on December 12, 1942, to an audience for the German School of Higher Learning and Institute of Free German Culture. It included his name in a list of patrons for the year. After his lecture titled "The Self-Government of Science," in which he criticized the imposition of Soviet ideology upon Soviet genetics, Polanyi received a letter from the institute's council criticizing his views at a time when the Soviet Union was allied with Great Britain in the war against Nazi Germany. Polanyi quickly concluded that the organization must be under Communist control. As for the Society for Cultural Relations, it sponsored the translation of Russian documents into English, many of which were valuable to Polanyi in his own work and to other scientists, scholars, and writers. The list of sponsors in England included well-respected writers such as Somerset Maugham, G. M. Trevelyan, and Graham Greene. Polanyi had withdrawn his membership in 1947 when some of the organization's statements were too pro-Soviet for him.[100]

By January 1953, the Liverpool office informed Polanyi that he might receive a visa under the revised McCarran Act, and he began to make plans for staying in Chicago in the winter of 1954 as a visiting professor.[101] He was incensed, however, as he reported in the *Bulletin for Atomic Scientists*, that his Liverpool interrogators also questioned him on the political opinions of his brother Karl, who was teaching at Columbia University. Karl had to make his permanent home in Canada, because the State Department would not give a visa to his wife Ilona Duczynska, who was a former member of the Hungarian Communist Party. At almost the same time, in 1951, Polanyi's good friend at Manchester, the physicist Patrick Blackett, was questioned by American authorities about his political views.[102] Polanyi's emigration from England to the United States had been prevented definitively, however. The economist Sir Henry Clay, an enthusiastic admirer of Polanyi's economic and political writings, wrote Willits that:

> Secretly . . . I was delighted . . . a complete transfer to America would not, I think, either improve his work or give him final satisfaction. His peculiar value is that his analytical mind has been reinforced by actual experience of life in Central Europe, in pre-war Germany (i.e. Western Europe) and in England. He ought not to lose contact with this experience. If he settled permanently in Chicago I should be afraid that either he would be continually irritated by the rather arid atmosphere of rationalist liberalism he would find himself in, or captured by it.[103]

## Jewish Identity and Internationalism in Science and Politics

Polanyi's experiences in central Europe and England not only structured the daily aspects of his scientific life and led him to grapple with economic and political questions of the time, but also forced upon him the necessity to reflect on his Jewish ancestry. Polanyi never practiced the Judaic religion. In his twenties he was attracted to what he called the Tolstoyan credo that denied the divinity of Jesus Christ, rejected belief in the reality of heaven and hell, and embraced the teachings of the Sermon on the Mount.[104] For practical reasons he obtained a baptismal certificate in the Roman Catholic Church upon leaving Hungary in 1919. His wife Magda was from a Roman Catholic background. Their marriage was civil ceremony, and she later told Polanyi's biographer, William T. Scott, that she did not know how long her husband remained a Catholic, because his interest in Christianity lay along the lines of Protestantism.[105] Once in England, their son George enrolled in a Quaker boarding school in York and their son John attended Manchester Grammar School. In Polanyi's later years, he adopted the habit of traveling with the Anglican Book of Common Prayer, and he occasionally went to services at Anglican, Presbyterian, and Methodist churches.[106]

Polanyi abhorred the Judaism of the ghetto, and he opposed Zionism insofar as it aimed to establish a nationalist ghetto in Palestine. When he was a child in Budapest, Jewish congregations were split along two main lines: the *neológ*, in the sense of Reform Judaism outside Hungary, and Orthodox Judaism, which included the truly conservative Hasidim. Nonobservant Jews, like Polanyi and his family, generally belonged to a higher economic stratum than the neológ and the Orthodox Jews, and in Budapest the Magyarized and nonobservant Jews comprised some 35 percent of gymnasium students.[107] Polanyi's Jewish association in Hungary was one of ancestry or what Vienna's chief rabbi of the late nineteenth century, Adolf Jellinek, articulated as *Stammesgenossen* or members of the tribe, an identification that Polanyi articulated in a speech in 1942 to a Manchester organization of British Jews, saying that "The Jews are not a nation, nor are they just Englishmen, Scotsmen, Frenchmen, etc., who happen to be of the Judaic religion. They are the descendants of a religious tribe."[108]

When he arrived in Manchester in the late summer of 1933, the influence of two leaders of British Zionism was strong in the Midlands city. Chaim Weizmann, who became the leader of the World Zionist Organization in 1920, was professor of chemistry at Manchester from 1904 until 1919. The Balfour Declaration of 1917, favoring the establishment in Palestine of a national home for the Jewish people, partly owed its origin to Weizmann's discussions in the early 1900s with Arthur Balfour, then a conservative MP from Manchester. One of Polanyi's faculty colleagues, Lewis Namier, who had become Manchester's professor of modern history in 1931, was spending most of each week in London working on behalf of the Jewish Agency. Namier's Russian Polish family had converted to Catholicism, and Namier had been educated as a Polish Catholic. It was a family "of Jewish descent, with strong Christian sympathies and Polish enthusiasms" before Namier's emigration to England in 1906. He and Polanyi became acquainted in 1933 and sometimes worked together on the resettlement of Jewish refugees.[109] Several Anglo-Jewish organizations created the Central British Fund for German Jewry, later the Council for Germany Jewry, which raised many thousands of pounds sterling to help refugees in Germany and Palestine. A Jewish Refugee Committee, later the German Jewish Aid Committee, raised money to cover admission monies for Jews to the UK, as well as maintenance, training and employment.[110]

In May 1934, Polanyi wrote Namier after reading Namier's introduction to Arthur Ruppin's *The Jews of the Modern World* and deciding he wanted to make clear his objections to Namier's call for a reintegration of assimilated Jews into Jewish nationalism. A handwritten draft of Polanyi's letter reads: "I am fundamentally opposed to your view . . . I will fight against a revival of the Ghetto spirit . . . New Palestine might yet make Jewish history in the future. God help it." It was the Jews, wrote Polanyi, who rejected Judaism, including the Jews who founded Christianity, who made history by daring to escape through the "half open door," even though treated with contempt by those left behind, and who attached their Jewish names to some of the noblest deeds of the human spirit.[111] Polanyi was not alone among Jews in opposing Zionism. Many Orthodox Jews regarded a return to Palestine before a return of the Messiah as religious heresy, and Reform and nonobservant Jews usually regarded it as ill-advised. The Board of Deputies of British Jews believed that Zionism endangered rights won by Jews and interfered with their loyalties as British subjects.[112]

In 1935 Polanyi received an invitation to join the Jewish Medical Society, He declined to join but accepted the group's invitation to speak in Liverpool in January 1936, preparing a lecture titled "On the Position of Jews." In his speech, Polanyi talked personally about some of his experiences as an assimilated Jew, struck, he said, by the delicacy with which the English try to avoid the very word "Jew" in the presence of a Jew, unlike their behavior with an Irishman or a Scotsman. Nonetheless, assimilation remains the best path, he counseled, not the exile and nationalism of Zionism. Establishing a Jewish state, concluded Polanyi, might give Jews a sense of a homeland, but it would not prevent a sense of alienation (*Galuth* or *galut*) among nations. Internationalism is the hope for the future, not the militant internationalism adopted by Jews who have turned to communism or socialism, but instead an international union without nation-states that could be established in what "will be left of Europe after Hitler has finished with it." Jews would do well to follow the example of Quakers who are detached from political life and express their faith in the works of Christianity.[113]

After the start of the war, Polanyi watched with alarm some manifestations of public anti-Semitism, especially insinuations that Jews were making great deals of money in the black market and shirking military service.[114] Toward the end of June 1942, he wrote a letter to the editor of *The New Statesman and Nation*, which appeared under the title "Anti-Semitism." Polanyi had harsh and damning words for Jews who might be guilty of black-market crimes, and he identified himself as a Jew criticizing such Jews.

> What matters to us Jews is to get rid of these swindlers—and quickly and radically at that . . . the law of the land has to be supplemented in this case by stern action by the Jewish community . . . As disloyal citizens of a great nation at war, these people are despicable; as Jews in a war against Hitler they are beneath contempt . . . I suggest the setting up of a Jewish Vigilance Council . . . to collect and publish records of Jews who have been convicted in the Courts. It should organize a life-long boycott of these persons as businessmen organizing local opinion. It will be a hard fight but a good fight.[115]

On September 24, Polanyi gave a speech to a Manchester trade group affiliated with the Board of Deputies of British Jews. The talk was published in the *Political Quarterly* in the spring of 1943. The gassing of Jews in Auschwitz had begun in June 1942, and the first deportations from the Warsaw Ghetto to concentration camps took place in July 1942. The Board of Deputies soon abandoned its anti-Zionism.

In his remarks, Polanyi expressed his horror of the ghetto life that many post-Ausgleich Jews imposed on themselves "speaking Yiddish, their men with forelocks, their married women with wigs over their shaven skulls. They recall the original mass of medieval degradation and mental narrowness to which most of our ancestors belonged about 150 years back."[116] The Jews lived for five hundred years in isolation because of religious persecution, he said, but they were mistaken to maintain their religion against their persecutors' Christian religion by adopting a newly defined ritual code of unsurpassed strictness, locking themselves from within even though no longer locked from without. Once nation-states began to develop in Europe, the Jews had been unable to form a nation because of their dispersal throughout Europe and, as a consequence, they could not achieve "westernization" and modernization by a national renaissance like Christians.

Modernization had to come from leaving the ghetto, Polanyi argued. It was accelerated by baptism and abandonment of distinctive Jewish traditions in the achievements of great Jewish figures such as Heinrich Heine, Felix Mendelssohn-Bartholdy, David Ricardo, Karl Marx, and Benjamin Disraeli. In the last decade, as opposition to the democratizing and liberalizing trends of the recent past acquired force and all Jews, baptized and unbaptized alike, were condemned by Hitler, it is not surprising that Zionism acquired new conviction. Still, Polanyi said, as he had done in the past, Jews have to be careful in moving toward Zionism and not abandon preparation for a future that will overcome nationalism. Faithful to views that he had begun expressing in his youth in Budapest, Polanyi concluded his talk, perhaps to the surprise of his audience, with a utopian vision of a post–World War II "Western Commonwealth" in which "all people of the West will have to undergo some assimilation towards a more uniform type of man." Its basis will be the "rule of law, equal citizenship and a religion rather similar to early Christianity with its admixture of Greek philosophy."[117]

Polanyi's relationship to Judaism was complex, not unlike many of his fellow scientists of Jewish ancestry. He confessed to Namier the feeling that he thought he shared with many assimilated Jews of the "shameful experience of being glad not to be taken for a Jew."[118] Szilárd later reminisced that when he converted to Calvinism, his mother counseled him that "it wouldn't be right to deny my origins. There was nothing to hide or be ashamed of."[119] Teller recalled that after he moved to Washington in 1935, an official of the Rockefeller Foundation gave him the advice, "You are Jewish. One terrible thing about Jews is that they have only Jewish friends. Don't do that."[120]

Among German Jews, James Franck was one who never converted: he viewed himself as a German of Jewish descent without membership in Jewish religious or social organizations. When he was decorated for bravery and selected to receive an officer's commission during World War I, one of the German officers asked Franck to do him the favor of getting baptized. "I smiled and asked whether he believed that I would be a better officer if against my own conviction I would be baptized to do him a favor. And he couldn't help to say he doesn't think so, but anyway it would show that I do belong to them. I said that I feel that I belong here, whether I am an officer or not. I have not asked for it. If they want to make me an officer, it is all right with me."[121]

Richard Willstätter never converted, he said, due to his rejection of the personal advantage that conversion would bring him. Otherwise, his statement of his views about Judaism and Christianity is remarkably similar to Polanyi's position:

> To me Judaism was history, an acknowledgment of my ancestry. It was not obedience to rules made thousands of years ago, therefore not fidelity to the Jewish law—not true, strict Judaism, certainly not a matter of faith in any sense. It seemed to me the weakness of obstinacy to cling, as if this were the essential point, to the old religious customs which in part have forfeited their validity in the cultural changes over the millennia, a weakness dangerous to the continued existence of Judaism in highly civilized countries, more dangerous than a liberal evolution. The retention and further development of the Hebrew language, too, seemed retrogressive to me for Central and Western Europeans, if perhaps not for the Eastern Jew.
>
> My faith has been that of Albert Einstein, who replied to a telegraphed inquiry from the Rabbi of Boston approximately as follows: "I believe in the God of Spinoza who reveals himself in the harmony and beauty of Nature, but not in a personal God who concerns himself with the fate of the individual . . ." I admire the ethic of the New Testament as humanity's greatest advance. The spiritual quality of Christianity, redemption through love, should encompass the common salvation of all, not the salvation of one church among many . . . if we [consider the history of Christianity] it becomes harder and harder to revere the thing into which the officials of the Church, "the guardians of God's secrets," have remodeled the tradition of

Jesus Christ . . . According to my conviction, faith and creed are private matters, no concern of the state, and never to be twisted to the ends of personal advantage. Conversion to Christianity was always out of the question for me because it entailed significant advantages, while remaining a Jew carried with it only civil disadvantages.[122]

The historian Fritz Stern wrote of German Jews that "the need to excel was instilled by tradition and nurtured by hostility" with hidden wounds inspiring visible achievement.[123] Of the doubly difficult situation of Hungarian Jews in Germany, Neumann said of his own achievement and that of his friends: "it was a coincidence of some cultural factors . . . an external pressure on the whole society of this part of Central Europe, a feeling of extreme insecurity in the individuals, and the necessity to produce the unusual or else face extinction."[124] Wigner reflected on his forced emigration: "Emigration can certainly be painful, but a young man with talent finds it stimulating. Outside your own nation, you lack a ready place. You need great ingenuity and effort just to find a niche."[125]

Michael Polanyi and his generation of refugee scientists dealt with many varieties of exclusion and alienation: They often were perceived as Jews, whether they were observant, nonobservant, or baptized into Christianity. They often were viewed as communists, whether they were Communists, Socialists, Liberals, or anti-Communists. They were labeled as Hungarians whether they lived in Imperial Austro-Hungary, Germany, Great Britain, the United States, or elsewhere. They were émigrés and aliens who were assimilated to new cultures in a post-nineteenth-century Europe marked by the horror of the Great War, the post-Versailles dismemberment and reconfiguration of Europe, the economic miseries of the early 1920s and the Great Depression, the shock of the Bolshevik Revolution, and the rise and fall of Communist revolutions in Germany and Hungary during 1918–1919. Scientific life in central Europe in the 1920s and 1930s was surrounded on the edges by waves of political violence and civil unrest that climaxed in the Nazi regime and the catastrophes of anti-Semitism and another World War.

While Jews often were labeled pejoratively by anti-Semites as "cosmopolitan" and "international," the cosmopolitan and international structure of the scientific community was the everyday experience of Polanyi and his fellow scientists both Jewish and non-Jewish. Jewish scientists such as Einstein were among those who publicly advocated a world government and international controls by the United Nations, but so were non-Jewish scientists and intellectuals, such as H. G. Wells. Scientists and nonscientists alike avidly read the fiction and nonfiction writings of Wells from the late 1890s to the 1940s. His *The World Set Free* (1914) not only predicted a new era of explosive weapons powered by the atom's energy, but warned that the only guarantee of world peace would be a benevolent and responsible world organization of government. Polanyi read the book shortly after it appeared and was a Wells enthusiast in his youth.[126]

The application of the organizational structure and governing values of the scientific community to political communities was a leading motif in Polanyi's political writings, which he began in earnest in the late 1930s. There is no doubt that the political experiences of his youth in Budapest, coupled with his émigré experiences and his reflections on the experiences of his fellow refugee scientists, especially his old friends from Budapest, led Polanyi to think deeply about the structure of scientific life as he had known it and the social conditions and behavioral values that make possible successful scientific careers and the advance of scientific knowledge. Had he never emigrated from one scientific center to another and then to another, in very different national and local cultural settings, he might never have posed the kinds of questions that he began to ask about the nature of scientific understanding.

During the middle years of Michael Polanyi's adulthood, as we will see in the next three chapters, his scientific work was his liberation and his preoccupation. While on vacation with his friend Hugh O'Neill and O'Neill's wife on the northwest coast of Wales in late June 1934, Polanyi read in the newspapers of Hitler's murder of more than a hundred of his possible opponents in the "Night of the Long Knives." Polanyi told O'Neill he could no longer remain with his hosts because he could not put aside his "chronic cares about a rotten world . . . the lab does me better than the sunny seaside."[127] In the next chapter we turn in detail to Polanyi's scientific life in Germany—his first and most satisfying experience as an émigré scientist in a new scientific culture—beginning with his early studies in Karlsruhe in 1912 and ending in Berlin in 1933. In chapters 3 and 4 we look at explicit details of the different kinds of experimental and theoretical scientific work that he accomplished in Germany and England, new difficulties occasioned for his research by the move from Berlin to Manchester, and some ways in which these professional experiences found their way into his later philosophical reflections.

# TWO

# Germany and Weimar Berlin
# as the City of Science

In 1920, Michael Polanyi launched what turned out to be a thirteen-year period of his career within the precincts of the Kaiser Wilhelm Gesellschaft (KWG) scientific institutes of the Berlin suburb of Dahlem. In the often tumultuous milieu of Weimar Berlin, he carved out a scientific life in what he and many scientists of his generation considered the scientific capital of the world. It was the city of Albert Einstein, Max Planck, Fritz Haber, Walther Nernst, and Lise Meitner, and it was the seat of the scientific institutions of the University of Berlin, the Berlin Technische Hochschule (Technical University), the Berlin Academy of Sciences, the Physikalisch Technische Reichsanstalt (PTR), and the new set of KWG research institutes that Kaiser Wilhelm II inaugurated before the outbreak of the Great War.

Arriving from his studies in Budapest and Karlruhe, Polanyi found himself in a milieu of laboratories, colleagues and coworkers, research seminars, and social groups which he came to think of as unrivaled anywhere on earth. The KWG was a new kind of scientific organization, devoted to research but providing opportunities for working with students at the university and at the Technische Hochschule. Funding for research at the KWG was precarious during the economic hardships of the early 1920s and the later Great Depression, but the KWG's structure of financial support was perhaps unique and certainly pioneering in the early 1920s. Its funds were a network of financial support from the German national state (Reich), the Prussian state (Land), German industry, philanthropy, and private foundations. With its multiple sources of support, the KWG maintained a professed mission devoted to academic values and fundamental science. Working within this institution, Polanyi was sheltered to some extent from pressures to make his own research conform either to pedagogical imperatives of university courses and examinations or to demands from employers or the state for immediately useful results.

Faced by the early 1930s with increasing violence, social unrest, and the broadening power of the National Socialist Party, Polanyi hesitated to leave his ideal city of science, even in the face of an unusually attractive offer from the University of Manchester. He resigned his position in Berlin only when faced with demands from the National Socialist government that non-Aryan scientists must be fired from their posts. Polanyi's daily career experiences in Berlin were a crucial foundation for his later writings on the nature of science and its everyday practice. So, too, was his immersion in a German academic tradition and rhetoric of pure and transcendent *Wissenschaft* that defied the realities of urban Berlin in the 1920s and 1930s. The ideals and realities of Weimar scientific culture and their impact on the economic, political, and philosophical writings of Polanyi are the subject of this chapter.

## First Steps in a German Scientific Life: Polanyi and Karlsruhe

Polanyi became interested in physical chemistry as a consequence of studying adsorption on colloidal gels in the physiologist Ferenc Tangl's laboratory while a medical student in Budapest. Polanyi learned that one of the most important centers for physical chemistry in Europe was the Institute for Electrochemistry and Physical Chemistry at Karlsruhe's Technische Hochschule. Fritz Haber, who had been a student at the Karlruhe Technische Hochschule in the 1890s, had become director of Karlsruhe's Physical Chemistry Institute in 1906. Just before Polanyi decided to matriculate at Karlsruhe in 1912, Haber accepted a position in Berlin to set up a similar institute at the newly founded Kaiser Wilhelm Gesellschaft.

Karlsruhe, a small city near the Rhine to the southwest of Frankfurt and the northeast of Strasbourg, has a distinguished place in the history of chemistry. The Technische Hochschule was founded as a polytechnic school in 1825 in emulation of the École Polytechnique in Paris, and it was the first technical institution of higher education in Germany. In 1860, Karlsruhe's professor of general chemistry, Carl Weltzien, hosted what became a famous chemistry conference which assembled 140 chemists for an international meeting to discuss standardization of chemical notation and chemical combining weights. The Italian chemist Stanislao Cannizzaro argued that the "atom" of the chemist and the atom of the physicist are the same substance in a debate that pitted his view against objections from August Kekulé. Dmitri Mendeleev and Lothar Meyer were among congress participants who were inspired by Cannizzaro, and each drew upon discussions at Karlsruhe to develop statements of the periodic table of the elements. In addition to Weltzien and Haber, faculty members at the Technische Hochschule during the decades of the 1860s to the 1910s included Lothar Meyer, Carl Engler, Hans Bunte, Hermann Staudinger, Max Le Blanc, Georg Bredig, and Kasimir Fajans.[1]

Polanyi arrived at Karlsruhe in April 1912 to enroll for the summer semester. One of his Budapest professors, Ignác Pfeifer, who was professor of chemical technology, made arrangements for Polanyi to have employment in Karlsruhe as a tutor in order to help with his expenses.[2] Polanyi attended courses during the summer semester in introductory physical chemistry and electrochemistry, contact (surface) chemistry, and catalysis, along with the first segment of a physical chemistry course. He developed a professional relationship with Georg Bredig, who succeeded Haber as director of the institute. Bredig had worked with Wilhelm Ostwald in Leipzig and had taught at Heidelberg. He was an expert on colloids and catalysis. Polanyi also got to know Bredig's assistant, Kasimir Fajans, who was four years older than Polanyi. Fajans was Polish, as was Bredig, and Fajans had taken his doctoral degree in 1909 with Bredig at Heidelberg. Fajans became a very prominent researcher in radioactivity, eventually working on the Manhattan Project. He was one of the founders of isotope theory along with Ernest Rutherford's early coworker Frederick Soddy, whose term "isotope" became the standard term for substances with differing atomic weights and the same chemical identity. Fajans learned radioactivity separation techniques in postdoctoral work with Rutherford at Manchester during 1910–1911, and he was present at the March 7, 1911, lecture where Rutherford announced his revolutionary conclusion that there is a small positively charged nucleus at the center of an atom.[3]

Polanyi had conversations with Bredig about thermodynamics, and they stayed in touch after Polanyi returned to Budapest to complete his medical degree. Polanyi came back to Karlsruhe in the fall of 1913 with his medical degree in hand and enrolled as a full-time student while continuing to work as a tutor. He had completed several papers on thermodynamics by the time the war broke out.[4]

To Polanyi's frustration he received an offer from Otto Warburg to join the physical chemistry group at the Kaiser Wilhelm Institute for Biology in late 1914, but he could not leave military service at the time. After passing the written examination for his doctorate in physical chemistry and learning that Professor Gustav Buchböck would accept as a dissertation a Hungarian version of his new and long article on adsorption on solids that appeared in the *Verhandlungen der deutschen physikalischen Gesellschft*, Polanyi wrote Bredig once again, this time to inquire about the possibility of doing a *habilitation* at Karlsruhe. The response was disappointing:

> We are compelled now, due to the War more than ever before, to take into account, as much as possible, the public opinion which urges us to fill in the available places for Dozenten by citizens of the Reich. Even though we like to treat the citizens of our Allies the same way as our own, you must have seen in my Institute that the situation was pushed so strongly in favor of them, that as of now, and more than ever before, I must attract more Imperial Germans.[5]

In August 1917, Polanyi received reserve status in the army, and he began working with László Berényi in the University of Budapest chemistry laboratory. Once his doctoral oral defense was finally scheduled for September 1918, he wrote Fajans, who had moved to Munich in a professorship, to inquire about doing a habilitation there. Polanyi outlined his current status and wrote that if religion were a concern, while he had formerly been a Jew, he presently had no church connection and he would join any Christian denomination that Fajans might suggest. Fajans was welcoming, but the military would not allow Polanyi to leave Hungary for Germany.[6] Just at the time that Polanyi officially received his PhD in July 1919, the political situation in Budapest became dire, and Polanyi was removed from his university post as an assistant to Georg von Hevesy. As Polanyi once more departed for Karlsruhe, in December 1919, he re-registered his family name change from Pollacsek to Polanyi and arranged through his Uncle Karl in Vienna for a copy of a certificate of baptism in the Roman Catholic Church. By the following spring, he had papers for Austrian citizenship. He also arranged to maintain his income as a consultant at Izzo, the United Incandescent Lamp and Electric Corporation's research laboratory in Budapest, which was directed by his former scientific mentor Ignác Pfeifer. His sister's husband, Sándor Striker, offered him financial help in Karlsruhe.[7]

Polanyi's friend Alfred Reis now had a position in physical chemistry at the Karlsruhe Technische Hochschule, and they would continue their relationship in Berlin when Reis received an appointment as professor of physical chemistry at the Berlin Technische Hochschule and became a research assistant (*Mitarbeiter*) at Haber's Institute. Imre Bródy, who had recently received his doctoral degree in chemistry in Budapest, was in the Hungarian group of twenty to thirty undergraduate, postgraduate and postdoctoral students who socialized together in Karlsruhe cafes. Bandi Kemeny and his sister Magda, whom Polanyi married in 1921, were among them. The Kemenys grew up in a Roman Catholic family of engineering and business in which their father managed finances for steam-operated grain mills in Hungary. Women could not finish a technical degree in Budapest, so Magda Kemeny was completing her degree in chemical engineering in Karlsruhe with the aim of going to Russia to help establish chemical industries there. In November 1920 she passed her preliminary examination for the doctoral degree.[8]

For the next six months or so, Polanyi immersed himself in studies of reaction kinetics, thermodynamics, and statistical mechanics, along with adsorption, mostly working in the library of the Karlsruhe Technische Hochschule, where he studied chemical literature and data from a theoretical point of view. He became interested in discrepancies between theoretical and experimental results in the investigations of hydrogen and bromine gases, a topic on which he would continue to work in the next two decades along with adsorption. In April 1920, he presented a paper on adsorption at the meeting of the Bunsengesellschaft in Halle, which he attended with Reis. On this occasion he again saw Fajans, who had moved to Munich, and Hevesy, who was in Copenhagen. Polanyi also met the young Viennese chemist Friedrich A. Paneth and Walther Nernst and Fritz Haber for the first time.

Although he had not met Paneth previously, Polanyi knew a great deal about his work through Fajans and Hevesy. Paneth, like Fajans and Hevesy, had briefly been a researcher in Rutherford's laboratory in Manchester.[9] After Manchester, Paneth spent the summer term of 1913 in Soddy's laboratory in Glasgow at the time that Soddy was coining the term "isotopes" for atoms having differing atomic mass and the same atomic number (the concept of atomic number having been just established by Henry Moseley). Collaborating in Vienna in 1913, Paneth and Hevesy took Soddy's position that different isotopes of the same atomic number are chemically identical rather than just very similar, exhibiting what Paneth and Hevesy called interchangeability (*Vertretbarkeit*) or the facility of replacing each other, in crystallization and in electrochemical reactions. Fajans, then in Karlsruhe and immersed in his own pioneering studies of isotopes, disagreed and included among his arguments against the chemical identity of isotopes some thermodynamic arguments that he got from an unpublished paper by Polanyi, predicting that two substances of different atomic weight would have different free energies.[10] The Paneth-Hevesy interpretation prevailed within a few years, however.[11]

By the summer of 1920, Polanyi decided to apply for a position in Berlin at Haber's Institute. He supported his application with letters of reference from Bredig, Fajans, Paneth, Hevesy, and others, presenting in his resume a publication record of almost twenty papers.[12] Among these papers was his work on Nernst's heat theorem and Einstein's quantum theory for specific heats, which are discussed in chapter 3.[13] This time Polanyi would be able to take up research at the KWG. Reginold O. Herzog, who had done his habilitation with Haber at Karlsruhe, was director of the Kaiser Wilhelm Institute for Fiber Chemistry, which was a division of Haber's Institute for Physical Chemistry and Electrochemistry. Herzog wrote to offer Polanyi a position in Berlin beginning in the fall of September 1920.

## The German Academic Tradition of *Bildung* and Transcendent Science

Polanyi entered Berlin scientific life at a time when radical change in scientific culture was in the offing following the "long nineteenth century" that ended in 1914 with the Great War.[14] In the past century, scientists writing about the history and philosophy of science hardly ever emphasized the practical or craft dimensions of their work. Physical scientists writing about the nature and value of chemistry and physics at the time were seeking to establish and expand their particular disciplinary specialties within the teaching programs of the universities. These physical scientists and their exponents walked a tightrope in their efforts to persuade university colleagues and administrators of the humanistic and philosophical value of studies of the natural world, despite the fact that instruments and equipment were necessary for their research and for their students' instruction. In 1840, the philosophical faculty at Berlin rejected Justus Liebig's idea for a reformation of German academic science by the teaching of chemistry in the philosophical faculty on the grounds that laboratory instruction has no place in the philosophical faculty. Chemistry, said the Berlin professors, was not a science of causes, and it was not a general theoretical science like mathematical physics. Chemistry was taught only in the medical faculty for a while longer.[15]

The German system of education by the mid-nineteenth century had produced a higher level of literacy and culture among a broader spectrum of its population than anywhere else in Europe. The German-born British engineer and philosopher John Merz argued around 1900 that Germany was far ahead of other European countries because Germany produced not only a scientific elite of the first rank but scientists of the second and third rank as well. Research had become the primary criterion for appointment in Prussian universities by 1840 under the earlier leadership of Wilhelm von Humboldt and Karl Altenstein. The University of Berlin was established in 1809, professional specialization became possible in the universities, and flexibility was emphasized in the broadening experience of following courses at several different universities rather than at one alone. The philosophy of education in the classical *gymnasia* and the educational approach within the universities emphasized the cultivation of what was called *Vielseitigkeit* or "many-sidedness" in teachers and students. Philosophy textbooks explained that *Bildung*, or, formation of ethical and moral character was the aim of education and that *Bildung* was acquired "in the stream of the world" through experience and systemization.[16]

In making arguments for the inclusion of their sciences in the educational mission of the universities, experimental physicists and chemists came to insist on the moral value, as well as the philosophical value, of their work for training students and scholars in self-discipline and analytical thinking, even as their laboratories filled with electrical generators, mercury vacuum pumps, and chemicals. Experimental science was said to have humanistic value fully comparable to the study of pure mathematics or classical languages because of the role of experiment in teaching a rigorous method of thinking.[17] Most scientists who reflected on the philosophy of science in the nineteenth and early twentieth centuries in Germany and elsewhere described the essence of scientific knowledge as knowledge of ideas and theories. They did not emphasize practical skills and laboratory routines. This was true of William Whewell in England and Pierre Duhem and Henri Poincaré in France, no less than Ernst Mach in Austria and Heinrich Hertz in Germany.[18]

The transformation of the medieval text and lecture-based curriculum into the modern science-based universities of the late nineteenth and twentieth centuries took place first in Germany. Everyone recognized this. By the late 1880s, half of all German universities had established physics seminars which, in turn, became kernels for teaching and research laboratories in well-funded university-based scientific institutes headed by single professors with subordinate staff. In 1863 the University of Tübingen became the first German university to establish a mathematics and natural philosophy faculty independent of the law, medicine, and philosophy faculties. This new arrangement in German universities, which had existed in French universities since the Napoleonic era, provided greater freedom to establish new scientific specialties without fighting objections by classicists and humanists. Physics and chemistry institutes proliferated in the late nineteenth century in Germany, evolving out of a combination of the seminar, the laboratory, and the colloquium. In England in 1865, Matthew Arnold counseled university reformers to learn from the Germans, saying, "The French university has no liberty, and the English universities have no science; the Germany universities have both."[19] In France, Adolphe Wurtz began to argue the need to emulate and imitate German science, citing its advantages of government subsidies, regional industrial support, and the building of fully equipped, modern laboratory facilities in research institutes associated with universities.[20] Scientists heralded the German model as the model for all of modern science.

During most of the eighteenth and early nineteenth centuries, university physicists and chemists used their own private funds to set up their laboratories. As remarked by Arnold and Wurtz, the modern model required government support of scientific laboratories and funds from wealthy patrons and local industrial groups. As science and medical student enrollments rapidly were expanding, science was becoming a larger enterprise. Much has been written about these developments in the history and sociology of science and education.[21] In the field of chemistry, for example, by 1900, Germany was far ahead of other countries in the numbers of chemists educated in the universities and in university-level Technische Hochschulen and other institutions, in the numbers of papers produced, and in paid positions held in industrial companies.[22] From 1840 to World War I, nearly eight hundred British and American students earned doctoral degrees in chemistry in German universities. In 1905 at Berlin, Emil Fischer's research group included some twenty-four to thirty assistants, advanced students, and guests. At his death in 1919, Fischer's collected works included six hundred experimental articles, of which a hundred and twenty were published under the sole authorship of his PhD students.[23] Research now was rarely the work of a single scientist working in isolation, although individual scientists such as Fischer and Hermann von Helmholtz, both heads of complex research facilities and acting as consultants to government and industry, were both lauded as exemplars of the heroic scientist. Despite the rhetoric of pure science, the disciplines of chemistry, physics, and engineering were gaining increasing cultural prestige in Germany by the end of the First World War because of their prominent roles in the building of the imperial German state. These were fields in which new jobs seemed assured, and there was among those who entered them an increasing number of German Jews after the relaxation of anti-Semitic prohibitions in some German states after 1848 and the promulgation of the law of civil and political equality under Bismarck in 1869.[24]

Paradoxically, or perhaps in reaction to what was happening as German universities were becoming more technical and complex in nature, the rhetoric of the "Humboldtian University" came increasingly to characterize the ideal of the German university after 1900. As Sylvia Paletschek and other scholars have noted, Humboldt's famous unfinished essay "Über die innere und äussere Organisation der höheren wissenschaftlichen Anstalten in Berlin" (On the internal and external organization of higher academic institutions in Berlin) was discovered only in the 1890s and published for the first time in 1903.[25] It received a great deal of attention. Humboldt wrote that loneliness and freedom are the driving principles of the pure idea of *Wissenschaft*, a view consistent with the burgeoning neo-idealism of the early 1900s.[26] The term "Humboldtian" became shorthand for a fundamental set of values: the unity of research and teaching, the function of the university as a research institution, the freedom from vested interests and pursuit of "pure" and objective knowledge, and the assumption that scientific and humanistic knowledge provides moral education (*sittliche Menschenbildung durch Wissenschaft*).[27] A calling to science was a calling to priesthood. In this tradition and quoting Arthur Schopenhauer, Einstein wrote in *Mein Weltbild* in 1934 that one of the strongest motives leading to art and science is a "flight . . . from everyday life with its painful rawness and desolate emptiness, away from the chains of one's own ever-changing desires."[28]

At the end of the Great War, in what became a famous lecture on "Science as a Vocation" at the University of Munich in 1918, the sociologist and political economist Max Weber decried the waning of asceticism and devotion in German scientific life (meaning both the human and natural sciences), saying that self-interested professionalism, rather than the pure pursuit of learning, was coming to characterize the new *Wissenschaft* in modern Germany. He also noted the continuing difficulties faced by Jewish intellectuals in the German academic world.

> Academic life is a mad hazard. If the young scholar asks for my advice with regard to Habilitation [the credential for teaching in the university], the responsibility of encouraging him can hardly be borne. If he is a Jew, of course one says "lasciate ogni speranza" [abandon all hope]. But one must ask every other man: Do you in all conscience believe that you can stand seeing mediocrity after mediocrity, year after year, climb beyond you, without becoming embittered and without coming to grief? Naturally, one always receives the

answer: "Of course, I live only for my 'calling.'" Yet, I have found that only a few men could endure this situation without coming to grief.[29]

Weber's public characterization of tension between the idealized tranquil tradition of noble calling and what he saw as the modern trend toward self-interested professional life was reiterated by Svante Arrhenius in a private letter to Jacobus van't Hoff following the suicide of the Berlin physicist Paul Drude in 1906. Drude had fallen victim to the Berliners' zeal, wrote Arrhenius: to "this notion that scientists necessarily have to be the most noble [people] in the world," an attitude which was straining people beyond their capacities. "This crazy system cannot continue forever," concluded Arrhenius.[30]

In the historian Russell McCormmach's novel *Night Thoughts of a Classical Physicist*, the fictional physicist Viktor Jacob similarly broods over Drude's fate in the realities of German scientific life: Drude's youthful exuberance, his accommodating reception of the new quantum theory, his editorship of the *Annalen der Physik*, and his fateful move in 1905 from the quiet of the small university town of Giessen to the grueling demands of the university and the institute of physics in urban Berlin.

> What courage it must have taken to walk through the entrance of the Berlin institute the day he became its master! . . . . He had to administer the institute and the instrument collections, deliver experimental physics lectures to huge audiences, conduct laboratory courses for beginners, pharmacists, and advanced students, direct the colloquium, and examine students from all over the university. All of this was only his official responsibility . . . Besides that, he had his editing and his work for the German Physical Society and, at the end, the Prussian Academy of Sciences. And he was permanently on call to answer all questions from ministries and faculties about physics and physicists and questions from all sides about optics. Staggering as all of this responsibility was, he had an even greater one: to do first-class research in a time of rapid advances in physics.[31]

In the end, Jacob concludes, Drude was defeated by the demands of the new modern science: "A sinister force had defeated Drude, broken his soul."[32] It was this hothouse of Berlin science into which Michael Polanyi entered in 1920 with his appointment at the Kaiser Wilhelm Gesellschaft in Berlin, a new kind of scientific research organization linked to universities, industry, and the Prussian and German states.

## The Scientific Culture of Berlin in the 1910s and 1920s

Henry Adams, whose grandfather and great-grandfather were presidents of the United States, visited Berlin in 1858 following his graduation from Harvard University. He attended some lectures in civil law at the university during the academic year, but he did not find Berlin much to his liking. Rather it was

> a poor, keen-witted, provincial town, simple, dirty, uncivilized, and in most respects disgusting. Life was primitive beyond what an American boy could have imagined. Overridden by military methods and bureaucratic pettiness, Prussia was only beginning to free her hands from internal bonds. Apart from discipline, activity scarcely existed.[33]

By the early 1920s, Berlin was entirely different. It reminded some visitors of an urban American city both in its aspect and in its population, full of life and uncertainty, novelty, and chaos. The well-traveled and French-born Count Harry Kessler, who was the son of a German banker and an Irish beauty, moved easily among European capital cities in the early decades of the 1900s as a diplomat, publisher, and supporter of aesthetic modernism. Writing in his diary in August 1922, he reported on the differences that he saw between Paris and Berlin. Paris was a city in which everyone understood his social function. In contrast, every form of social function was in a state of evolution in Germany—especially in Berlin. Berlin "looks like a swarm of separate entities," wrote Kessler, where one cannot slip easily "into the long-made bed and the long-settled posture of his social function" as in Paris.[34] The historian István Deák wrote that Weimar Berlin was a cultural center of central and Eastern Europe with a rapidly expanding population of exiles escaping the Bolshevik Revolution, white terror, poverty in Vienna, and Ukrainian pogroms, joined by cultural and criminal refugees hungering for the "new, the sensational, and the extreme."[35] The Baedeker guide to Berlin published in 1923 warned the traveler that "the loss of the Great War has effected vast changes in the social composition of Berlin. The brilliance of the imperial court has disappeared. New classes of society with new aspirations have risen to commercial power, while the former calm based on assured prosperity has given way to a restless self-indulgence."[36] Shortly after Polanyi arrived in Berlin in the fall of 1920, Alfred Reis wrote from Karlsruhe to warn Polanyi that he was entering a "jungle."[37]

As described by Eric Weitz in *Weimar Germany*, by Brian Ladd in *Ghosts of Berlin*, and by many historians and witnesses, Berlin in the 1920s became a city of new architecture (Walter Gropius, Ludwig Mies van der Rohe, Erich Mendelsohn), new theater (Max Reinhardt, Erwin Piscator, Bertolt Brecht), new painting (Max Beckmann, George Grosz, Otto Dix), new music and entertainment (Arnold Schoenberg, Paul Hindemith, Kurt Weill), new cinema (Fritz Lang, F. W. Murnau, Marlene Dietrich), and the New Physics, the latter epitomized by the giant figure of Einstein.[38] Kessler wrote of a jubilee banker's banquet which he and Einstein attended in 1924 and remarked on "the ironical (*narquois*) trait in Einstein's expression, the *Pierrot Lunaire* quality, the smiling and pain-ridden skepticism that plays about his eyes."[39] Berlin could be a life of constant stimulation in the avant-garde. Tea with Virginia and Leonard Woolf, an evening concert by the young Yehudi Menuhin, a first-night performance at the Staatsoper of La Scala's *Falstaff* conducted by Toscanini, Max Reinhardt's new production at the Deutsche Oper of *Die Fledermaus*, and a visit from Erich Maria Remarque were on Kessler's calendar of engagements in the first half of 1929 when he happened to be in the city.[40] Metropolitan and cosmopolitan Berlin supported some 120 newspapers, 40 theaters, 200 chamber groups and more than 600 choruses in 20 concert halls and numerous churches.[41]

The political and economic situation of Weimar Berlin in the 1920s and early 1930s oscillated between difficulty and disaster. The German war effort had collapsed in late autumn of 1918 with more than nine million military deaths within the warring armies, approximately two million of them German. Soldiers' and workers' soviets, or councils, sprang up in German cities with demands for the abdication of Kaiser Wilhelm II. As the emperor fled to the Netherlands, the moderate Social Democrat Philipp Scheidemann proclaimed a German Republic from the window of the Reichstag on November 9, 1918, two hours before Spartacist League leader Karl Liebknecht called for a Free Socialist Republic from the balcony of the Berlin Stadtschloss. Two days later, on November 11, German civilian leaders signed an armistice, which veterans' groups, Bolsheviks, Socialists, liberals, conservatives, nationalists, and extremists on both sides of the political spectrum came to criticize during the next fifteen years as they struggled with each other for political power. The Weimar Republic was established with support from the army—but not from many of its soldiers—and with suppression of the January 1919 Spartacist uprising in Berlin. An assembly in Weimar finished a new constitution, which Max Weber helped draft, on August 11, 1919, one year after the armistice. Eight years later, Kessler reflected on the observance of Constitution Day, August 11, 1927, noting that he could see flags flying on government buildings, buses, trams, and the subway, but hardly at all on big business houses, department stores, hotels, and banks. The flags confirmed Kessler's opinion that an increasing proportion of the citizenry had come to accept the republic for the time being, but the great majority of the "'captains of industry,' the powerful financiers, the civil service, the Reichswehr, the bench, the large and medium-size landowners (*Junkers*), and university professors and students" had not.[42]

Many conservative members of the university regretted the passing of the monarchy and many others simply kept their distance, expressing allegiance to the ideal of objective teaching and scholarship independent from political party. University professors became involved in politics only when it was a matter of protecting internal university traditions from direct state interference. Their behavior reflected what historian Fritz Ringer has called a "mandarin" attitude in reference to the politically neutral protocols observed by Chinese civil servants over thousands of years. It was an attitude that betrayed many members of the German academic community in the 1930s.[43]

Within the academic elite, natural scientists for the most part joined colleagues in withdrawing from active participation in the political debates of Weimar Germany. Immediately following the Great War, they hoped to be able to teach and to pursue researches that were abandoned during the long war and military service. By the beginning of the Weimar period, some of the giant scientific figures of the early twentieth century had passed away, foremost among them Emil Fischer, who committed suicide in July 1919 in despair over his son's wartime death and the end of imperial Germany. The Berlin leadership triumvirate of Max Planck, Walther Nernst, and Fritz Haber had been valiant German patriots during the war who now hoped for a restoration of stability and calm. Planck had lost one of his sons and Nernst lost both of his sons to the war. These three scientists were among the 93 German intellectuals who signed the October 1914 "Appeal to the Cultured Peoples of the World" supporting the German army's invasion of neutral Belgium and denying charges of German atrocities in Belgium and France. Planck later had profound regrets about signing the political document.

In contrast, Albert Einstein earned the unrelenting enmity of German chauvinists and anti-Semites because of his signature on a counter-manifesto to the "Appeal," which Einstein and three other scholars signed in an appeal for cooperation and peace.[44] Despite his apparent distractedness and naïveté, Einstein became adept at dealing with the press once he was world famous, and he was politically courageous in his outspoken commitment to individual freedom, internationalism (including the idea of a European or world government), and Zionism (based on the conception of a cultural Jewish nation but not necessarily a political Jewish state). Einstein, who adopted Swiss citizenship and maintained it after settling in Berlin in 1914, was not representative of the German academic mandarinate.[45]

In the early 1920s, Planck, Nernst, Einstein, and Haber could often be found among the sixty or so people at the weekly Wednesday afternoon colloquium of the University's Physical Institute, which was held from 5 p.m. to 7 p.m. in the small lecture hall on Reichtagsufer. A good-sized classroom was arrayed with three sets of connected wooden chairs. Until his death in 1922, Heinrich Rubens led the colloquium along with Max von Laue. Laue had been appointed to a new associate professorship in theoretical physics under Planck in 1918. So anxious was Laue to return to Berlin from Frankfurt that he gave up a full professorship in order to accept the Berlin position. The Wednesday colloquium room often featured four or five Nobel laureates in the early 1920s. Laue had received the award in Physics for 1914, Planck for 1918, and Einstein for 1921; Haber received the award in Chemistry for 1918, and Nernst for 1920. Gustav Hertz's sharing of the Nobel Prize in Physics with James Franck in 1925 brought the number to six Nobelists at the Laue colloquium on Wednesdays. Erwin Schrödinger, a 1933 Nobelist, joined the colloquium in 1927 when he succeeded Planck at the university. Lise Meitner and Otto Hahn were other regulars, and younger, soon-to-be-famous physicists, such as Werner Heisenberg and Wolfgang Pauli, attended when in town.[46]

The ostensible purpose of the colloquium was the review of important new physics papers. Laue made assignments to mostly junior physicists, such as Eugene Wigner or Rudolf Ladenburg. In Wigner's view, the arrangement "allowed the major physicists, who were absorbed in their own work, to still follow other branches of physics."[47] Wigner later reflected that it was the Wednesday discussions that made him realize the meaning of theoretical physics. What he saw as its purity appealed to him. "Science was a monastic occupation then, with a monastic spirit," he later wrote nostalgically.[48] At a talk in Chicago in 1950, Michael Polanyi recalled the weekly Berlin Wednesday colloquia and informal discussions as "still the most glorious intellectual memory of my life."[49] For these scientists, Berlin's culture of theoretical physical science was a welcome escape from some of the darker realities of Berlin.

The Wednesday colloquium was not the only game in town. In addition to the ubiquitous coffeehouse conversations around large communal tables, Wigner attended Max Volmer's colloquium at the Technische Hochschule in Charlottenberg, where Ferdinand Kurlbaum was head of physics and Volmer directed the Institute for Physical Chemistry and Electrical Chemistry. The German Physical Society's colloquium met in the university's Physical Institute on Fridays. The other principal venue was Fritz Haber's biweekly Monday colloquium at his Institute for Physical Chemistry and Electrochemistry, some five miles from central Berlin in Dahlem. Most of the KWG colloquia were closed discussions for coworkers in a single institute, but Haber's was different, because there were speakers from the different departments in Haber's large institute and from other KWG institutes. The regular participants included workers in Herzog's Fiber Chemistry Institute and in Hahn's Chemistry Institute. Nernst, Einstein, and other central Berlin physicists and chemists sometimes took the subway out to Dahlem for the meetings. Haber delegated the planning of the programs and introduction of the speakers to Herbert Freundlich, who was head of the colloid chemistry subdivision, but Haber reserved for himself the right to interrupt the colloquium speaker at any time. Every lecture was followed by passionate discussion stirred up by Haber, who liked to see ideas fly. Wilhelm Jost, who was preparing for his habilitation in the late 1920s with Max Bodenstein, wrote, "There were so many independent scientists in Berlin that the atmosphere for work and the contacts and connections among them were so free . . . that each young man had a chance of being successful with his own ideas. There was much poverty here, but that was really no limitation for truly creative young people."[50]

The star organization in the network of Weimar Berlin's scientific research centers was the Kaiser Wilhelm Gesellschaft enclave in Dahlem where Haber's colloquium took place. The KWG was the realization of the plan of Friedrich Althoff, the retired director of the Prussian Ministry of Culture and Education, to establish a German Oxford in the undeveloped suburb of Dahlem on the southwestern side of Berlin. After Althoff's death in 1908, the historian and theologian Adolf von Harnack prepared a white paper for the kaiser in which he proposed the creation of an organization for fundamental research in which scientists could be independent of "clique and wealth" and of obligations for teaching. He noted the beneficial role elsewhere, particularly in the United States, of private endowments for furthering science, and he argued in good measure that "military power and science are the twin pillars of Germany's greatness."[51] Harnack proposed that the Prussian government provide land in Dahlem for research institutes that could be built one-by-one as funds and directors became available. The society's private contributing members would pay for facilities and operating expenses. Primary researchers would become state employees associated with the universities, but with no obligations to teach and with their KWG salaries paid by the state or private funds. The kaiser agreed to the plan as a celebration of the hundred-year jubilee of the founding of the University of Berlin.[52]

The KWG was meant to be a new and different kind of institution from the Physikalisch Technische Reichsanstalt (PTR), the major nonuniversity and nonindustrial organization for research in Berlin around 1900. The PTR had been built in 1887 in the west-central Berlin suburb of Charlottenburg with funding from the wealthy industrialist Werner Siemens and the state. Siemens's son-in-law Hermann von Helmholtz became its first director, and Helmholtz's immediate successors were Friedrich Kohlrausch, Emil Warburg, Walther Nernst, and Friedrich Paschen. The PTR consisted of ten buildings, five for a scientific section devoted to fundamental research and five for a technical section focused on industry-related research. The PTR became a model for later state-funded research laboratories, such as the National Physical Laboratory in England, established in 1900, and the National Bureau of Standards in the United States, founded in 1901.[53]

The first two KWG institute buildings were the Kaiser Wilhelm Institute for Physical Chemistry and Electrochemistry and the Institute for Chemistry. This outcome resulted in large part from the role of Leopold Koppel, the Saxon Privy Councilor for Commerce and a banker, industrialist, and entrepreneur whose enterprises included the Auer Gas Light Company, the Hotel Bristol, and the Wintergarten Theater. He promised 700,000 marks (approximately US$167,000 at the time) for the building and equipment of an Institute for Physical Chemistry, provided that its first director would be Fritz Haber. Svante Arrhenius and others also strongly recommended Haber. From his Institute at Karlsruhe, Haber demanded a professorship at the University of Berlin and election to the Prussian Academy of Sciences, along with the directorship of the new Physical Chemistry Institute. At Haber's request, Koppel provided another 300,000 marks and a ten-year operating budget of 35,000 marks a year for the Physical Chemistry Institute, provided that the state would contribute the director's salary of 15,000 marks, a housing allowance of 5,000 marks, and another 35,000 marks for expenses. Koppel's private foundation and the state directly provided the funds and oversight for Haber's institute until 1923 when it formally came under the administration of the KWG. The first general assembly of the KWG was held in the machine room of Haber's new institute on October 12, 1912.[54]

Haber's ambitions for his institute immediately extended to the recruit-

Figure 2.1. Fritz Haber and Albert Einstein at the Haber Institute, ca. 1915.
(Courtesy of Archiv der Max-Planck-Gesellschaft, Berlin-Dahlem.)

ment of Einstein from Zurich to Berlin, an idea discussed sometime in late 1912 with Friedrich Schmidt-Ott, the ministerial director in the Prussian Ministry of Culture and Education, and his assistant Hugo Andres Krüss. Haber proposed that he could provide space for Einstein on the upper floor of his institute and that Koppel would help with Einstein's salary. By the summer of 1913, Planck and Nernst had journeyed to Zurich to recruit Einstein. The Koppel Stiftung donated the salary that Einstein was paid at the Berlin Academy of Sciences, and Koppel agreed to partially finance a separate physics institute specifically for Einstein, as proposed by Harnack and supported in an application to the ministry by Haber. When Planck,

Nernst, and Laue successfully persuaded Einstein to leave his professorship at the Zurich Polytechnic Institute and come to Berlin in 1914, they courted him with the offer of election to a special salaried membership in the Berlin Academy of Sciences, a professorship with no teaching duties at the University of Berlin, and directorship of the not-yet-founded Kaiser Wilhelm Institute for Physics.[55]

Among potential institutes for the KWG, it is hardly surprising that the field of chemistry was a prime target for successful funding among the society's philanthropist members. The economic and military value of chemistry to the German state was taken for granted. The KWG's contributing members, with the exception of a few scientists, were drawn from industrialists and bankers who were part of Prussia's super-rich elite. Some 35 percent of the original contributions came from Jewish philanthropists, Koppel among them. At the time, Jews made up less than 1 percent of the population of Berlin, a proportion that increased to 4 percent in the 1920s. Large numbers of Jews were prominent not only in finance and industry but also in journalism and publishing, the garment business, and department stores. The crown jewel was the Wertheim Department Store, at the corner of Leipziger Strasse and Leipziger Platz, with its glass-roofed atrium, ten thousand light bulbs, and eighty-three elevators.[56] Jews played an important role in chemical industry as entrepreneurs, employees, and consultants, particularly in the fast-moving dye, pharmaceutical, and lighting industries. Koppel's Auer Gas Light Company (Deutsche Gasglühlichtgesellschaft-Aktiengesellschaft; DGA) was among these businesses, competing in the rapidly expanding lighting industry, which was revolutionized in 1913 by the invention of the inert-gas-filled tungsten lamp by Nernst's American student Irving Langmuir. Fritz Haber was one of the expert consultants for DGA.

Haber was in fact an outstanding exemplar of the successful religiously converted Jewish chemist and intellectual who was at ease in the worlds of both the university and industry. While working in his laboratory at Technische Hochschule in Karlsruhe, he signed a five-year contract in 1908 with the Baden Aniline and Soda Factory (Badische Anilin- und Soda-Fabrik; BASF) to "support the interests of BASF insofar as his official and scientific work allows it" and place his research results at their disposal, although the work would remain his literary property for publication. The commercial exploitation of Haber's research belonged to BASF. In return, Haber received 6,000 marks annually at a time when his professor's salary was 4,000 marks (with a 1,200-mark housing allowance), and he was promised 10 percent of any commercial net profits. Working with his English collaborator Robert Le Rossignol and his doctoral student assistants, Haber in 1909 solved the century-old problem of synthesizing ammonia from the simple raw ingredients of free nitrogen and hydrogen gases, using the very high pressure of 200 atmospheres, a temperature of 200°C, and a catalyst composed of the powder of the rare metal osmium. Nitrogen is unlimited in the air and hydrogen is abundant in coal gas, so the discovery had unlimited potential for agricultural fertilizers and for explosives, although Haber's process required an expensive catalyst.

Recognizing the tremendous profits now likely at hand and regretting the terms of his consulting agreement with BASF, Haber threatened to begin working for Koppel's company. In response, Haber received a new contract from BASF for 23,000 marks, including 8,000 marks for instrumentation and assistants, as well as permission to advise Koppel's company DGA, but only in its lighting division. The first commercial production of ammonia at BASF began in 1913 in Oppau, northwest of Ludwigshafen, with modifications made to a cheap iron catalyst by BASF's salaried chemists Carl Bosch and Alwin Mittasch. Bosch received the Nobel Prize in Chemistry in 1931 for his research on high-pressure chemical processes, and he succeeded Max Planck as president of the KWG in 1937 following a career as director at BASF and IG Farben.[57]

On October 23, 1913, the kaiser presided over the opening of the first two chemical institutes of the KWG. Their design and construction had been speedy, in part because the Institute of Chemistry had funds at its disposal from an earlier scheme of Fischer, Nernst, and Carl Duisberg, the managing director of the Bayer Company (Farbenfabriken Bayer), to establish a national chemical institute as a sister institution to the Physikalisch Technische Reichsanstalt. They already had collected some 1,000,000 marks of industrial monies into a Verein Chemische Reichsanstalt (Society for a National Chemical Institute; VCR). The VCR worked out an agreement with the KWG in late 1911 for an institute completely devoted to fundamental science. Ernst Beckmann at Leipzig became the first director of the Chemical Institute's three independent departments. Beckmann himself directed a section for inorganic and physical chemistry, Richard Willstätter directed organic chemistry, and Otto Hahn directed radioactivity, with Hahn taking charge of chemistry and Meitner of physics in the radioactivity department. Haber and Fischer persuaded Willstätter to leave ETH in Zurich for Berlin. Fischer promised Willstätter, "You will be completely independent. Nobody will bother you or interfere with you. You can go walking in the Grunewald for a few years or, if you wish, think up some beautiful new things." Haber helped Willstätter buy a plot of land where a house was built for him on

Faradayweg, next door to Haber's own official residence, with the two villas separated by a large garden. Willstätter, who was doing pioneering studies in the molecular structures of plant alkaloids (such as cocaine), hemoglobin, and chlorophyll, became the KWG's first Nobel Prize winner in 1915 for his description of the structure of chlorophyll. He left in 1916 to succeed his mentor Adolf von Baeyer in Munich.[58]

Willstätter was among Hahn's colleagues and staff who devoted most of the institute's research time to war work during 1914 to 1918. The kaiser appointed Haber directly to the military rank of captain despite the War Ministry's opposition to making a Jew an officer. Haber, Fischer, and Walther Rathenau, who was head of Allgemeine Elektrizitätswerke Gesellschaft (AEG), were among the first to foresee the huge wartime shortages for materials such as toluene, nitric acid, sulfuric acid, petroleum, and rubber. Fischer turned his attention to the manufacturing of alternatives for sulfates, and Haber focused on nitrates. Haber was responsible for the first use of chemical weapons on the western front after persuading the German General Staff that gases, such as chlorine or phosgene, could be used to clear out enemy trenches. He directed what became a huge military weapons project at his institute. By the final years of the war, some 150 scientific coworkers and 2,000 staff members were working on the production of nitric acid for explosives and fertilizers and on the preparation of poisonous gases. The 10 department heads in the umbrella Group for Offensive and Defensive Installations for Gas Warfare included Herbert Freundlich, Reginald Herzog, the organic chemist Heinrich Wieland, and Willstätter. Willstätter worked with Freundlich on a filter for gas masks before Willstätter left for Munich. The filter was produced by one of Koppel's companies. The gases, including mustard gas (toward the end of the war), were produced by Bayer under the management of Duisberg.[59] Haber personally supervised the first use of chlorine gas against Franco-Algerian troops at Ypres on April 22, 1915, and, along with Fischer, Nernst, and three other chemists, served as director of a government-funded Kaiser Wilhelm Foundation for War Technology. At war's end, Haber fled to Switzerland briefly until his name was removed from an Allied list of war criminals. The suicide of his wife Clara in 1915, a few days after Haber's return from Ypres, was widely attributed to her disgust with Haber's work. [60]

The decision by the Swedish Academy of Sciences in 1919 to award the 1918 Nobel Prize in Chemistry to Haber, as well as the 1918 and 1919 Nobel Prizes in Physics to the German scientists Max Planck and Johannes Stark, was met with outrage in Allied countries—particularly in France and Belgium. Not only had Haber, by his own account, pressed for the development of chemical weapons, but his industrial process for producing nitrates had made possible Germany's long commitment to the war despite Allied blockades of Germany's resources. Haber continued to advise the German government on its secret research on chemical weapons after the war but focused mainly on administration and research in his own institute. Following up on wartime development and agricultural applications of pesticides, Haber supported the postwar development of improved pesticides in his institute. From 1919 to 1920, Ferdinand Flury's department in Haber's institute developed a preparation based in hydrocyanic acid which was turned over to the firm of Degesch in Frankfurt, where Haber was an adviser. Horribly, a version of this pesticide called Zyklon B was used in Auschwitz and other extermination camps during the Second World War.[61]

## Routines of Scientific Life at Haber's Institute in the 1920s and 1930s

At Polanyi's arrival in Berlin in the fall of 1920, there were seven institutes in Dahlem, including the Institute for Fiber Chemistry directed by Reginald Herzog.[62] The wartime barracks and other temporary buildings had been torn down and the institute had been demilitarized by an Allied military commission. Flury's department was closed in 1920 due to financial problems caused by inflation, and the department for textile research, established in 1919 under Herzog's directorship, had become the separate Institute for Fiber Chemistry. The Institute for Physics, directed by Einstein, still had no funds for a building. The KWG faced a financial shortfall, and the Prussian state and the Weimar federal government agreed to share expenses for covering this deficit, which amounted to three times the society's income. In return, the government acquired the former kaiser's right to choose half the KWG Senate. Industrialists stepped in with more funds, and by 1926 representatives of Krupp, I. G. Farben, and other companies held five places on the seven-member executive committee of the KWG. One of Haber's immediate ventures to help fund the society was to establish a research group to develop a method for extraction of gold from seawater, but a commercial process proved unfeasible. Haber contributed his own patent royalties for a while to the budget of his institute, and Koppel agreed to support Haber's institute with 15,000 gold marks annually. The arrangement of the society with Prussia and the federal government persisted over the next two decades so that about half the society's operating expenses would come from government funds.[63]

In 1920, in addition to Haber's research group in his Institute for Physical Chemistry and Electrochemistry, which focused on gold extraction, chemiluminescence, and reaction kinetics, Freundlich directed a research section for colloid chemistry and James Franck briefly headed a section on atomic physics. Franck's assistant was Hertha Sponer, who in 1925 became the first woman after Lise Meitner to qualify to teach physics in German universities (*habilitation*). Just before Polanyi arrived, Franck (and Sponer) left Berlin for Göttingen, where he was teaching when he and Gustav Herz received the Nobel Prize in Physics in 1925 for their work on forces between atoms and electrons. In 1924 Rudolf Ladenburg was appointed to Franck's former position in Haber's institute.[64] The Fiber Chemistry Institute was formally independent from Haber's institute and specifically had as part of its mission a responsibility to address problems of interest to industrial firms and the Prussian state. The Fiber Chemistry Institute consisted in 1920 of a few small rooms in Haber's institute and equipment scattered around the KWG premises, including a microscope in the Institute for Biology and photographic equipment in the Materials Testing Laboratory a few blocks away. The basement of Haber's private villa next door to the Physical Chemistry Institute contained X-ray equipment.[65]

When Polanyi arrived in Berlin, he had left his fiancé Magda Kemeny in Karlsruhe to complete her preliminary examinations for the doctorate in chemistry. They married in Budapest in a civil ceremony in February 1921 and moved into two furnished rooms in Dahlem without cooking facilities and within a long walk of Polanyi's laboratory. She enrolled at the Berlin Technische Hochschule the following year while working on a dissertation but gave up her efforts following the birth of their son George in October 1922 and the temporary closing down of the Technische Hochschule for financial reasons that same fall. After George's birth, the Polanyis lived in a little flat rented originally by Magda's mother, and they were able by 1924 to build a small two-story house on Waltraudstrasse with a kitchen and a room for live-in help in the basement. Their son John was born in 1929.

Michael and Magda Polanyi were part of a close network of Hungarian family members and of old and new friends who often met for coffee and conversation at the Romanische Café on the Kurfürstendamm not far from rooms rented by Polanyi's Hungarian friend Leo Szilard, who worked at the Fiber Chemistry Institute from 1922 to 1924 and taught courses at the University of Berlin as Privatdozent.[66] Wigner completed his doctoral thesis formally under Polanyi in 1928 and also collaborated in Berlin with their mutual Hungarian friend John von Neumann, then a Privatdozent at the university.[67] Along with his colleagues, Polanyi regularly went across town from Dahlem to the Mitte to take in Laue's weekly colloquia at

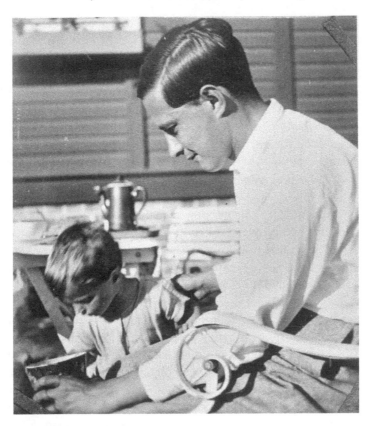

Figure 2.2. Michael Polanyi with a young John Polanyi in 1931 at the Polanyi home in Berlin. (Courtesy of Dudley R. Herschbach and John C. Polanyi.)

the University. Decades later Wolf Berg told William T. Scott that Berg and other physics students knew Polanyi by sight as the "remarkably good looking" young physical chemist from Hungary. Berg admired Polanyi's ability to explain ideas that Berg found difficult in a simple way and to clarify disagreements. He reported that Polanyi sometimes took the colloquium podium and delivered lucid summaries of what had been said.[68]

The Fiber Chemistry Institute moved into its own buildings on Faradayweg up the street from Haber's institute in September 1922. The Fiber Chemistry Institute's researchers had an understanding that they could focus on basic research in response to problems posed by industry without concern for immediately applicable results. In addition to Herzog's research group in the Fiber Chemistry Institute, where Polanyi worked initially as a research assistant, there were departments for organic chemistry and for technology when he arrived. The organic chemistry department quickly lost its first two department heads, when, first, Max Bergmann and then his successor Burckhardt Helferich left for other jobs. The department closed in 1922. Hermann Mark came to the institute in 1922 to succeed A. Geiger as head of the technology department, but Mark himself left in 1926 for other positions, first at the society's Institute for Silicate Research and then at I. G. Farben in Ludwigshafen. In January 1922 Polanyi was named head of a section for physical chemistry at the Fiber Chemistry Institute and a member (*Wissenschaftlichen Mitglied*) of the institute. The institute membership gave him membership in the Kaiser Wilhelm Society. Now he was set.[69]

In the spring of 1923 Polanyi completed the habilitation using his work on the structure and properties of crystals. He received an appointment as Privatdozent at the Berlin Technische Hochschule in the Faculty of Materials Technology, where he became associate (*Extraordinarius*) professor in 1926.[70] Haber was anxious to transfer Polanyi from Fiber Chemistry into his own institute and, after negotiations over funding, the transfer became official in September 1923 when Polanyi became Haber's director of the department of basic research in physical chemistry. Haber made the request for Polanyi's transfer at the same time that he was recruiting Ladenburg into Franck's vacant position in atomic physics. In rationalizing Polanyi's transfer, Haber wrote Harnack that Polanyi's theoretical talents could not be fulfilled in the Fiber Chemistry Institute and that theoretical development is central to national scientific progress. As for the salary funds for Polanyi, Haber informed Harnack that he could supply them through the currently vacant state-funded salaries of one and a half research assistants for the Physical Chemistry Institute.[71] In September 1923 Polanyi moved from Herzog's building at Faradayweg 16 to Haber's building at Faradayweg 4, where he occupied several laboratory rooms on the third floor.[72]

By the spring of 1923 the financial situation in Germany was acute. In May 1921, a Versailles Treaty commission that had been appointed to determine the exact amount of reparations announced a total sum in the amount of 132 billion gold marks. The first annual payment was due in August 1921. From a value of 60 marks to the dollar in early 1921, the German mark fell rapidly to 9,000 marks to the dollar by November 1922. As anticipated by the British economist John Maynard Keynes, who resigned in disgust from his post representing the British Treasury at the Paris Peace Conference of 1919, the Versailles policy designed by the French prime minister Georges Clemenceau had the aim of destroying the German economic system, which had been the basis of German military might.[73] The German chancellor Joseph Wirth brought into the government as foreign minister the Jewish financier and political visionary Walter Rathenau, who was director of Allgemeine Elektrizitätswerke Gesellschaft (AEG), a director of dozens of local and foreign companies, and an author of texts of politics and economics. Rathenau had been head of the Board of Wartime Raw Materials under the Ministry of War. Some of his friends, including Einstein, had misgivings about Rathenau's accepting the position. Vilified as a Jew and as a Communist sympathizer after the signing of a diplomatic treaty with the Soviet Union, Rathenau was shot dead in his car in June 1922. Much of Berlin, as Count Harry Kessler put it, was "thunderstruck."[74]

Things only got worse during most of 1923. In January of that year French and Belgian troops occupied the industrial, coal- and iron-rich areas of Germany in the Ruhr Valley, fueling renewed popular anger and resentment in Germany at the Versailles Treaty and at the Weimar government. On November 8, in anticipation of the upcoming anniversaries of the proclamation of the Weimar Republic and the armistice, Adolf Hitler and his brown-shirted storm troopers tried to mount a coup in Munich. The Bavarian state and the Weimar Republic survived the overthrow attempt, and the financial situation stabilized after the appointment in the late summer of Gustav Stresemann as chancellor. Stresemann was a leader in the right-center German People's Party (*Deutsche Volkspartei*, or DVP) and he proved willing as chancellor and later as foreign minister to work with parties of the left and center. His government introduced a new currency, the Rentenmark, in November 1923 and implemented a severe reduction in government spending which, along with an injection of American loans into the economy, promised relief to the battered economy.[75] Einstein, who regularly received death threats during the early 1920s, took seriously a threat in early November 1923 and fled to the Netherlands, returning only after news of the failure of Hitler's brown shirts and pleas from Planck to return to Germany. Weimar Berlin still remained an intellectual climate in which he could thrive.[76]

There was a great deal of financial stress and worry about funds for salaries, equipment, materials, and other operating costs during Polanyi's first few years in Berlin. The economy further improved after the Allies agreed in 1924 to end their occupation of the Ruhr and offered a more realistic schedule for reparations. Only a few of the KWG positions, such as department heads, were fully salaried with benefits such as a house or a housing allowance and support for at least one paid assistant. KWG salaries were supplemented by university salaries for those who were appointed as professors or associate professors at the university or Technische Hochschule. Industrial consulting contracts or patent income could provide additional sources of revenue for KWG members or staff.

Polanyi's multiple sources of funds illustrate how this worked. From the time of his arrival in Berlin, Polanyi had a regular income from Izzo in Budapest, and his agreement with Haber in 1923 guaranteed that he could keep this industrial contract.[77] Other sources of funding could be carefully crafted together. One source was the Notgemeinschaft der deutschen Wissenschaft. Established in October 1920, it was an organization supported by the central government, industry, the Rockefeller Foundation, and other sources and it had the mandate of distributing funds to worthy scientific projects. Under the leadership of Planck, Haber, Nernst, and Laue, the distribution of funds was tightly targeted to friends of the elite scientific leadership. As Planck wrote Arnold Sommerfeld, "Tell me what you need, and I will see that you get it all."[78]

Polanyi quickly became an astute observer and participant in the funding game, drawing upon industry and eventually upon the Rockefeller Foundation. Industrial monies could be used not only as a personal stipend, but to support research staff. Polanyi learned how this was done when Herzog set up what was called a *Studiengesellschaft* at the Fiber Chemistry Institute in 1922. This study-society was an in-house consulting group which contracted with industrial firms to conduct specific pieces of research in the institute. In 1922, Ignác Pfeifer approached Polanyi about his returning to Budapest. Like Polanyi, Pfeifer had been forced to retire from the University of Budapest under the Horthy regime, in Pfeifer's case because he had been a member of Bela Kun's government. Pfeifer got a position as head of the new research laboratory of the United Incandescent Lamp and Electric Corporation (Izzo), which developed the krypton bulb and carried out basic research in experimental physics. Pfeifer had employed Polanyi as a consultant since 1920, and in 1922 he offered him a full-time position in Budapest.[79] Polanyi's mother begged him to take it and Magda thought it a "fabulous" job.[80] Polanyi instead worked out an arrangement that he would devote 30 percent of his time in the Fiber Chemistry Institute to projects of interest to Izzo and he would provide research experience to assistants, who were paid by Izzo and sent from Budapest to Berlin. In return, Izzo was to underwrite 30 percent of Polanyi's regular stipend.[81] Polanyi signed subsequent contracts with Siemens Electric Works, Osram Lamp Works, and Philips Lamp Works.[82]

Patents were another source of income, although not as lucrative for Polanyi as consulting. Haber's policy was that 30 percent of all patent income generated from work done at his institute must be returned to it, an agreement that Polanyi signed in 1923 while he continued to work as a consultant and develop patent claims in the next years.[83] Some of these patents came from work in the late 1920s with the Russian physical chemist Stefan Bogdandy in another *Studiengesellschaft* arrangement. One of Polanyi's professional friends was Abram F. Joffe, who headed the Physical Technical Institute in St. Petersburg. Joffe knew Germany well after studying physics in Munich where he was Wilhelm Röntgen's assistant. Joffe and Polanyi set up a *Studiengesellschaft* in Polanyi's research department which employed Bogdandy to do work for Siemens and Halske and for AEG. This arrangement provided research assistance for Polanyi and resulted in some patent income.[84]

Another source of funding was the Rockefeller Foundation, established by John D. Rockefeller in 1913 to "promote the well-being of mankind" worldwide through education, science, and medicine. The foundation consisted of five divisions for scientific research, in health, medical sciences, natural sciences, social sciences, and the humanities. It set up a European division at the war's end and began an aid program to European centers that initially included Copenhagen, Paris, Vienna, London, Uppsala, and Göttingen. By 1930, the Kaiser Wilhelm Institutes were attracting strong interest from Lauder Jones, one of the foundation's officers for physical sciences in Paris, and by 1932 the Rockefeller Foundation had allocated or pledged about $2 million to the KWG at a time when the exchange rate was roughly 3.25 Reichsmarks to the dollar. (The Reichmark gradually strengthened against the dollar from 4.25 in 1932 to 2.50 in 1938.)[85]

In addition to awarding grants to foreign guest researchers to come to scientific centers such as the KWG, the Rockefeller Foundation helped fund equipment costs. Some of the Rockefeller money was matching money. A new centrifuge for Herbert Freundlich, for example, received $7,000 in 1930, contingent upon the society's providing $1500. Some grants provided full costs for equipment, for example, $25,000 in 1932 for a liquid air machine, as part of a total of $55,000 given to the society for equipment. After meeting in Berlin with Lauder Jones, Polanyi got equipment for his work on atomic reactions and free radicals, and his colleague Hartmut Kallmann received high-voltage apparatus for experiments on the disintegration of atoms.[86] A much bigger grant resulted from Jones's learning that the society had an option to buy one of the last large parcels of land in Dahlem near the Institute of Biology and the Institute of Chemistry. The foundation committed $655,000 for a new Institute for Cell Physiology for the Biology Institute's leading biochemist, Otto Warburg, and for a brick-and-mortar Institute of Physics for Einstein.[87]

The crash on the New York Stock Exchange in late October 1929 and the descent into the Great Depression began to take a toll in the early 1930s at Haber's institute and other scientific centers in Berlin. At the end of his first decade in Berlin, however, partly because of offers of positions elsewhere and largely because of his scientific work, Polanyi substantially improved his institute contract. In 1928 he received the offer of a full professorship at the German University in Prague and the directorship of its Physical Chemistry Institute. In May and June of 1929, the Hungarian minister of education tried to lure him back to Hungary with a full professorship at the University at Szeged, south of Budapest, and Izzo renewed its offer of a full-time position. It was also rumored that Polanyi had received an offer from Harvard. Negotiating with Haber and Harnack, Polanyi received a new contract at the KWG, which gave him a lifetime membership in the society and a generous retirement plan. His contract specified a salary of 10,500 marks, along with a housing allowance, a child subsidy, and a bonus that brought his total contract income from the society in 1929 to over 14,000 marks.[88] In addition he had a salary at the Technische Hochschule. By comparison, when Eugene Wigner was offered what he considered a spectacular six-month contract by Princeton University in 1930, the sum offered was $4,000 or, approximately 16,000 marks.[89]

By 1930, Polanyi was working in an environment that seemed to him unparalleled as a scientific community, just as it seemed unrivalled to Einstein and to so many Berlin scientists who tried as best they could to ignore the uglier side of politics in the Weimar Republic. Haber's institute was a mecca for physical chemists; during the 1920s and early 1930s, coworkers and guests included James Franck, Hertha Sponer, Rudolf Ladenburg, Hermann Mark, Hans Beutler, Karl Friedrich Bonhoeffer, Henry Eyring, Ladislaus Farkas, Paul Harteck, Hartmut Kallmann, Paul Knipping, Fritz London, Eugene Rabinowitch, Karl Söllner, Eugene Wigner, Karl Weissenberg, and Erika Cremer.

The biweekly Haber colloquium moved in late 1929 to the newly built Harnack Haus, which was dedicated on Harnack's seventy-eighth birthday, just before his death. Polanyi took over scheduling for the Haber seminar in 1930 from Freundlich, who had managed it since 1920. Weekday lunches were served at Harnack Haus under a vaulted ceiling in a dining room adjacent to a chair-lined hall facing the terrace and gardens. Three tennis courts, a brightly lit indoor gym, and shower rooms provided opportunities for exercise and relaxation. Otto Hahn was one of Polanyi's partners for gymnastics exercises and tennis matches, recreations which supplemented Polanyi's weekend swimming and sailing on the Wannsee and his Sunday family walks in the Grunewald. Polanyi's coworker Ervin von Gomperz sailed with him, and Erich Schmid joined him for short skiing trips. Out-of-town visitors could stay in guest rooms at Harnack Haus and join KWG members in a library of over a hundred magazines and journals and German and foreign newspapers. There were rooms for large and small gatherings. KWG members, like Polanyi, could schedule evening dinners or discussion groups, as Polanyi did for an economics dining club that included his longtime Hungarian friends Eugene Wigner, Leo Szilard, and John von Neumann.[90]

In 1930, Polanyi had a staff of nine in his division. His budget included one assistant and two technicians, Martin Schmalz and Kurt Hauschild. The rest of his coworkers were mostly foreign and short-term, either covering their own expenses or receiving fellowships or industrial support. Polanyi had five small rooms on the third floor for his research groups. Most of their apparatus was called in for immediate use from a common storeroom, and Hauschild did glass-blowing as needed on site in each of the research rooms. There was also a large machine shop for complex and high-precision work.[91]

In the spring of 1930, the *Manchester Guardian* journalist James G. Crowther visited Berlin. He was politically left and sympathetic with the efforts in the Soviet Union to modernize the formerly czarist country quickly through the expansion of science and technology. Crowther talked with Hermann Mark and asked him what had prompted his study of the chemistry and physics of fibers and materials. Mark pointed to the postwar shortage of raw materials in the textile industry and to industrialists' need for scientists to help them. This, Mark emphasized, led to fundamental research on fibers at the Fiber Chemistry Institute and elsewhere.[92] Crowther also talked with Haber in Haber's villa adjacent to the Physical Chemistry Institute, asking him about the current relationship between German industrial firms and academic chemists, Haber told Crowther that "there [is] no split between the scientific and commercial side," that this first had become true in chemistry and was becoming characteristic of metallurgy and other fields. Crowther attended Polanyi's economics discussion group at Harnack Haus and marveled at the conviviality and high intellectual level of what he heard.[93]

Indeed, Crowther was riveted by Dahlem in 1930. "The first principle of the KWG," he wrote in an article for the *Manchester Guardian*, was "to search for the newest developments in science and encourage them, and to employ the best managerial ability for the administrative side of scientific organization and [to] relieve geniuses of all possible distractions."[94] In Crowther's view, the dream had been realized for establishing a German Oxford in the quiet suburb of Berlin for scientists who would be "independent of clique and wealth."[95] Crowther likely noted, too, although he did not mention it, that the Dahlem scientists generally were quite free of students and exams, unlike in Oxford, although KWG members often had formal appointments at the Berlin University or Techische Hochschule. Polanyi himself was horrified by teaching loads at American universities when he visited the United States in 1929.[96] Asked in 1928 by the editor of a Hungarian newspaper about his experiences in Berlin, Polanyi replied:

> [Here in Berlin] the professors grab with great enthusiasm the hands of students who are thought to be gifted. They are like art collectors whose main passion is to discover talent . . . They educated me; they placed me where I can do my utmost. They provide me with everything and do not ask for anything. They trust that the man who is aware of the joy of science will not leave it for the rest of his life.[97]

Polanyi's longtime friend Eugene Wigner later said of him, "I doubt he was ever again as happy as he had been in Berlin."[98]

Crowther had some misgivings, however. He was British, and he was a left-leaning journalist who would become a partisan in the 1930s of what came to be called Bernalism and the scientists-for-social-responsibility movement in England. There was something about Berlin, Crowther wrote in 1930, that was "a little frightening." "I was left with the impression that the brilliant scientific efflorescence . . . had an intellectual life of its own, above that of industry and the people, in spite of the integration of the scientific research with industry." "This division of the high intellectual life from the brutal rumblings underneath was one of the most striking features of the Weimar Republic."[99]

In 1930, following the stock market crash, American financiers began withdrawing loans from Germany and businesses were failing. Unemployment rose to almost three million in this country of sixty-six million. Unemployment would increase to six million by 1933, in part due to the deflationary economic policies of Chancellor Heinrich Brüning and his opposition to implementation of public works projects. The September 1930 election for the Reichstag gave the National Socialists more than 18 percent of the popular vote, increasing their representation in parliament from twelve seats to a hundred and seven and making them the second strongest party in Parliament. Following the opening of the Reichstag on October 13, 1930, masses of demonstrators who were organized by the National Socialists smashed the windows of the Wertheim and other department stores on Leipziger Strasse and assembled in Potsdamer Platz shouting "Germany Awake!" "Death to Judah!" and "Heil Hitler!"[100] Polanyi's mother, Cecile, wrote her Viennese sister-in-law from Berlin in 1931:

> The times in Berlin are beginning to be frightful. Unemployment, privation, disheveled economic, political and emotional life. Riots, brawls . . . in short, from one side, Fascism, from the other Bolshevism! In the middle Democracy, starved, beaten, demolished, . . . behind all [the artistic happenings] the question: how much longer? One says the worst will come in January, the other in February . . . but that it will come they all believe.[101]

The KWG's administration called upon the national and Prussian governments for increased subsidies. Memoranda argued that German scientific development was essential to economic recovery, that catastrophe would ensue without further government subsidies, and that these monies must be granted with no brakes upon the scientific freedom that is necessary for researchers. In January 1932, Wilbur E. Tisdale, a former physicist working for the Rockefeller Foundation, wrote Lauder Jones that he had found the situation in Haber's institute to be more critical that at other KWG laboratories. Haber's personnel had been formally reduced from sixty-five to forty-two during the previous year and would be cut further to thirty-six in April 1932. No personnel had been discharged, but only thirty-six were receiving their customary stipends. The twenty-nine unpaid personnel would be denied admission to the institute in April because there were no funds to maintain their equipment. Tisdale also reported that Haber had advised Ladenburg to accept a permanent appointment in Princeton and that Haber would close Ladenburg's department.[102]

By comparison to 1930–1931, national funding was cut by 30 percent to the society for the 1932–1933 year.[103] The ratio of national to Prussian funds was approximately four to one. All through the Weimar period, Harnack and then Planck had enjoyed success in convincing the national government that the society's institutes should remain internally autonomous and not subject to governmental control, even though they depended a great deal on government funding, because they also were supported by private donors and industries. Brigitte Schroeder-Gudehus notes that that were no major limitations on the autonomy of the Gesellschaft (nor of the Notgemeinschaft der deutschen Wissenschaft, which dispensed funds to individual researchers) before 1933. Haber, like Harnack and Planck, made the argument that scientific research was essential to the domain of the national economy, public health and social welfare, but that no strings should be attached to research funding.[104] This was an argument that Polanyi strongly endorsed as worries increased about the commitment of the state to the support of fundamental research in a time of financial stress. In an article in the May 1930 issue of *Der deutsche Volkswirt* (The German economist), Polanyi insisted on the need for government support of science even when practical benefits might not be immediately obvious.[105]

## The Darkening of Berlin and the Dismantlement of Haber's Institute

During the 1920s and early 1930s there were rivalries, frustrations, and missed opportunities for Polanyi and his colleagues in Dahlem. Despite traveling widely, he never seems to have spotted a professional position that he preferred despite continual worries about his personal income in Berlin. In January 1933, Polanyi ended ten months of discussions by declining an exceptionally generous offer of a chair and a laboratory in physical chemistry at the University of Manchester, which was known for the international caliber of its physics and chemistry departments.[106] He made his decision to stay in Berlin in early 1933 despite the fact that Haber encouraged him to accept the lucrative Manchester position, just as he had advised Ladenburg to move to Princeton because of the institute's increasing financial problems and the uncertainty in Haber's view of the future in Germany for those of Jewish descent.[107]

Polanyi's discussions with Manchester had begun in March of 1932 after he received a letter from one of Manchester's senior chemists, Arthur Lapworth. The organic chemist Robert Robinson, who would receive the Nobel Prize in 1947, left Manchester for Oxford in 1928 and, after some deliberations, the university offered the vacant chemistry professorship to Hugh S. Taylor, who had taught physical chemistry at Princeton University since 1914. Taylor, who was one year older than Polanyi, had worked with Arrhenius and Bodenstein before taking his doctoral degree at Liverpool in 1914. He was British and he never relinquished his British citizenship. After lecturing in Manchester during October and December of 1931, Taylor declined the permanent position in Manchester on the grounds of his recent ill health in England and his need for the clear and crisp winters in Princeton.[108]

One of the names that came up in renewed discussions about the chair was Polanyi. He had given some lectures in the summer of 1931 at King's College in London, and the King's Professor of Chemistry Arthur John Allmand suggested Polanyi to Lapworth.[109] The Manchester search committee was a university senate committee that included the physicist William Lawrence Bragg and the chemist Bernard Mouat Jones (who was head of Manchester Municipal College of Technology). The procedure for new appointments was one of deciding on an invitation to a prospective candidate rather than publicly advertising an open position. The committee members identified possible candidates, including W. E. Garner at Bristol, Samuel Sugden at Birkbeck College, and both F. Philip Bowden and R. G. W. Norrish at Cambridge. Norrish would receive a share of the Nobel Prize in 1967 for his work in photochemistry and chemical kinetics.

In a meeting in late February 1932 the senate search committee under the chairmanship of Vice-Chancellor Walter B. Moberly reviewed outside letters of recommendation, which included Robinson's opinion that Polanyi should be appointed, while also recommending Garner and Norrish. Bragg read aloud letters from Hermann Mark at I. G. Farben and Paul Peter Ewald at Stuttgart in favor of Polanyi. A letter solicited from Richard Willstätter came to the vice-chancellor shortly after the committee decided to approach Polanyi. Willstätter's candid view was that Polanyi was a person of much importance, scientific productivity and originality with a position and salary at Berlin that was very good. He was a lovely man, but the pressure of too many physical chemists working together in a close space had made for tensions in the last few years. Polanyi likely would benefit from different circumstances, wrote Willstätter.[110]

Lapworth's letter of March 1, 1932, inquired whether Polanyi would seriously entertain a position at Manchester. The salary would be about £1,500 a year, which, Lapworth added, is a very high salary in Great Britain (equivalent to about US$7,500, or 22,000 German marks). Regarding the subject of Manchester as a place to live, explained Lapworth about a clearly touchy subject, the winters are never very cold and the nights in July and August are never oppressive. Rents are much lower than in London and music is better than any British town except London. Polanyi would have plenty of independence in the Chemistry Department, he could restrict his lecture courses to advanced students, and it was expected that he would make his main contribution to the university in his research work.[111]

Polanyi's next actions demonstrated his extreme reluctance to leave Berlin and the Kaiser Wilhelm Gesellschaft. The previous fall he had declined an offer of a position at Joffe's institute in Leningrad.[112] A couple of weeks after receiving Lapworth's letter, he drafted a reply in German and then sent

Figure 2.3. Research Staff of the Kaiser Wilhelm Institute for Physical Chemistry and Electrochemistry, with some of their family members, in the spring of 1933. Michael Polanyi and Fritz Haber are seated in the second row, second and third from the right. Herbert Freundlich is seated two seats to the left of Haber. (Courtesy of Archiv der Max-Planck-Gesellschaft, Berlin-Dahlem.)

a revision of the letter, in German, in which he offered to visit Manchester in early May. The final version of Polanyi's letter to Lapworth omitted a paragraph in which he explained how hard it would be for him to leave Berlin. "Although I first arrived in Germany in my later years, I nonetheless am rooted here with the greater part of my being [*meines Wesens*]. Even if I wanted to leave here, in order to secure greater latitude in my professional work, this decision would be especially difficult for me at the present moment when Germany endures such hard times. One would reluctantly give up a community [*Gemeinschaft*] which finds itself in such difficulties, when one has shared earlier in the good times."[113]

Before embarking for a previously scheduled trip to Holland, to be followed by the visit to Manchester, Polanyi scheduled an appointment to talk with Friedrich Glum, KWG's director-general, about the offer from Manchester. Polanyi drafted a handwritten account for Haber of what happened at the meeting with Glum, although he perhaps never sent the account to Haber. Polanyi asked Glum whether there was hope for assurance from the society that his position in Haber's institute was secure. The reply was disappointing. As a foreigner and a Jew (*Ausländer und Juden*), said Glum, Polanyi's position was not secure against possible changes that might be demanded on political grounds in the society's senate. If, for example, either Philipp Lenard or Johannes Stark was made president of the society there would be real danger to his position.[114] Lenard and Stark, both Nobel laureates in physics, were well-known as pro-Nazi anti-Semites who proselytized for the Aryanization of German science and who sought to influence control of German scientific institutions.[115]

Polanyi's May visit to Manchester was pleasant, although a pall of smoke hung over the city. Polanyi laid out some terms for his accepting the chair at Manchester—terms that initially shocked his hosts. He wanted a new physical laboratory to be built, estimated at a cost of £20,000 to £25,000, with an initial fund of £10,000 for apparatus and a sum of £1,000 per year for eight to ten personal research coworkers. Hit with news of these terms, the prominent physical chemist Frederick Donnan at University of College London wrote Lauder Jones to inquire whether the Rockefeller Foundation might help with funding for the laboratory, and Donnan encouraged Polanyi to request Lapworth and Bragg to do likewise.[116] The initiative to Jones did not go well. Jones had just been in Berlin in April, where he had met with Freundlich, Polanyi, and Kallmann about their needs in Berlin. By June, Lauder was in London where he had dinner with Donnan on June 10 after receiving a letter from Vice-Chancellor Moberly in the morning. Jones told Donnan that there was no possibility that the Rockefeller Foundation would fund a building project of this magnitude, largely in order to increase the prestige of English science, given the present financial condition in the United States.[117]

In late June 1932, Haber responded to a letter from Polanyi about the Manchester offer. In this matter, wrote Haber, "I have made clear in full candor that if I received this call [*Ruf*] in your situation, which I see as exceptionally honorable [*ehrenvoll*] and advantageous, I would accept it." "My character is one of the pessimist," continued Haber, "and I think it is conceivable that public [*öffentliche*] changes in Germany can make it impossible that your contract with the Society will be binding." Further, Haber soberly warned, "the limits of my defense if you are threatened lie naturally in the question of whether the very existence of the Institute is in danger, and I am convinced that the same is true for the President of the Society" (namely Planck).[118] Lauder Jones wrote Haber in July to ask for any news about Polanyi's possible appointment at Manchester, and Haber replied that he was not sure whether Polanyi would accept it. Jones also contacted Freundlich, who said that neither he nor Haber should interfere with Polanyi's decision but that Polanyi would likely remain in Berlin.[119]

A decision on what to do next in Manchester was put off by the senate committee until the new academic term in November. The vice-chancellor wrote Robinson about the dilemma that they faced: "in view of the financial requirements we have to make up our minds not only *whether* Polanyi is the best man for our purpose, but *how much* better he is than any of the English chemists who might conceivably be willing to come (e.g. Garner of Bristol or Allmand of King's College)."[120] Indeed, Moberly and Lapworth now found themselves faced with increasing outside knowledge of the offer to Polanyi and protests from some of the older generation of English chemists against the hiring of a foreigner instead of an Englishman. William J. Pope at Cambridge wrote a letter that Henry J. Armstrong, the crotchety old lion of chemistry at the City and Guilds College in South Kensington, sent on to Moberly, in which Pope not only complained that there were plenty of young English physical chemists to fill the post but that Polanyi's work had only been possible due to the vast resources at his command in Berlin. Robinson encouraged Moberly to stay the course with Polanyi but added that he would not favor appointing a foreigner if it meant that postwar efforts to build up British schools of chemistry could then logically be characterized as failures. Still, the important thing was to resuscitate the "Manchester School" of chemistry.[121]

In late November Moberly reopened negotiations with Polanyi and suggested that he visit Manchester again. The university promised Polanyi almost all that he had asked, with only the issue of the paid research assistants and mechanics up in the air. In the meantime, Polanyi had become more optimistic about the future in Berlin. In the elections of the past July, the National Socialist Party had won 230 seats, making them the largest party in the Reichstag. The elections of November 1932, however, had resulted in a slight decrease in National Socialist representation down to 196 seats, with the Social Democrats at 121 seats. Perhaps the National Socialists were past their peak. Arriving in Manchester in early December, Polanyi met with the chemistry staff and the senate committee that constituted the search committee. Thirty-year old Norman Burkhardt, the organic chemist in the department, remembered, "He was very good-looking, a very lively face, very dark—almost Indian—tremendous vitality. Straight away my reaction was, I hope we get him."[122] Some concerns had been expressed in the department about the extent of Polanyi's interest in the university in general and about his commitment to teaching, but these fears were largely muted by his presence.[123] After his return to Berlin, Polanyi received a letter reiterating Manchester's offer and promising to make arrangements to expedite his permanent residence. An addendum addressed the possibility that he might not be able to adapt to English life and that there would be no reproach if he were to resign his Manchester chair after only a short tenure. Another letter arrived with the news that tenants on the construction site for the new physical chemistry laboratory had been given a deadline of March 25 to leave their accommodations.[124]

In late December, Polanyi wrote a letter alerting Allmand that he would decline the Manchester offer. On January 13, 1933 Polanyi wrote to Lapworth and Moberly of his decision, giving as an explanation a weeklong bout of rheumatism following his December visit to Manchester when there had been a thick layer of industrial fog and soot in the city. He could not support a continually damp climate, he wrote in German, and indeed he and his wife had taken a long time to adjust to the darker skies of Berlin after growing up in Hungary. Faced with the difficulties of the move and the uncertainties about whether it would work, he had come to this negative decision.[125] He wrote Donnan, too, on January 17, that he had decided that he could not live in Manchester because the climate was a danger (*Gefahr*) for him.[126] It was not a very compelling explanation, although ironically it was similar to the reason given by Taylor for staying in Princeton. There was consternation and embarrassment in Manchester. On January 30, 1933, however, a greater danger than Manchester weather presented itself to Polanyi when the unthinkable (which Haber had thought imaginable) happened: Paul von Hindenburg, who had defeated Hitler in the presidential election of 1932 and had dissolved Parliament twice during the year, appointed Hitler as chancellor.

The next month, when the Catholic Center Party refused to join a coalition of the National Socialists and the German National Peoples' Party, Hitler persuaded Hindenburg to call elections for March 5, 1933. A week before the elections, the Reichstag building was set on fire, providing Hitler with a rationale for the declaration of a state of emergency, the suspension of civil liberties, and the arrest of Communists and other opposition leaders. At its opening on March 23, 1933, the newly constituted Reichstag, with a membership of approximately 44 percent from the National Socialist Party and 8 percent from the National Peoples' Party, and with many members arrested or in hiding, passed an Enabling Act that gave Hitler the power to govern by decree. Leo Szilard later reminisced that he visited Polanyi right after the Reichstag fire on February 27 and told Polanyi that he thought the government itself was responsible for the fire: "He looked at me and said, 'Do you really mean to say that you think that the secretary of the interior had anything to do with this?' and I said, 'Yes, that is precisely what I mean,' and he just looked at me with incredulous eyes."[127]

Einstein was in the United States in early March 1933 and he made a statement that was widely reported around the world. He would not return to a country, he said, that no longer enjoyed "political liberty, tolerance, and equality of all citizens before the law."[128] Planck immediately wrote Einstein that he feared that Einstein's public action would only cause more difficulties for Einstein's "racial and religious brethren."[129] Haber, too, worried about the effect of Einstein's public protests, which included Einstein's resignation from the Berlin Academy of Sciences before the academy had the opportunity to expel him on the grounds of public slander. Haber wrote to Einstein, "The outcome of this business for you is 'many enemies, much honor,' but we have to carry the worst part."[130] In mid-March Polanyi met with Donnan and asked if there were still a possibility for him to come to Manchester, and Donnan wrote Moberly with a caution that Moberly should not write Polanyi directly but communicate through Donnan or Allmand. Moberly set to work to see if he could revive an offer to Polanyi, but no decision seemed likely for a couple of months.[131]

From the beginning, National Socialist literature had focused on what was called "the Jewish problem" and preached the reversal of Jewish assimilation into German culture. The Nazis called for a nationwide boycott of Jewish businesses on April 1, 1933. Gangs of hoodlums led by Hitler's storm troopers roamed the streets and beat up those who appeared to be Jewish. Erwin Schrödinger, who had succeeded Planck in 1927 in the University of Berlin's chair of theoretical physics, happened to be downtown in front of the Wertheim Department Store. Only the intervention of one of his students, who was wearing Nazi insignia, prevented violence against Schrödinger, who was not Jewish, when he protested against what he was witnessing.[132] A week later the Law for the Restoration of the German Civil Service was enacted. It demanded dismissal of political enemies and non-Aryans who were employees of the state. Non-Aryans were defined as all persons with at least one Jewish grandparent, irrespective of religion. Exemptions were made for World War I Jewish frontline soldiers, at least until the Nuremberg laws of September 1935. All university personnel were among those subject to the law, as well as staff at the KWG institutes, for which at least 50 percent of the funding came from the state.[133] A questionnaire had to be completed giving dates of employment, type of war service, religion, and the "racial membership of four grandparents" (*Rassezugehörigkeit der 4 Grosseltern*). Previously it had been possible for those of Jewish descent to record on employment forms only their religion, which might be Roman Catholic, Lutheran, or Calvinist among assimilated Jews. Now the word *nichtarisch* had to be recorded.[134]

By mid-April 1933, Haber was informed by the Prussian Ministry of

Education that his institute could not reopen in its current organization after the holidays. At least a dozen of his forty-nine member staff could be affected, including the department heads Polanyi and Freundlich, although both, like Haber, appeared technically exempt because of their service during the last war. Haber's coworkers and research assistants Fritz Epstein, Paul Goldfinger, Ladislaus Farkas, Leopold Frommer, Hartmut Kallmann, Karl Söllner, Ernst Simon, and Joseph Weiss all were vulnerable. The technician Martin Schmalz fell under the new law, as did Haber's private secretary, Rita Cracauer, and another secretary, Irene Sackur. Irene was the daughter of Otto Sackur who had died testing military explosives in Haber's institute in 1914.[135]

Paul Harteck and Karl-Friedrich Bonhoeffer were among those who had worked at Haber's institute and were trying to assess the situation. Bonhoeffer was a professor at the University of Frankfurt, and Harteck was spending a year at the Cavendish Laboratory in Cambridge. Neither joined the Nazi Party. Bonhoeffer was the brother of the pastor and theologian Dietrich Bonhoeffer, who was an outspoken critic of the Nazi regime and was executed in April 1945 in the Flossenbürg concentration camp for his connections with plots to assassinate Hitler. Karl-Friedrich Bonhoeffer wrote Harteck just after implementation of the civil service law: "I would not have thought that the anti-Semitism would take on such vulgar forms, even though I have often gotten annoyed about the Jews. These methods are somehow shameful for *us*." Harteck reported in May, "In London the Jews, the one-half and one-quarter Jews of Germany, are gathering. If these people have ever had a liking for Germany, it can only have been a very superficial one, because now you really don't notice anything of it."[136] By June the opportunism of Nazi-implemented anti-Semitism showed up in the two men's correspondence, with Bonhoeffer writing Harteck that there are many openings in Germany now, and "you should get one of them, actually." Harteck received the vacant position of the refugee Jewish physicist Otto Stern at Hamburg in 1934.[137] Szilard wrote later of Polanyi's astonishment at what had happened:

> They all thought that civilized Germans would not stand for anything really rough happening. The reason that I took the opposite position was based on observations of rather small and insignificant things. What I noticed was that the Germans always took a utilitarian point of view. They asked, 'Well suppose I would oppose this, what good would I do? I wouldn't do very much good, I would just lose my influence. Then why should I oppose it?' You see, the moral point of view was completely absent, or very weak . . . And on this

basis did I reach the conclusion in 1931 that Hitler would get into power, not because the forces of Nazi revolution were so strong, but rather because I thought that there would be no resistance whatsoever.[138]

Szilard packed two suitcases in his rooms at Harnack Haus after Hitler's appointment as chancellor and he left for Vienna by the end of March.[139]

On April 19, Polanyi drafted a letter that he appears not to have sent to Haber. It was two days after Franck, director of the Second Physics Institute in Göttingen, submitted his resignation, even though he was exempt from the law because of his war service, in protest: "We Germans of Jewish descent are being treated as aliens and enemies of the Fatherland."[140]

In fact, Franck had been in discussions since 1932 with Planck, Haber, and other officials in the KWG and the Ministry of Culture about the possibility of moving from Göttingen to Berlin, where the Rockefeller Foundation would be ever more likely to fund a building for the Institute for Physics with Franck as its potential director. Now Franck publicly opposed the government's policy toward Jews, although he expressed his wish to remain scientifically active in Germany if possible.[141]

Polanyi praised Franck's courage for resigning his chair. Franck himself had written Haber in mid-April about his decision, saying that he would not "make use of the scrap of charity which the government offers veterans of the Jewish race."[142] Polanyi's unsent letter asked Haber to take some kind of action: "A hundred thousand will breathe easier if Haber stands up against the rising oppression," wrote Polanyi. "A right which one permits to be taken away without resistance is lost . . . Today's insanity will pass. But what will remain is the damage to their reputation caused by the Jews' own attitude." Three days after writing these words, although Polanyi was a war veteran and had kept his Austrian citizenship, he handed Haber his formal retirement application.[143]

Since Max Planck was on vacation in Italy in April, Haber met with Friedrich Schmidt-Ott, the vice president of KWG. In the second of these meetings Haber handed Schmidt-Ott the letters of resignation of his department heads Freundlich and Polanyi. Both had resigned rather than participate in the dismissal of their Jewish subordinates, even though their own resignations were not required.[144] Haber requested that the phasing out of employment should take place over the next five months rather than immediately. After further soul-searching and discussions with Willstätter, Laue, and Meitner, Haber submitted his own resignation to Planck and to Bernhard Rust, the Minister of Education for the Reich and for Prussia. His resignation read in part:

My decision to request retirement derives from the contrast between the research tradition in which I have lived up to now and the changed views which you, Minister, and your ministry advocate as representatives of the current large national movement. My tradition requires that in a scientific post, when choosing coworkers, I consider only the professional and personal characteristics of applicants, without considering their racial make-up. From a man in the sixty-fifth year of his life you will expect no change in the thinking that has directed him for the past thirty-nine years of his university life, and you will understand that the pride with which he served his German homeland all his life now stipulates this request for retirement.[145]

Faced with the terrible situation that included Haber's resignation, Planck had an interview with Hitler on May 16 in which Planck later said that he aimed to explain the dangers to German science of the forced emigration of Jewish scientists and in particular to say a word in favor of Haber whose war work had been crucial to the German effort. The interview failed miserably, with Hitler's replies including his exclamation that Jews are all Communists. Still, Planck encouraged Laue and Heisenberg to think that some of their senior and most distinguished colleagues might be saved for German science.[146] Otto Hahn, who had been in the United States during the spring, suggested to Planck upon his return to Berlin that a letter of protest signed by a group of distinguished "Aryan" professors should be sent to Rust. Planck's reply was fatalistic: "If today thirty professors get up and protest against the government's actions, by tomorrow there will be 150 individuals declaring their solidarity with Hitler, simply because they're after the jobs."[147] Bonhoeffer, who encouraged Harteck to go after one of those jobs, wrote Haber early in May 1933 of his sorrow at hearing of his former mentor's resignation, saying that he would always proudly and gratefully acknowledge that Haber had been his teacher.[148]

It was hard to know the best course of action to take. Planck counseled fellow non-Jewish scientists to stick with their posts and to do the best they could by their colleagues and German science. Nernst, who had succeeded Heinrich Rubens in 1924 as professor of experimental physics and head of the Physical Institute at the University of Berlin, retired from academic duties as he had earlier planned to do in the fall of 1932. He opposed National Socialism and the unjust treatment of his Jewish colleagues, giving up his Berlin residence and living southeast of the city. He had lost his two sons in the last war, and two of his three daughters had Jewish husbands. They left Germany.[149]

When Frederick Lindemann showed up in Erwin Schrödinger's Berlin office in mid-April to explain that he had offered a post at Oxford to Schrödinger's assistant, Fritz London, but that London had asked for time to think about it, Schrödinger astonished Lindemann by saying that he would take the post if London declined it. Both London and Schrödinger were in Oxford by October—London because he had been dismissed and Schrödinger because Lindemann saw an opportunity he could not miss. Schrödinger's letter of resignation made no protest against the new regime, but his biographer Walter Moore reported that he was entered into Nazi records as a "politically unreliable" Austrian. The Berlin *Deutsche Zeitung* carried a report in late October of the "severe loss" to "the German world of learning" due to Schrödinger's call to Oxford and Hermann Weyl's "call" to the Institute for Advanced Study in Princeton. Weyl's wife was Jewish.[150] Wilhelm Schlenk, who had succeeded Emil Fischer at the University of Berlin, spoke out and defended Haber. He was forced out of his professorship in 1935.[151]

The response of Bodenstein was not untypical of non-Jewish colleagues who sympathized with their Jewish friends and worried about the future of German science. He paid a visit to Lauder Jones's Paris office in May, reporting that Haber's institute was collapsing, but that his own Institute for Physical Chemistry at the university would not be affected. The trouble, in Bodenstein's view, was coming from the younger students who were calling for these dismissals. Bodenstein was forced to retire from the university in 1936 and died in Berlin in 1942.[152] Friedrich Paschen, who was president of the Physikalisch-Technische Reichsanstalt, suffered the fate of many who were not Nazi members and who stood in the way of Nazi scientists. He was compelled to retire from the PTR in 1933 and was succeeded by Stark. Planck soldiered on at the KWG until 1937, when he retired under exhaustion and pressure. He was replaced by Carl Bosch who was not a Nazi but was not an outspoken opponent either.[153]

Those who were Jewish had the chance to resign before their dismissal. Herzog apparently criticized Franck, Polanyi, and Freundlich in late April for taking the "comfortable" path of resignation instead of resisting the National Socialist movement from within their posts. Months earlier, Polanyi had suggested to Herzog and other colleagues, including Haber and Bonhoefer, that they should all resign if Hitler came to power.[154] Herzog was forced to retire, and he accepted a position at the University of Istanbul. He committed suicide during a vacation in Zurich in 1935.[155] On April 25, 1933, newspapers published a list of dismissed civil servants, including Max Born and Richard Courant at Göttingen. "Though we expected this," Born wrote to Einstein, it hit us hard. All I had built up in Göttingen during twelve years' hard work, was shattered."[156] Willstätter, who never had converted to Christianity and who had resigned in 1924 at the age of fifty-two from the University of Munich in protest against anti-Semitism, remained in Munich until 1938 when he had to flee to Switzerland from the Gestapo. He wrote, "It was my resolve to hold out in Munich as long as decently possible."[157] Lise Meitner, who had Austrian citizenship and baptism in the Evangelical Church, declined an offer for a year at Bohr's Institute in Copenhagen after Planck encouraged her to stay in Berlin. Although she was dismissed from her university position in 1933, she remained at the Kaiser Wilhelm Institute of Chemistry, where Max Delbrück was her assistant, and she continued to work with Otto Hahn and Fritz Strassmann in studies of radioactivity. She had to flee in 1938, following the Anschluss, after she was denounced by one of her colleagues, Kurt Hess, who lived in the apartment next to hers in Dahlem. She later wrote Hahn in June of 1948 that it had became clear to her that "I committed a great moral wrong by not leaving in '33 because staying had the result of supporting Hitlerism."[158]

In mid-July 1933, the Board of Trustees for the KWG accepted the final reports of the three administrators who were retiring from Haber's institute, with Planck presiding over the session. Freundlich reported on colloid chemistry, Haber on chain reactions and spectroscopy, and Polanyi on reaction kinetics. Planck thanked them for their contributions to German science, and Glum reported the decision of the minister that no exceptions were to be made in the employment of non-Aryans in the institutes.[159] Polanyi had heard in late April that he would be offered a position at Manchester, although no longer with the promise of a new physical chemistry laboratory and a munificent budget. He kept the information private for a while, until an announcement appeared on July 1 in the *Deutsche Arbeiter Zeitung*, following a report a week earlier in *Nature*.[160] Frederick Donnan helped arrange a position for Freundlich at University College in London, and Freundlich moved to the University of Minnesota in 1938. Karl Söllner was able to go to Minnesota after a first stop at Cornell University.[161] Ladislaus Farkas received a temporary position at Cambridge University and moved to the Hebrew University in Jerusalem in 1935. Ernst Simon got an offer at Chaim Weizmann's new Daniel Sieff Institute in Rehovot.[162]

Ute Deichmann provides accounts of others who were assistants and Privatdozents. Among them, Hans Beutler was able to work for Bodenstein until Bodenstein's retirement in 1935 when Beutler emigrated from Germany.[163] Hartmut Kallmann was employed at I. G. Farben and returned in 1945 to the KWG Institute for Physical Chemistry as its director before moving to New York University in 1948.[164] Mark, who had been advised at I. G. Farben in 1932 that his future at the company might be in jeopardy because of his Jewish wife, left his position at the University in Vienna in 1938. After interrogation by the Gestapo, he escaped through Switzerland and in 1940 began teaching at Brooklyn Polytechnic. Polanyi's old friend Alfred Reis, who had an informal arrangement as *Mitarbeiter* at Haber's institute while teaching at the Berlin Technische Hochschule, left for France in 1933 and eventually took a position at Cooper Union Institute of Technology in New York.[165] Leopold Koppel, whose philanthropy had made possible the very existence of the Institute for Physical Chemistry and other KWG institutes, died in August 1933 after losing his membership in the KWG Senate. His banking house and his electrical company, Auergesellschaft, were confiscated under the Nazi policy of the Aryanization of German businesses. The German company Degussa took control of Auergesellschaft and later provided uranium for research on a uranium machine at the Kaiser Wilhelm Institute for Physics, headed in the early 1940s by Werner Heisenberg.[166]

Under Planck's presidency, the KWG maintained alignment (*Gleichschaltung*) with the regime by taking Nazis into the Senate, flying the swastika, and ending correspondence with the customary "Heil Hitler!" Hahn temporarily directed Haber's institute, carrying through on the required dismissals, until Rust appointed Gerhart Jander, who had done research on chemical weapons. Jander quickly was replaced by Nazi Party member Peter Adolf Thiessen, and Rudolf Mentzel became head of a technical-chemical department at Haber's old institute. Mentzel showed up at the KWG Senate in an SS uniform and revolver.[167] In 1939 Haber's former institute received the title "National Socialist Model Plant."[168]

Haber left Dahlem in August 1933 shortly before Polanyi's departure for Manchester. Frederick Donnan and William Pope were determined to get Haber to settle in England, while the former Manchester University chemist and Zionist leader Chaim Weizmann tried to persuade Haber, just as Weizmann had tried to persuade Willstätter, to come to the new Daniel Sieff Institute in Palestine. While in Switzerland, before going to Cambridge, Haber collapsed in Brig and spent time recuperating at a Swiss sanatorium near Lake Constance. When he arrived in Cambridge in November, Pope had arranged laboratory facilities for him as well as passage and visas for Rita Cracauer, Joseph Weiss, and Haber's sister Else Freyhahn. Haber met with Freundlich, Polanyi, Farkas, Kallmann, and Harteck in Oxford and London before going to Basel at the end of January, where he was met by his son Hermann and his Berlin physician Rudolf Stern. They were shocked at his appearance. Haber died of a heart attack not long after, on January 29, 1934. James Franck wrote from Chicago to Hermann Haber and Rudolf Stern that "in my whole life I will not be able to forgive forcing a man like Fritz Haber, who did more for his country than is almost imaginable, to end his life in exile."[169]

Haber's death briefly focused a world spotlight on the Kaiser Wilhelm Gesellschaft. Planck became determined to organize an official memorial service on the occasion of the first anniversary of Haber's death. Planck surely had in mind the views of foreign colleagues as well as the opinion of the Rockefeller Foundation, from whom Planck still was seeking funding for a building for the Physics Institute, now formally headed by Peter Debye. In mid-January 1935, Rust forbade all state employees from attending the Haber ceremony scheduled for January 29 at Harnack Haus on the grounds of Haber's resistance to Nazi policy. Reminded by Planck of international attention already addressed to the event, Rust relented a bit by specifying that the ceremony could only be a private event with no spectators or news coverage.

Hahn, Bonhoeffer, and the retired colonel Joseph Koeth, who had directed the raw materials division of the Prussian War Ministry from 1915 to 1918, had agreed to give speeches, but after Rust explicitly forbade Bonhoeffer to attend, his speech was read by Hahn. Only a few members of the KWG defied Rust's prohibition, namely Meitner, Strassmann, and Delbrück. Most of the participants were the wives of Berlin professors and KWG staff. Among others who braved Nazi hostility to memorials for Haber and gave tributes elsewhere or published memorial notices were Carl Neuberg, Bonhoeffer, Schlenk, Willstätter, Laue, and Bosch. In a personal conversation with Hitler, Bosch had unsuccessfully warned of the damage that would be inflicted on German chemistry and physics by dismissals of non-Aryan scientists. In remarks at the Berlin Academy of Sciences in June 1934, Bodenstein expressed appreciation for Haber's scientific contributions and those of his Jewish colleagues Polanyi, Freundlich, Beutler, Kallmann, Wigner, and Franck, noting that it was conflict with the state and not Haber's death that had ended this research group.[170] As Dieter Hoffmann and Mark Walker write, the Haber memorial service was the high point of public dissent by scientists against National Socialist policies.[171]

The 1935 ceremony was a strange and tragic shadow of the earlier ceremony in Dahlem on the occasion of Haber's sixtieth birthday celebration on December 2, 1928. Among those who were present and photographed in 1928 at the planting of the Haber Linden in the courtyard of Haber's institute were the later Nazi members or sympathizers Kurt Hess and Adolf

Figure 2.4. Planting the Haber Linden in celebration of Fritz Haber's
sixtieth birthday on December 2, 1928. Max Planck and Michael Polanyi are sixth
and seventh from the left side of the photograph; Lise Meitner is near the center.
(Courtesy of Archiv der Max Planck-Gesellschaft, Berlin-Dahlem.)

Kühn, along with Glum, Planck, Polanyi, Laue, Kallmann, Hahn, Ladenburg, Weissenberg, Meitner, Harteck, Söllner, Freundlich, and Farkas.
Among the tributes offered Haber on that joyous occasion was a speech
by Polanyi. After the linden tree was planted in honor of Haber's steady
leadership, Polanyi told the audience that there have been two kinds of
great leaders in science: innovators and traditionalists, revolutionaries and

conservatives, *Zerstörer* and *Erhalter*. Einstein, Planck, and Rutherford are among the revolutionaries, said Polanyi. Their works are victories of the great heretics. In contrast, Fritz Haber is a conservative, who believes in the true correctness (*wirkliche Richtigkeit*) of the scientific picture. He is a leader who recognizes that science can not be considered as just pure enlightenment, but that its pursuit and stewardship require decisive action. Scientists must take it upon themselves to control the life of their science as a family of researchers who know how to rule themselves while working within the fabric of the state and the economy.[172] In his description of the aims and structure of the Kaiser Wilhelm Institute for Physical Chemistry, Polanyi extolled in 1928 what he later called a "city" or "republic" of science with Haber as its chief legislator.[173]

By the time of his forced departure from Berlin in 1933, Polanyi had developed a more critical attitude toward Haber—the disappointment, perhaps, of the son in the father who could not protect his family from harm or in the master who failed the standards of the guild or in the elected leader who abandoned his citizens to unjust and immoral laws. Haber had been Polanyi's hero, and Haber's institute became a blueprint for Polanyi's ideal of the scientific community. Until Polanyi confronted the brutal ugliness and moral abyss of the new German state and its destruction of his beloved city of science, Polanyi could not accept the notion of leaving Berlin. Nor was he alone in this commitment to scientific life in Weimar Berlin in the 1920s and early 1930s. His attitude was shared by many scientists of Polanyi's generation. It all was destroyed when a new German elite abrogated the old contract for politically neutral scientific knowledge. As a National Socialist leader declared at the opening of the university in Leipzig for the 1933–1934 academic year, "The university in the new Germany *will be political*, an educational institution for political persons who place their knowledge and their abilities in service to the nation."[174]

Michael Polanyi's thirteen years in Berlin gave him the experience of what he later transformed into an idealized vision of the scientific research community. Polanyi's later reflection on the world he had lost propelled him into an intellectual exposition and defense of the social and institutional preconditions that are necessary for the flourishing of modern science. The freedom of research that he had experienced in a tightly networked community of world-class colleagues within the tree-lined precincts of Dahlem became an induplicable but idealized memory that formed the foundation for his later writings on the nature of scientific life and scientific achievement. The intensity of his scientific life in Berlin infused the passion of his later sociologically inflected philosophy of science. The loss of his Berlin scientific community and gradually of his own scientific productivity led to later reflections on the social conditions of scientific work and on the difficulty of transplanting established traditions in new terrain. He was hardly alone in these reflections among refugee scientists. Erwin Chargaff, for example, wrote in his autobiography of the limits on transferring from place to place scientific modes of thought and practice which "live in the womb of a particular language and civilization."[175] The next two chapters will look carefully at the specific lines of research and development of scientific theories that structured Polanyi's scientific life and career in Berlin and in Manchester. Like the social and institutional arrangements that were the fabric of his scientific life, the daily routines of doing science were also the stuff of which his philosophy of science was to be made.

# THREE

# Origins of a Social Perspective:
# Doing Physical Chemistry in Weimar Berlin

In his writings on the nature of scientific inquiry and the meaning of science, Michael Polanyi drew both explicitly and implicitly upon his experiences in physical chemistry over a period of some thirty-five years. His first scientific publication squarely in physical chemistry dates from 1913, and the last appeared in 1949.[1] Polanyi's first philosophical paper, in which he developed the notion of an essential tension in scientific work between tradition and innovation, might be said to be his 1928 speech delivered on the occasion of Fritz Haber's sixtieth birthday, as discussed in the previous chapter, but it was only in the 1940s that a flurry of philosophical writings began to appear as Polanyi's interests turned away from physical chemistry toward economic, political, and philosophical writings. He drew upon memories of the everyday routines and ups and downs of his own scientific career in order to develop what he intended to be a novel and even controversial description of science as a community of dogmatic traditions and social practices rather than a march of revolutionary ideas and individual genius. Some of his first arguments along these lines were delivered in 1946 in the form of the Riddell Lectures at the University of Durham. The lectures received a good deal of attention when they were published in Great Britain and the United States as the volume *Science, Faith and Society*, and many of the themes in the Riddell Lectures were repeated and extended in Polanyi's writings in the 1950s and 1960s, including *Personal Knowledge* (1958).

Polanyi's career at the Kaiser Wilhelm Institute in Berlin provided a deep reservoir of detailed experiences and impressions from which he drew in his reflections on scientific practice. The general applicability of his sociological and philosophical conclusions was ensured by the wide range of his scientific researches in the overlapping fields of thermodynamics,

X-ray crystallography, chemical kinetics, reaction mechanisms, and quantum theory. Polanyi had many scientific triumphs over his career, especially in the fields of chemical kinetics and chemical dynamics, but he also experienced some disappointments, for example in his work in surface chemistry and X-ray diffraction studies of fibers and metals. Physical chemists judged his surface theory of adsorption inadequate and expressed strong preference in the 1930s for the very different theory of American chemist Irving Langmuir despite Polanyi's attempt to improve and generalize his own approach. Polanyi's pioneering work in X-ray diffraction met resistance from colloid chemists, who objected to his interpretation of the molecular structure of cellulose, and it suffered indifference from physicists, who did not see any fundamental theoretical significance at the time to his investigations of the strength of materials. Polanyi saw his own successes and setbacks as typical of scientists' experiences , although he underestimated the privileged position of his membership in one of the world's scientific centers, as well as the esteem in which he was held in the broad scientific community.

This chapter examines Polanyi's work in surface chemistry and X-ray diffraction by way of analyzing how he later drew upon his experiences in order to develop the notion of the "typical" or ordinary scientist who is at the heart of everyday scientific practice (what Thomas Kuhn more famously called "normal science"). Polanyi's investigations in these two fields generated greater skepticism and less recognition from his colleagues than his results in chemical kinetics and reaction dynamics, which are the subject of the next chapter. Polanyi's reflections on resistance to his work turned him to sociological explanation, rather than logical explanation, for the mechanism by which scientific priority and recognition are accorded within the structure of scientific authority. The chapter concludes with an analysis of how he reinterpreted his work on surface chemistry and X-ray diffraction within a sociological framework in two essays published in the early 1960s.

## Debate in Surface Chemistry: Polanyi's Potential Theory versus Langmuir's Valence Theory

Polanyi's interest in thermodynamics and the adsorption of gases on solid surfaces originated in his studies with Ferenc Tangl in Budapest and accelerated after his coursework in Karlsruhe. His youthful ambition to establish a reputation in thermodynamics is demonstrated in his attempts to interest

Walther Nernst and Albert Einstein in some of his earliest researches. When he returned to Budapest from Karlsruhe in 1912, Polanyi wrote a theoretical paper on Nernst's heat theorem, a law which requires changes in entropy (a measure of a molecular system's disorder) to become zero at the temperature of absolute zero. Nernst's work correlated experimental results from thermodynamics and thermochemistry, using concepts of energy and entropy in order to successfully predict specific heats of solids at low temperatures. (Specific heat is the heat required to raise the temperature of one gram of mass one degree Centigrade, and experimental discrepancies from its theoretical values had been noted since the 1860s.) In his original treatment of the problem, Nernst did not address the behavior of individual molecules, but in 1907 Einstein used the new quantum theory of energy and applied it to individual particles in order to predict the behavior of solids as the temperature approaches absolute zero.

Polanyi extended Nernst's theorem to high pressures at ordinary temperatures using an application of Einstein's quantum theory for specific heats. Polanyi predicted that at infinitely high pressure the entropy would go to zero, because the decreasing volume would have an effect on energy frequencies similar to that of decreasing temperature.[2] Writing from Budapest to his friend Alfred Reis at the Karlsruhe Technische Hochschule, Polanyi asked Reis to transmit this manuscript to Georg Bredig. When Bredig replied that he could not judge the manuscript, Polanyi requested him to send the work to Albert Einstein in Zurich. Bredig soon advised Polanyi that Einstein's opinion was favorable, and Bredig himself offered some stylistic suggestions and advised Polanyi to contact Karl Scheel in Berlin about publishing the work in the German Physical Society's *Verhandlungen*.[3] An ecstatic Polanyi wrote his sister Mausi that "I gambled and I won." He later reminisced, "Bang! I was created a scientist."[4] As he worked further on the heat theorem problem, Polanyi wrote Einstein, who now was in Berlin, and asked directly for advice, initiating a correspondence around December 1914 that lasted until the following July. Polanyi failed, however, to convince Einstein of some of his arguments. In a second article published in 1915 on the derivation of Nernst's heat theorem, Polanyi acknowledged Einstein's help.[5]

Nernst was less encouraging than Einstein when Polanyi sent him a letter at the University of Berlin in the summer of 1913 with a copy of his first paper on the derivation of the theorem. Nernst especially objected to Polanyi's historical introduction to the paper in which Polanyi attributed to Einstein the prediction that specific heats will tend toward zero as temperature declines toward absolute zero. Nernst wrote Polanyi that he had pointed out this possibility in the heat theorem two years before Einstein and that this behavior of specific heats was self-evident, given the decrease of entropy to zero at absolute zero. Nernst also questioned the clarity of Polanyi's discussion of cyclic processes in his manuscript and broke off correspondence in the fall of 1913 with the comment that the only way to solve their points of disagreement was by speaking personally. At least two behavioral norms that Polanyi learned from this correspondence were the scientist's preoccupation with priority and the significance of personal contact in resolving controversy.[6]

Polanyi's doctoral thesis, completed in Budapest during the war, was built on publications of 1914 and 1916 that focused on the adsorption of gases on the surface of colloidal droplets and on porous solids and used data from published literature on adsorption of $CO_2$ by charcoal.[7] Adsorption is a process that occurs when gas or liquid particles bind to the surface of a solid or a liquid (adsorbent), forming a layer (adsorbate) of molecules or atoms. The binding to the surface is usually weak and reversible. Adsorption is different from absorption in which a substance diffuses into a liquid or solid.

Polanyi's aim was to develop a theory of the adsorption of gases by solids that leads to quantitative predictions of the effects of changes in temperature and pressure on the quantity of gas adsorbed. In his thesis, Polanyi conceived of the forces of adsorption as working through a potential energy gradient, along the same lines as Arnold Eucken, who introduced the term "adsorption potential" in 1914. Eucken assumed that adsorption forces are independent of temperature, that adsorption of any one gas molecule is independent of other already adsorbed molecules, and that molecules in an adsorbed layer obey the same nonideal equation of state (i.e., the 1873 Van der Waals equation) as when they are not adsorbed. Eucken also proposed a specific general formula for the adsorption force.[8] Questioning the validity of the formula, Polanyi sought to explain adsorption data through the derivation of an adsorption isotherm, in which the volume of gas adsorbed (adsorbate) per gram of adsorbing material (adsorbent) is plotted against the equilibrium pressure at a constant temperature.[9] Polanyi assumed that there are long-range intermolecular attractive forces of a Van der Waals type acting between the adsorbing solid and the atoms or molecules of gas that locate themselves in layers at the surface of the solid. Van der Waals forces are weak attractive forces between atoms or nonpolar molecules. In addition to deriving the adsorption isotherm, which is the representational curve or contour line for adsorption at a single temperature as pressure varies, Polanyi described the adsorption potential in a simple functional equation:

$$\varepsilon = f(\varphi)$$

In the equation, $\varepsilon$ is the "adsorption potential," or energy of affinity of the adsorbate for the adsorbent, and $\varphi$ is "the space enclosed by the level having this potential," that is, the volume through which the attractive force is effective.[10] Polanyi did not offer a generalized equation for the isotherm but rather an experimental isotherm from which other isotherms could be calculated at different temperatures.[11]

Polanyi defined the adsorption potential at a point near the adsorbent as the work done by the adsorption forces in bringing a molecule from the gas phase to that point. The work was conceived as work of compression, in analogy with vapor condensation, with the effect that the first layer of adsorbate is under greatest compression, the second layer under lesser compression, etc., until the density decreases to that of the surrounding gas. Thus there is a gradient of potential surfaces from $\varepsilon_0$, $\varepsilon_1$, $\varepsilon_2$ . . . to $\varepsilon_i$ and the adsorbed layer resembles the atmosphere of a planet with the layer's density decreasing outward from the surface of the solid.

Polanyi assumed that the potential is independent of the temperature of the adsorbent and that the pressure exerted by particles of the adsorbate on its immediate neighborhood is the same as that which it would exert, at the same density and temperature, if it were in the free state.[12] Polanyi's approach lay thoroughly within the framework of nineteenth-century classical thermodynamics and adhered to common assumptions about the process of adsorption. In short, he was working within accepted practices in physical chemistry, but at a time when there was no good explanation for the Van der Waals attractions.

In 1915, Irving Langmuir, who had taken his doctoral degree in Berlin with Nernst in 1906, published a paper that, like Polanyi's work, applied to gas adsorption in general and to catalysis, in particular, where catalysis is the acceleration of a chemical reaction by addition of a substance that does not participate in the reaction. A common system in catalysis was the use of minute amounts of a solid material to speed up a reaction involving a gas. Langmuir rejected the usual assumption that gas adsorption and heterogeneous catalysis occur in thick layers of adsorbed gases. Langmuir postulated instead that adsorption happens in a single molecular layer on a metal or oxide surface.

Langmuir's investigations on surfaces were part of his research at the General Electric Laboratory in Schenectady, New York on the physics and chemistry of incandescent light bulbs, including the adsorption of various gases on different filaments, for example, oxygen on tungsten and hydrogen on platinum. Langmuir proposed the action of electrostatic forces in adsorption and began framing his work within the new electron-pair theory of chemical valence, which was first clearly articulated in print by G. N. Lewis in 1916.[13] Langmuir argued that the force that retains the adsorbed gas particles on the solid surface results from the electrical valence forces of the atoms or molecules in the outermost layer of the solid. Adsorbed gas particles fit into a single layer at the solid's surface, as if on a chessboard where each square, or hole, can only be occupied by a single gas particle. Thus, adsorption ceases when the surface is fully occupied.[14]

Langmuir's theory constituted a bold break with classical thermodynamic theory and with the prevalent assumption that a gas adsorbate is relatively thick, becoming less dense as the distance from the adsorbent surface increases. Langmuir's approach was innovative in its use of the language of the electrical, or electron-pair, chemical bond, a theory to which Langmuir soon tied his name by writing a series of landmark papers in 1919 in which he introduced the terms "covalent" and "electrovalent" to describe nonpolar and polar bonds.[15] In October 1921, Langmuir lectured at a joint session in Edinburgh of the British Association for the Advancement of Science on what would become known, to Lewis's dismay, as the "Lewis-Langmuir" theory of the electron bond.[16] That fall Fritz Haber invited Polanyi to give a full account of adsorption theory at Haber's biweekly colloquium which brought together members of Kaiser Wilhelm institutes across Berlin-Dahlem.[17] The result was considerable criticism from Haber and Einstein, who both faulted Polanyi for ignoring the new electrical theories of the structure of matter. Polanyi later said, "Professionally, I survived the occasion only by the skin of my teeth."[18] It was a sobering experience for him.

Herbert Freundlich, who headed the colloid department in Haber's Physical Chemistry Institute, had been studying gas adsorption for well over a decade. During the war Freundlich's research centered on the adsorption of gases used in chemical warfare and on the properties of charcoal and other adsorbents in gas mask canisters.[19] His textbook *Kapillarchemie* (1909) was one of only two references cited in Polanyi's paper on adsorption in 1914, and Polanyi expected Freundlich to be sympathetic to the potential theory. An empirically based "classical" adsorption isotherm became known as the "Freundlich equation," even though Freundlich was not its originator.[20]

Freundlich gave an account of both Polanyi's and Lewis's adsorption theories in subsequent editions of *Kapillarchemie* in 1922 and 1923, and he was ambivalent about strongly choosing one theory over the other.[21] He told Polanyi, "I am heavily committed now to your theory myself; I hope it is correct."[22] In the textbook Freundlich wrote:

> The arguments vary considerably according as we assume the adsorption layer to consist of several layers of molecules or only of one. As I am at present unable to choose definitely between the two possibilities, they may both be dealt with at length: as an example of a theory postulating several layers of molecules, Polanyi's will be taken; and as an example of one depending upon only one layer, that of Langmuir.[23]

Freundlich noted, however, that Polanyi's theory, "as is well known, is more readily explained on the assumption of a multimolecular layer, *although it does not exclude a unimolecular one.*"[24] Hermann Mark later recalled that most organic chemists were not then much interested in electrons or in the new physics in contrast to physicists and physical chemists, and the organic chemists found Polanyi's theory perfectly satisfactory.[25]

From 1914 to 1922, Polanyi wrote a total of twelve papers on adsorption, but after the October 1921 Haber colloquium, he did little work in the field for six years, focusing instead on X-ray diffraction studies of fibers, crystals, and metals, and on chemical reaction rates. Given the preoccupation among many physical chemists and most theoretical physicists in the 1920s with atomic theories and quantum mechanics, Polanyi began to doubt that he could even have published his original classical potential theory if he had first tried to publish it in 1921 rather than during 1914 to 1917.[26] In 1926, however, Arthur S. Coolidge at the GE Laboratory in Schenectady, New York, published data supporting Polanyi's theory, as did Homer H. Lowry and P. S. Olmsted at Bell Telephone Laboratories in New Jersey in 1927. Polanyi's interest in adsorption revived, coinciding with the arrival in 1928 in Berlin of Fritz London, who took up appointments at the Kaiser Wilhelm Institute for Physics and at the University of Berlin's Institute for Theoretical Physics.[27] London and Walter Heitler had just published a revolutionary paper on the chemical bond and binding energy, in which they extended Werner Heisenberg's theory of quantum-mechanical resonance and exchange energy from the two-electron helium atom to the hydrogen molecule.[28]

In 1928, Polanyi began a series of laboratory experiments on adsorption rather than continuing to rely only on published data; he collaborated with F. Goldmann, K. Welke, and Walter Heyne and substituted for his original adsorption theory a picture that covered both unimolecular and multimolecular adsorption.[29] This new approach employed two-dimensional equipotential lines, instead of three-dimensional equipotential surfaces, using an interpretation of "islands" of compressed gas or liquid that grow in areas of lower potential on the adsorbent surface. The work with Goldmann treated each adsorbed molecule as having its own attractive potential at a spot on the surface. As pressure increases, new islands form, and some of them flow together until the whole surface becomes covered with adsorbed liquid when the vapor pressure is reached. Using this model, Polanyi and Goldmann concluded that adsorption of vapors on charcoal was not unimolecular, although the data obeys the Langmuir equation.[30]

Polanyi agreed to organize an all-day colloquium on heterogeneous catalysis and surface chemistry during the four-day meeting of the Deutsche Bunsen Gesellschaft in Berlin in May 1929. He scheduled seven speakers, including Max Bodenstein, Nernst's successor at the University of Berlin. Other speakers included the Princeton University physical chemist Hugh S. Taylor (who was Bodenstein's former student), Fritz London, and Polanyi himself. At the beginning of the meeting, Haber delivered a ringing introductory speech in which he extolled the usefulness of heterogeneous catalysis and other applications of the adsorption phenomenon for German industry, as well as its theoretical interest in connection with the new wave mechanics.[31]

Polanyi had committed his paper to memory the evening before the Friday colloquium.[32] In his remarks, he discussed how Langmuir's adsorption isotherm could be derived as a special case from his own revised potential theory. He attempted to account for some observational differences between Langmuir's and his theories, allowing that there could be monomolecular layer adsorption on some solid surfaces, but not on all. Experimental information was accumulating, he noted, which suggested that Langmuir's theory was useful at higher temperatures and lower pressures on smooth surfaces, while Polanyi's was more suitable for lower temperatures and higher pressures on porous surfaces. Oddly, likely because Polanyi had made concessions to the applicability of Langmuir's theory under some conditions, Haber concluded from Polanyi's remarks that he had given up his original theory. Eugene Wigner later recalled that he, too, thought Polanyi now was leaning toward Langmuir's theory.[33] Rather, Polanyi was trying to make the case that Langmuir's formula represents an idealization which is not obeyed in all cases, and that he was amending his own theory so that Langmuir's isotherm could be derived as a special case. Polanyi's claim was that his amended theory now was the best general theory.[34]

This was a theme repeated by Polanyi in a paper sent in June 1929 to the *Zeitschrift für Elektrochemie*, where he responded to criticisms of his work by Heino Zeise, who claimed that data from experiments reported by three sets of investigators earlier in 1910 and 1917 better fit isotherms calculated on the basis of Langmuir's theory than those calculated from Polanyi's potential theory.[35] Since Zeise's method of calculating the isotherms clearly favored Langmuir's theory over Polanyi's, Polanyi had good reason to question the results.[36] Researchers in the adsorption field mostly continued to favor Langmuir's theory, partly because it appeared that the most active sites of adsorption and the most energetic ones, which especially figure in catalysis, follow the Langmuir equation and isotherm.[37]

If Polanyi's theory had seemed old-fashioned to Haber and Einstein in 1921, Polanyi thought he could overcome this objection and persuade researchers of the power of his theory after London's work began to intersect with his own. On his arrival in Berlin, London frequented both Haber's biweekly seminars at the Physical Chemistry Institute and Max von Laue's weekly Wednesday physics colloquia at the University of Berlin.[38] London applied himself to the study of Van der Waals forces between atoms and molecules, forces which, unlike valence forces between atoms, are additive and relatively unaffected when a third molecule is brought in the vicinity of the two molecules. These are the kinds of forces Polanyi had proposed as acting across his adsorption potential gradient.

By analogy to dispersion of light, London in 1930 applied the name "dispersion forces" to the long-range intermolecular forces that exist between nonpolar molecules, giving them a theoretical foundation which they previously had not enjoyed.[39] In a paper published in *Zeitschrift für physikalische Chemie*, he explicitly mentioned the heat of adsorption as a phenomenon that could be explained by physical dispersion forces and referred in the article to a note coauthored with Polanyi that would soon appear in *Naturwissenschaften*. Dispersion forces provided a common theoretical basis, London argued, to the phenomena of Van der Waals deviations of real gases from ideal gases, heats of evaporation, and surface adsorption.[40] In their paper, Polanyi and London demonstrated that the adsorption potential of an adsorbent decreases with the distance from the adsorbent wall just as Polanyi had first argued in 1914. The potential gradient now had a firm theoretical basis in the newest version of quantum mechanics.[41] In January 1932, the Faraday Society sponsored a symposium at Oxford on the subject of adsorption of gases. The three invited keynote speakers were Freundlich, Polanyi, and Eric Rideal of Cambridge. As it turned out, Freundlich and Polanyi had to cancel their trip at the last minute due to the strained financial situation at the Berlin institutes during the winter of 1931–1932.[42]

Taylor, who was on leave from Princeton at Manchester University, introduced the agenda for the Oxford symposium. Two rival theories about surfaces had been struggling for supremacy, he said, one assuming thick compressed films and long-range forces of attraction extending outwards from solid surfaces, and the other emphasizing extremely short-range interatomic and molecular forces resulting in a unimolecular adsorption layer. Now, Taylor claimed, "As to adsorption, one can summarize the situation by saying that the thick compressed film has during the last decade become progressively thinner until now the tendency is to reinterpret the ideas of the compressed film in terms of the unimolecular layer."[43] Taylor highlighted recent results on activation energies and their applicability to adsorption, along with Polanyi's work with London on resonance energies between atoms of the adsorbent and adsorbate. Langmuir's chemical approach to adsorption, however, clearly appealed to Taylor more than Polanyi's, perhaps because of Taylor's own interest in chemically activated sites and catalysis.[44]

As in his concluding remarks at the Berlin colloquium in 1929, Freundlich's paper, circulated for the Faraday symposium, suggested that current experimental results did not permit a clear decision between the rival theories. "Monomolecular layers are perhaps the rule at low pressures, polymolecular ones at higher pressures, especially near to the saturation pressure . . . theories using an adsorption potential lead to other formulae correlating the amount adsorbed with the equilibrium pressure." Freundlich reiterated the usefulness of the empirically based equation relating the volume of adsorbed material to the equilibrium pressure known as the "Freundlich equation."[45]

Polanyi's Faraday Society paper emphasized the role of the newly formulated cohesion or dispersion forces as an explanation of adsorption. It also reiterated his old claims that 1) the adsorption potential is not essentially dependent on temperature, 2) it is independent of the chemical nature of the molecules of adsorbent and adsorbate, and 3) molecules in the adsorbed state exert approximately the same forces on one another as they do when they are free (which is not the case if any chemical forces are acting).[46] According to Polanyi, forces exist both of chemical and physical adsorption, forces which later were to be characterized under the clear rubrics chemisorption and physisorption. In general, Polanyi's theory worked best for microporous and heterogeneous adsorbate surfaces at higher pressures and Langmuir's for other systems. In Polanyi's view, Langmuir was partly right and so was Polanyi.[47]

If the symposium of January 1932 was inconclusive, deliberations of the Swedish Academy of Sciences were not. In the fall of 1932 it was announced that Langmuir would receive the 1932 Nobel Prize in Chemistry. The prize was awarded "for his discoveries and investigations in surface chemistry," including his experiments on monomolecular layers of oily substances on water. His Nobel lecture in December made specific mention of only a few colleagues—among them Hugh Taylor, but not Polanyi—although Langmuir did discuss cases of adsorption in which a surface saturated with a molecular layer of adsorbed molecules might attract and adsorb a second layer. He also explicitly noted that there are different types of adsorption and that at sufficiently low temperatures Van der Waals forces alone suffice to cause adsorption. Langmuir added that he had made the distinction in 1918 between Van der Waals (physical) adsorption and activated (chemical) adsorption.[48]

Langmuir had been nominated earlier than 1932 for a Nobel Prize, including nominations in physics from Niels Bohr in 1928 and 1929. Bohr nominated Langmuir along with the British physicist O. W. Richardson for their work on electronics and the theory of emission of electrons from the surface of hot metals. Richardson alone received the Nobel Prize in Physics for 1928, which was reserved and awarded in 1929.[49] Herbert Freundlich, who himself had been nominated in 1928 for chemistry for his work in colloid chemistry, was among those who nominated Langmuir in 1928 for the chemistry prize on the grounds of Langmuir's work in surface chemistry and adsorption. Freundlich particularly drew attention to Langmuir's studies of the behavior of thin films of organic materials in water, allowing calculation of the size of the cross section of the organic molecule. Freundlich's nomination arrived too late to be taken into account for the 1928 award, but went into the file for later consideration.[50] In late September 1928, the five-member Chemistry Committee for the Swedish Academy of Sciences completed their report on scientists nominated for that year's prize. At that time Langmuir had two nominations that were considered in the deliberations, one from John Bayard Bates and another from Leo Heindrik Baekeland, the American inventor and manufacturer of the plastic known as Bakelite. The committee also knew of Bohr's nomination of Langmuir in physics.

An appendix to the committee's report was written by the theoretical physicist Carl Wilhelm Oseen and committee member Ludwig Ramberg, who was a physical organic chemist. They organized Langmuir's research and contributions in chemistry into three areas: gas-filled incandescent lamps, surface chemistry and thin molecular layers, and the discovery of atomic hydrogen and the atomic hydrogen welding process. The welding process uses an electric arc between tungsten electrodes in a hydrogen gas environment to produce temperatures up to 4,000°C, which is higher than reached by an acetylene torch. In the research area of surface chemistry, Oseen and Ramberg divided Langmuir's researches into the categories of adsorption, surface tension, and catalysis. Under adsorption, they discussed what they termed "Eucken and Polanyi's" theory, mentioning briefly, too, Polanyi's work with László Berényi in Budapest in 1919. Langmuir's approach, they said, was entirely different from the Eucken-Polanyi approach in that Langmuir broke with the old theory and attributed adsorption forces to chemical valence.[51]

The Chemistry Committee again filed reports on Langmuir for the 1931 and the 1932 chemistry prizes. The determining nomination likely was the one in 1931 by Theodor Svedberg, the prominent physical chemist and member of the committee whose work on molecular dimensions helped establish what came to be called "molecular reality" and the value of the kinetic-mechanical picture in physics and chemistry.[52] Langmuir's approach fit perfectly with Svedberg's.[53] Another member of the Chemistry Committee who nominated Langmuir was Hans von Euler, with a nomination in January 1932.[54] Svedberg himself had received the Nobel Prize in Chemistry in 1926, as had Euler in 1929, so their opinions especially mattered, and Svedberg wrote the official resume of Langmuir's work for the 1932 prize (with no mention of Polanyi among recent contributors to surface chemistry).[55] An additional factor in the award to Langmuir was the visit to Stockholm to receive his Nobel Prize in 1930 by Karl Landsteiner of the Rockefeller University in New York. Landsteiner's award in physiology or medicine recognized his work on the typology of blood groups. He told Svedberg and probably others such as Euler, all of whom received funds for their research from the Rockefeller Foundation, of resentment in the United States that T. W. Richards had been the only American Nobel Prize winner in Chemistry (1915). In a letter of January 9, 1931, Landsteiner proposed Langmuir for a Nobel Prize on the basis of his work on surface films. Historian Robert Marc Friedman notes the interest in chemistry committee members for this kind of work because of its significance in biology as well as in chemistry and physics.[56]

At the Nobel ceremony in December 1932, Hendrik Gustaf Söderbaum, who was an organic chemist and secretary of the Swedish Academy of

Sciences, made the presentation speech. He began speaking about Langmuir's award with a pun: "Superficiality is a quality which has always had a bad reputation, not least within the field of science." He then explained to the audience that opinions were divided about the forces operating in adsorption and that the general conception had been that the gas at the boundary surface with a solid is in a condensed state with the density decreasing outwards in the same way as the density of the earth's atmosphere decreases with distance from the earth's solid crust. "This year's Nobel prize-winner in chemistry has advanced an entirely conflicting theory," stated Söderbaum, "and one which at first sight seems to be particularly bold." Unsaturated valences in surface atoms exert forces effective at fixed points and at fixed distances, he explained, so that gas particles are adsorbed as if occupying squares in a single-layer chessboard. This bold theory was supported by evidence, concluded Söderbaum, and it met the aims of the Nobel Foundation to recognize useful discoveries and inventions because Langmuir's results were valuable for understanding chemical reactions that are important both in industry and biology.[57]

After 1933, Polanyi published only one more article on adsorption, a review article published in 1935.[58] His correspondence and personal papers show that he felt frustration about the award to Langmuir and concern about ever receiving just due for his work. Erika Cremer later recalled Polanyi saying to her, "Whose fate is better, mine or Langmuir's? My theory is absolutely right but not accepted. Langmuir's theory is wrong but he is very famous . . . Langmuir is better off!"[59] Polanyi had laid out in print in 1929 an argument that his adsorption theory had been *original* in 1916 in comparison both to Eucken and Langmuir. As he saw it, Polanyi claimed, his contribution lay in establishing the principle that adsorption forces include cohesion (after 1930, dispersion) forces which act between adsorbed molecules and that were neglected by Langmuir.[60] Yet Polanyi decided not to teach his theory to undergraduates in physical chemistry classes after he moved to the University of Manchester in 1933 because he thought that students' examination results might be compromised in their grading by an external examiner and by younger staff members at Manchester who were familiar with Langmuir's approach and the electron-valence theory of chemisorption.[61]

Some textbooks from the 1940s do not refer at all to Polanyi's theory and mention that theories different from Langmuir's had been almost completely discarded. In Frank Henry Macdougall's *Physical Chemistry* (1943), the Eucken-Polanyi theory is described briefly as an alternative to Langmuir's, with the still-applicable warning that "at the present time, the experimental evidence is perhaps not sufficiently complete or conclusive to enable us to decide which of the two theories gives a more correct picture of the adsorption process. We have given preference to Langmuir's hypothesis mainly because of its greater simplicity and consequent greater ease of application."[62] Langmuir's appeal, as with Svedberg, partly had to do with its visual atomistic mechanical model. As Taylor wrote, "With its thermodynamic nature, Polanyi's theory does not provide so detailed a *picture* of the mechanism of adsorption as does Langmuir's."[63]

One 1940s monograph on physical adsorption of gases and vapors does give attention to Polanyi's experimental and theoretical work, assigning him a prominent place alongside Langmuir, but then goes on to develop a generalized adsorption theory on the basis of Langmuir's equation rather than Polanyi's isotherm.[64] *The Adsorption of Gases and Vapours* (1944) was written by Stephen Brunauer, who collaborated with Paul Emmett at Johns Hopkins University and Edward Teller at George Washington University. Beginning in 1935, Brunauer coauthored a series of papers with Emmett of which the best known is a much-cited 1938 paper in the *Journal of the American Chemical Society* in which Brunauer, Emmett, and Teller laid out what became known as the "BET theory."[65]

The BET theory, like Brunauer's 1944 book, takes the Langmuir equation to be the "most important single equation in the field of adsorption" and the starting point for derivations of both unimolecular and multimolecular adsorption, including the working out of five types of adsorption isotherms. The BET theory assumes, like Langmuir's, that an adsorbent surface possesses uniform, localized sites and that adsorption at one site does not affect adsorption at neighboring sites. The theory allows, however, the adsorption of molecules in second, third, etc., layers. Brunauer aimed at a general theory with equations and he faulted Polanyi's potential theory for failing to provide any analytical formulation for the adsorption isotherm.[66]

In contrast to Brunauer, the Russian chemist Mikhail Moiseyevich Dubinin and his collaborators argued in the 1940s that the adsorption potential is a valuable general theory but that it must be recognized, in a departure from Polanyi, that adsorption varies according to the nature of the adsorbate as well as the solid adsorbent, specifically with respect to the polarizability of the adsorbed molecule. Thus Dubinin kept Polanyi's principle of the temperature independence of adsorption but rejected Polanyi's principles of the independence of adsorption from chemical and valence forces. Dubinin introduced an "affinity coefficient," and he treated the adsorption process as volume-filling of micropores rather than layer-by-layer adsorption on the pore walls. Dubinin's reputation was considerable in the field of surface chemistry, and he served as head of the Chemical Division of the Soviet Academy of Sciences from 1948 to 1963.[67]

In a brief history of these developments, the physical chemist Ljubisa R. Radovic wrote that a general theory of adsorption still eludes scientists for explaining why and to what extent one adsorbent performs better than another. Practitioners who need to quantify differences among adsorbent properties generally prefer the Dubinin approach, and theorists reluctantly prefer the BET theory.[68] We can only conclude that in the 1920s and 1930s, clear preference for Langmuir's approach, whether in the deliberation of the Swedish Academy of Sciences or among Polanyi's close colleagues, was rooted in Langmuir's novel invocation of molecular dimensions and chemical reactivity at a time when Van der Waals forces were of less immediate interest.

## X-Ray Diffraction Studies of Fibers and Molecules

At the same time that he was pursuing the problem of adsorption, Polanyi began collaborating in the Fiber Chemistry Institute's new program to develop X-ray diffraction studies of the structure of crystals, fibers, and other materials. This work was a program of fundamental research that was intended to have practical applications for the German textile industry. In 1920, the institute's space consisted of a few small rooms in Haber's Physical Chemistry Institute with equipment for the X-ray investigations scattered around the KWG premises. A microscope could be found in the Institute for Biology and photographic equipment in the Materials Testing Laboratory a few blocks away. An X-ray apparatus was set up in the basement of Haber's private villa next to the Physical Chemistry Institute.[69] Mark later reflected on the remarkable progress made by their group. At the beginning of the 1920s, there were no X-ray tubes in Berlin that could be operated at high intensities over long periods without careful supervision. Research groups had no precision cameras in which crystals or crystalline objects could be conveniently mounted for irradiation in all possible orientations, nor did they have instruments for working at very low or very high temperatures or at other extreme conditions. Over the course of the decade, a well-equipped and world-class X-ray crystallography center was created at Berlin-Dahlem.[70]

At the beginning, Polanyi had no experience with X-ray crystallography. His friend Alfred Reis was working in this field at Karlsruhe and encouraged Polanyi to have confidence in his ability to do innovative work, giving him a pep talk that he was, after all, a "decisive, consequential, and rigorous-thinking man."[71] The theory and practice of X-ray crystallography had just been highlighted during the summer in Max von Laue's long-delayed Nobel lecture in Stockholm for the physics award of 1914. Laue was the associate director of the Kaiser Wilhelm Institute for Physics, where Einstein was nominally the director. Using the hypothesis that X-rays have tiny wavelengths comparable to the distance between atoms in a crystal lattice, Laue predicted in 1912 that diffraction phenomena resulting from the interference of x-radiation waves with one another would be produced by passing x-radiation through the natural three-dimensional diffraction grating of a crystal. Two doctoral students at Munich, Walter Friedrich and Paul Knipping, confirmed this prediction in Arnold Sommerfeld's laboratory using polychromatic, or "white," x-radiation and a crystal of copper sulfate.[72] At the time, Laue and his colleagues mainly were interested in investigating the production of secondary fluorescent X-rays produced by oscillating atoms in the crystal.[73] Both physicists and crystallographers paid attention to these results. Physicists saw in the photographs a demonstration that X-rays are electromagnetic radiations, not particles or pulses. Crystallographers recognized evidence for the space group theory of crystallography that modeled crystal structure as ordered three-dimensional arrays of point particles.[74]

Contrasting with Laue's approach was the work of the University of Leeds physicist William H. Bragg and his son William Lawrence Bragg (better known as Lawrence Bragg) during 1913–1914. The younger Bragg preferred to think that incident radiation was being reflected from planes in the crystal, with the reflected beam flashing up only at a particular angle of incidence and this angle becoming greater the closer together the planes are. Thus each spot in the X-ray diffraction photograph corresponds to a particular set of planes and the spots farthest out in the photographic pattern correspond to the most closely spaced sets of planes. The angle $\theta$ for a spot is determined by the wavelength $\lambda$ of the x rays and by the interplanar spacing d in the crystal, which was expressed in a formulation that became known as the Bragg Law.

$$n\lambda = 2d \sin \theta$$

The Braggs oversaw the building of an X-ray spectrometer that relied on ionization to register the intensity of reflection of the X-rays rather than using the blackening of photographic film. Their instrument opened up the exploration of X-ray spectra emitted by different elements, as well as the analysis of crystal structure. Their joint work gained father and son a shared Nobel Prize in Physics in record time in 1915.[75]

A surprising development occurred that same year in Max Born's laboratory in Göttingen, when Peter Debye obtained with Paul Scherrer a set of characteristic diffraction lines from a fine-grained powder—not a crystal—of lithium fluoride. They used monochromatic x-radiation and a cylindrical diffraction camera. The powder sample, in rotation at the center of the camera, was bathed in radiation, and a cylindrical film around the circumference of the camera recorded the diffracted images.[76] Albert Hull at the General Electric research laboratory used a high-vacuum X-ray tube developed by his GE colleague William D. Coolidge and independently published a theory of powder crystal analysis in 1917 as part of an investigation of the crystal structure of iron.[77] It was the use of newer and more powerfully evacuated X-ray tubes that led to the recognition of the crystallinity of substances previously held to be amorphous.[78] Herzog's assistant Willi Jancke learned how to build the new tubes from Scherrer and was able to produce powder patterns from a pulverized salt of cellulose after he arrived in Berlin. He began irradiating a compressed bundle of cellulose fibers in ramie with monochromatic radiation in the summer of 1920 and obtained photographs of twenty-six smeared points arranged symmetrically in groups of four around two mirror-planes, one plane passing through the primary beam and the axis of the fibers, and the other oriented to the normal.[79] Herzog gave the photograph to Polanyi to interpret and within a couple of weeks Polanyi concluded that the four-point image was due to a group of parallel crystals within the fiber arranged around the axis of the fiber, and that the fiber as a whole has rotational symmetry around its axis. His results appeared in a joint publication with Herzog and Jancke.[80]

In the early spring of 1921 Polanyi was asked to speak at the Haber colloquium scheduled for March 7 and to evaluate X-ray data for cellulose within a theoretical framework of hypotheses about the structure of colloidal substances of high-molecular weight such as cellulose, rubber, and resins. From the X-ray data he derived the dimensions of the periodic unit cell in cellulose as a rhombic cell with edges 7.9, 8.45, and 10.2 Angstroms, which is close to the modern value. (An Angstrom is one ten-millionth of a meter, or $1 \times 10^{-10}$ meter.) From the volume of the cell and the known density of pure cellulose, Polanyi calculated that each unit cell consists of the glucose anhydride $C_6H_{10}O_5$ with a molecular weight of 162, which corresponds to the chemically determined empirical formula for cellulose that had been known since the early nineteenth century.[81] He suggested that subsidiary or secondary valences (*Nebenvalenzen*) might hold the unit cells together into larger aggregates within cellulose, an idea favored by Herzog and by Kurt Hess, who was head of the department of cellulose chemistry at the Kaiser Wilhelm Institute for Chemistry. Hess specifically proposed the notion that cellulose was formed by an aggregation of cyclic molecules composed of five glucose anhydride units.[82] As Polanyi played with this idea, Herzog wrote Polanyi a note to clarify the history of priorities for the idea of secondary valences which, Herzog said, had been established within Johannes Thiele's school and further developed by himself and Hess. Herzog also drew Polanyi's attention to a recent summary on the subject in a new paper on polymerization and long-chain molecules by Hermann Staudinger at the Zurich Polytechnic Institute.[83]

Staudinger's research was striking to Polanyi because it was entirely different from Herzog's approach. At this time Staudinger and his students were using traditional methods of organic synthesis and they only began to employ X-ray diffraction studies in the late 1920s. In the early 1920s, Staudinger pressed forward the hypothesis that the molecular units of glucose or amino acids in cellulose or protein, respectively, are polymerized into very long-chained molecules, adding up to molecular weight for cellulose and protein in the tens of thousands. In contrast, the aggregate theory argued that colloidal substances, such as cellulose, rubber, starch, proteins, and resins are physical aggregations of relatively small molecules held together by weak secondary valences that do not withstand mild heating. Staudinger's paper of 1921 proposed long-chain structures for polystyrene, polyoxymethylene, and rubber. Karl Freudenberg at Heidelberg had reported in 1920 that the yield of cellobiose during cellulose degradation favored long-chain structure, but Polanyi appears not to have known these results at the time.[84]

The prejudice against huge molecular weights was rooted in the past work of Germany's perhaps most eminent chemist, Emil Fischer, who had held the University of Berlin's chair of chemistry from 1892 until his death in 1919. Fischer thought it impossible that natural proteins have molecular weight higher than around 4,000. If evidence—for example, from freezing point depression—seemed to point to higher weights, the explanation must lie in the "aggregate theory" of colloids accounting for "pseudo-high molecular weight." Colloidal chemists in Freundlich's department for colloid chemistry in Haber's institute favored this explanation, as did Herzog and Hess.[85] The vehemence of the prejudice against very high molecular weights can be seen in the attitude of Heinrich Wieland, who had worked under Haber at the KWG in Berlin during the war and would receive the

Nobel Prize in Chemistry in 1927 for his studies of the structure of bile acids. He told Staudinger to "drop the idea of large molecules; organic molecules with a molecular weight higher than 5000 do not exist. Purify your products, such as rubber, then they will crystallize and prove to be low molecular compounds."[86]

In the final version of his talk delivered at the Haber colloquium, Polanyi reported that his analysis of the cellulose diffraction patterns allowed two different interpretations: the molecular structure was *either* an aggregate of hexobiose-anhydrides *or* it was one giant molecule composed of a single file of chemically linked hexoses. Both structures, he said, were compatible with the symmetry and size of the elementary cell, the repeat unit within the three-dimensional space lattice of the crystal.[87] After his remarks, there ensued what Polanyi later called "a storm of protest from all sides."[88] His critics drew the conclusion from Polanyi's work that the small size of the unit cell within crystalline cellulose proved that cellulose is a chemical substance of low molecular weight on the grounds that the molecule cannot be larger than the unit cell. Cellulose could only appear to be of higher molecular weight because physical forces, or secondary valences, are weakly holding the molecules together.[89] Polanyi, who was a physical chemist and an outsider in the field of organic chemistry, deferred to his elders. He later concluded that this was a missed opportunity: "I failed to see the importance of the problem."[90] In contrast Hermann Mark, later known as a pioneer of polymer chemistry, followed up in the next decade on the problem of the structure of cellulose and large molecules, coming to remarkably important conclusions such as the flexibility and spiral coiling of the chains.[91]

In 1926, two scientists at the University of California at Berkeley, the botanist L. O. Sponsler and the X-ray crystallographer W. H. Dore, published a detailed structure for a long-chain cellulose molecule. They proposed an alternating 1,1 and 4,4 bonding of the glucose units in the chain, rejecting a 1,4 bonding despite the fact that there was irrefutable chemical evidence of a 1,4 glucosidic bond in cellobiose, the cellulose degradation product that is two linked glucose units. Not surprisingly, many organic chemists renewed their objections to claims of X-ray crystallographers trying to solve problems of molecular structure while ignoring chemical evidence.[92] When Polanyi's attention was drawn to Sponsler and Dore's interpretation, he responded that their structure was impossible, given his own early diffraction data and the geometry of the four-point system.[93]

The objection to macromolecules on the grounds of the size of the unit cell continued to have influence until 1927 or so. Staudinger moved from

Zurich to Freiburg in 1926, succeeding Wieland in the chair of organic chemistry on the basis of Wieland's recommendation. Although Wieland did not like Staudinger's macromolecule or long-chain molecule hypothesis, he highly respected Staudinger's work on ketenes and other organic substances of low molecular weight.[94] Shortly after Staudinger's return to Germany, a symposium on polymers was scheduled at the annual meeting of the Gesellschaft Deutscher Naturforscher und Ärzte (Society of German Natural Scientists and Physicians) in Düsseldorf, with Haber and Richard Willstätter among the organizers. Before the meeting, Haber called Mark into his office and told him that he wanted Mark to give a lecture on X-ray structure analysis of organic substances and especially on the question of whether a small crystallographic elementary cell excludes the presence of a very large molecule. Mark later recalled:

> Haber was, as usual, right to the point and had understood the situation very clear[ly]. Most of the arguments . . . were qualitative . . . On the other hand, the small volume of the elementary cells of cellulose and silk had been clearly established; each of them could only contain four units. Now, if it were true that a molecule could not be larger than the elementary cell established by crystallographic investigation there would be a clear-cut and objective argument against the existence of very large molecules.[95]

Mark's paper may have disappointed Haber, since Mark allowed the possibility that the molecule could be larger than the unit cell, with lattice-like forces at work within the molecule, referring to work by his KWG colleagues and by Reis. Staudinger's presentation at the meeting impressed Willstätter, but failed to persuade other participants at the time. What was more telling was research in X-ray studies that Staudinger directed when he returned to Freiburg which resulted in evidence that the molecular size of high molecular compounds cannot be determined by X-ray measurements.[96]

Later in 1926, Mark left Berlin to become director of the I. G. Farbenindustrie research laboratory in Ludwigshafen, a town midway between Karlsruhe and Frankfurt. He continued research there on the crystal structure of cellulose, collaborating with Kurt Meyer. Shortly thereafter, in 1928, they proposed a compromise theory in which long-chain, high-molecular weight molecules are held together by special Van der Waals–type forces in "micelles" or "banks" as flexible coils. This theory is essentially the present interpretation.[97] By the mid-1930s, Staudinger's original hypothesis that natural and synthetic polymeric substances are macromolecules of extremely high molecular weight had support from new molecular weight determinations in Svedberg's ultracentrifuge researches at Uppsala. Staudinger received the 1953 Nobel Prize in Chemistry for his work. As Max Perutz later reminisced in a tribute to Lawrence Bragg, few chemists paid attention to insights from X-ray crystallography into molecular structure until the late 1930s. It is hardly surprising that Polanyi's alternative proposal of high molecular weight in 1922, based on the new X-ray diffraction technique and very little evidence, was rejected at the time.[98]

Polanyi's research group was one of three groups in Berlin in the X-ray field during the period of the 1920s and 1930s. Paul Becker and Jancke were working with Herzog; J. Böhm and H. Zocher with Freundlich; and Mark, Karl Weissenberg, and Ervin von Gomperz with Polanyi.[99] Mark had taken his degree in chemistry with Wilhelm Schlenk in Vienna and came to Berlin with Schlenk when Schlenk succeeded Emil Fischer in the University of Berlin's professorship in chemistry. Haber invited Mark to head the Fiber Chemistry Institute's technology department in 1922. Weissenberg, a talented Viennese mathematician, arrived in Berlin in 1920 as assistant to the applied mathematician Richard von Mises and joined the fiber chemistry group in 1921. He later held a chair of physics at the University of Berlin. Gomperz was a Hungarian whom Polanyi had tutored in Budapest, and Wigner originally was a member of Polanyi's X-ray group at the Physical Chemistry Institute, but switched in 1925 to a focus on chemical reaction rates as the subject of his dissertation.[100]

Following up on a paper published by Polanyi in May 1921 on his discovery of "layer lines" in X-ray diffraction photographs, his group developed important new methods in studying all kinds of crystalline substances. Polanyi had found that all but two dots forming a fiber diagram, such as the diagram for cellulose, lie on a series of symmetrical hyperbolas when a flat photographic film is used as detector, with each hyperbola composed of dots reflected by planes having identical indices with respect to the crystal axis parallel to the fiber. He concluded that the formula for the series of these hyperbolas is a function of the identity period (the distance between the identical diffracting centers) parallel to the axis of the fiber. Circles, rather than hyperbolas, resulted when the film was wrapped cylindrically around the fiber bundle or crystal rotation axis. The separation between any two successive curved or "layer" lines could give a direct measure of the size of the unit cell.[101]

The "fiber diagram," which the group was producing resulted from monochromatic radiation directed at a stationary fiber. The fiber consists of many single-crystal subfibers, all with parallel long axes. The subfibers have random orientation about the common axis, so that together they are composed of many identical crystals, each in a different orientation about the common axis. Ernst Schiebold earlier had studied the use of a different kind of experimental arrangement at Leipzig in which a single crystal is mounted with an axis normal to the X-ray beam and rotated. He recognized the equivalence of the fiber photograph and a rotating-crystal photograph. Soon Schiebold, Polanyi, and Weissenberg began further developing the rotating crystal method, particularly in the study of metals.[102] Starting with Polanyi's simple geometrical formula for determining the identity period from layer lines, Weissenberg derived a mathematical treatment of the layer-line relationship in which the distances between the layer lines provided direct information about the identity period of the lattice under investigation so that the size of the repeating unit could be calculated.[103] Mark directed the first experiments with the rotating crystal method, and Polanyi published with Weissenberg and Schiebold a detailed account of the method in 1924. Weissenberg also developed in 1923 what would become known as the Weissenberg camera, or X-ray goniometer, in which the photographic film is rotated with the crystal so that each layer of lines corresponds to a particular setting of the crystal around the axis, providing a powerful analytic tool for determining three-dimensional crystal structure.[104]

Using both the powder and rotating crystal methods, with the improvements made by Weissenberg, Polanyi and his coworkers undertook a program of research on materials in the solid state. They constructed a variety of machines for treating fibers, particularly cellulose and silk, by swelling, stretching, relaxing, and drying them. Mark and Polanyi noted that the increase in the modulus of cellulose fibers on stretching seemed similar to the reinforcement of metal wires during cold-drawing, leading them to a systematic investigation of the changes accompanying the cold-drawing of a zinc wire. Thus, their studies of rigidity, extensibility, elasticity, melting, softening, and swelling expanded from fibers to metal wires. Work in support of the research of a University of Berlin PhD candidate, Margarete Ettisch, showed how the faces in the cubic crystals of the metal tended to line up during the drawing process and how a cross-section of the wire would appear when examined under the microscope. Investigations by Mark, Erich Schmid, and Polanyi included the stretching of single wire crystals of zinc to five or six times their original lengths, while also measuring the force applied in the process and the degree of elongation. Polanyi supervised construction of a new instrument, a stretching apparatus (*Dehnungsapparat*), on which workers from several departments collaborated.[105]

Together Polanyi and Mark discovered the slip properties of single-crystal tin, while Schmid worked out a law for the shear stress component along the slip direction in a slip plane. They were studying why crystal planes slide readily across each other in some cases and what happens in shear hardening, where resistance increases without bending and twisting.[106] These studies had clear industrial as well as theoretical interest, and Polanyi continued with them throughout the 1920s, with a particular interest in the phenomenon of crystal weakness and an explanation at the molecular level of how solids rupture and break. Following up on some work of Ludwig Prandtl in 1928, Polanyi developed by 1932 his own theoretical concept of dislocation (*Versetzung*) in crystals.[107] Every process that destroys the ideal structure of crystals increases material strength: in a particular instance of dislocation, for example, "ten atoms on one side are opposed by eleven atoms on the other side of the line."[108]

Polanyi presented his theory of dislocation in April 1932 to members of Abraham Joffe's research institute in Leningrad, a few months after he was unable to attend the Faraday Society meeting on adsorption. Polanyi's dislocation theory noted that observed diffraction lines from powdered samples of cold-worked metals are diffuse and that diffraction spots in single-crystal photographs are elongated. While a material like diamond is a reasonably perfect crystal, metallic crystals were now proposed to consist of small units that are slightly out of alignment with one another. Thus, many properties of metals depend on imperfection of the crystal structures, resulting, for example, in reduction of the actual tensile strength of rock salt 200-fold from its theoretical values.[109]

After returning to Berlin from Leningrad, Polanyi learned that a similar idea on dislocation was to appear in a thesis by the Hungarian engineer Egon Orowan under the direction of Richard Becker at the University of Berlin. Orowan knew of Polanyi's work on the subject and suggested that they write a joint paper, but Polanyi instead published separately, in *Zeitschrift für Physik*. In 1934, the physicist Geoffrey I. Taylor, the Royal Society Professor at Cambridge University, published another version of the theory of dislocation.[110] Gradually, dislocation—including the concept of the point defect, consisting of missing, misplaced, or foreign atoms, ions or molecules—came to be well understood, especially through publications of Robert W. Pohl's research group in Göttingen during the period 1920 to 1940.[111]

Polanyi's solid-state work, carried out in an institute with significant ties to commerce and industry, had clear industrial applications, and the strength of solids was a technical problem assigned to Polanyi's group.[112] Yet partly on account of the direct industrial link, Polanyi found the reactions of many of his colleagues to his work disappointing. Most organic chemists were not only largely uninterested in the application of the tool of X-ray diffraction to molecular structure, but also ignored his group's work on the mechanical behavior of organic solids. Hermann Mark later expressed frustration that so many organic chemists missed the significance of the work Polanyi's group was doing in the early 1920s.

> We, in our fiber research, were interested in . . . strength, elasticity, water absorption, and abrasion resistance. Until this point, for solid organic compounds, interest was focused on melting point, solubility, color, surface activity . . . but never on mechanical behavior. We soon recognized that the solution viscosity of polymeric systems is not their most important property, but the enormous influence of polymer chain length on all physical and mechanical characteristics is . . . [important, as can be seen in] . . . the hardness of ivory and ebony, the elasticity of kangaroo tails, the toughness of alligator skins, and the softness of cashmere and vicuna wool.[113]

In publications in the 1930s, Staudinger also linked the long-chained macromolecular structure of substances to their physical properties, noting that it is because substances such as cellulose are linear macromolecular substances that they are fibrous and tough and dissolve with considerable swelling.[114]

Even more disturbing to Polanyi than the views of organic chemists was the response of his colleagues in physics and physical chemistry to the work in X-ray diffraction studies. Einstein and other Berlin physicists sometimes visited Polanyi's laboratory in the 1920s, but one of their main purposes was to ask if Polanyi's X-ray research group could find the increase in wavelength of secondary X-rays scattered from free electrons which Arthur H. Compton had discovered in 1922. Compton's discovery was a confirmation of Einstein's particle or photon theory of radiation first proposed in 1905, which predicted a loss of energy in reflected radiation after the energy of its photons was imparted to free electrons in a target. In some of the first follow-up work to Compton's announcement of his discovery, William Duane had not been able to duplicate Compton's measurements at Harvard University, but Mark and Harmut Kallman were able to do so in Berlin in 1924.[115] As for Mark's investigations on the strength of materials:

> Truthfully, the results of our studies failed to impress the leading members of the scientific community in the Kaiser Wilhelm Institute, including Max von Laue, Fritz Haber, O. Hahn, Lise Meitner, James Franck, K. F. Bonhoeffer,

and others who were preoccupied with radioactivity, atomic and molecular quantum phenomena, and catalysis.[116]

In correspondence, Polanyi wrote despondently that the scientific problems of the strength of materials (*Festigkeit*), magnetism, electrical conduction, viscosity, and so forth, were areas in which no physicist wanted to study unless they had some relation to atomic physics.[117] "Dirt physics" was the term of opprobrium used by Wolfgang Pauli for the study of processes in real solids.[118] Orowan, by original training an electrical engineer, later recalled of his turn from electrical engineering to materials science, "plasticity was a prosaic and even humiliating proposition in the age of De Broglie, Heisenberg and Schrödinger, but it was better than computing my sixtieth transformer."[119]

The state of the materials science field changed dramatically after the Second World War when the chemistry and physics of semiconductors, transistors, and integrated circuits on silicon chips revolutionized fundamental and technological sciences. The defining text of modern solid-state science, *Modern Theory of Solids*, was written by Wigner's Princeton graduate student Frederick Seitz during 1936 to 1940. Wigner, Seitz, and Pohl became pioneers of the solid-state field, rather than Polanyi, largely because of their mastery of the new quantum mechanics, in which he never was at ease. The ascendancy of solid-state physics and materials science was signaled in 1970 when the *Physical Review* split into two separate journals, with *Physical Review B* devoted to solids and containing more articles than could be found in *Physical Review A* on nuclear and high-energy physics.[120] Polanyi's work in "dirt physics" became overshadowed by the revolutionary quantum-mechanical understanding of materials.

## Recognition and Authority in the Scientific Community

In February 1962, Thomas Kuhn formally interviewed Polanyi for the History of Quantum Physics archival project. On that occasion Polanyi brought up his old adsorption theory. He told Kuhn that his adsorption theory had been dismissed by Einstein and others in the 1920s because it did not fit in with the new electron theories. Polanyi's view was that once the adsorption theory was discarded at its inception by influential scientists, his subsequent attempts to update the theory and his demonstration at the Faraday Society meeting in 1932 of the consistency of evidence with his new theory had attracted no interest.[121]

In an article that appeared in *Science* in September 1963, Polanyi used the history of his potential theory of adsorption as the centerpiece of an argument about the ways in which scientific authority is constructed and recognition is conferred in science. He claimed that recent surveys of work on adsorption by Dubinin and by R. S. Hansen and C. A. Smolders indicated that his own "correct theory of adsorption," which had been "delayed by almost half a century," now was gaining acceptance. Presumably, by his "correct theory," Polanyi meant his principle that adsorption forces include cohesion forces, which he had highlighted in 1929 as his original discovery.[122] In fact Polanyi's statement in *Science* was an oversimplification of the complexities still plaguing theories of adsorption in the 1960s, but as a result of reading Polanyi's article, the molecular biologist Gunther Stent highlighted adsorption theory in an essay on prematurity and uniqueness in scientific discoveries published in 1972 in *Scientific American*.[123] Relying on Polanyi's account in *Science*, Stent wrote:

> Despite the fact that Polanyi was able to provide strong experimental evidence in favor of his theory, it was generally rejected. Not only was the theory rejected, but it was considered . . . ridiculous by the leading authorities of the time. . . . [At] the very time he put it forward the role of electrical forces in the architecture of matter had just been discovered . . . [a] point of view . . . irreconcilable with Polanyi's basic assumption of the mutual independence of individual gas molecules in the adsorption process . . . It was only in the 1930s after F. London developed his new theory of cohesive molecular forces . . . that it became conceivable that gas molecules *could* behave in the way which Polanyi's experiments indicated . . . Meanwhile, Langmuir's theory had become so well-established, and Polanyi's had been consigned so authoritatively to the ash can of crackpot ideas, that Polanyi's theory was rediscovered only in the 1950s.[124]

In fact, as we have seen, Polanyi's theory had not been considered either crackpot or ridiculous, and the theory had figured in discussions by leading figures in the field since the 1920s, although it consistently was judged inadequate as a general theory or inferior to alternatives for particular purposes, especially in early comparisons with Langmuir's theory.

In an essay in 1962 Polanyi discussed another of his disappointments—the negative reaction in 1921 to his macromolecular chain hypothesis for cellulose, in which organic chemists, colloid chemists, and even Haber had dismissed his proposal of an alternative to the generally favored aggregate theory. Polanyi also was disappointed to learn from an assistant who was asked to survey X-ray crystallography textbooks published through the

1940s, that most books failed to mention his name in connection with the rotating crystal method of X-ray diffraction. Bragg, Hull, Shiebold, J. D. Bernal and others were cited, but not Polanyi.[125]

With much perspicacity and undue humility, Polanyi suggested that his disappointments had taught him that:

> The example of great scientists is the light which guides all workers in science . . . There has been too much talk about the flash of discovery, and this had tended to obscure . . . the intellectual situation in which scientists find themselves . . . It is easier to see this for the kind of work that I had done than it is for major discoveries.[126]

Thus, Polanyi described many of his scientific investigations as ordinary or typical of science—what Kuhn would call "normal science."

Polanyi addressed the intellectual situation in which scientists find themselves by explaining the reception of his adsorption theory as a consequence of the way in which evidence is assessed and authority is exercised in science:

> My verification could make no impression on minds convinced that it was bound to be specious . . . There must be at all times a predominantly accepted scientific view of the nature of things, in the light of which research is jointly conducted by members of the community of scientists. A strong presumption that any evidence which contradicts this view is invalid must prevail . . . in the hope that it will eventually turn out to be false or irrelevant . . . I am making, therefore, no complaint about the suppression of my theory for reasons which must have seemed well founded at the time, though they have now been proved false.[127]

Here Polanyi took a novel tack in explaining the relationship between theories and evidence: good evidence is often ignored when a community of opinion favors a competing theory which scientists assume will eventually successfully explain originally anomalous data.

Thus, the popular notion of a straightforward relationship between empirical data and scientific discovery or verification is rooted in a misunderstanding of just how science works. The simple prescriptions of nineteenth-century positivism or updated logical empiricism are wrong. Polanyi used statements in Bertrand Russell's *Impact of Science on Society* as examples of these pernicious wrong ideas about how science works. Russell wrote, for example, that "the triumphs of science are due to the substitution of observation and inference for authority. Every attempt to revive authority in intellectual matters is a retrograde step."[128] Nothing could be further from the truth, argued Polanyi, based in his own experiences of a scientific career and authority structures in science.

Polanyi's later critical reflections on the nature of science very clearly were reflections on the successes and disappointments of his own career. He came to portray himself as an ordinary researcher in a scientific discipline in which, in fact, he had been regarded as a master, as reflected in his relationships with students, collaborators, and peers who were part of the Berlin scientific culture. Further, he was honored time and again by the authority network of the scientific community in professional awards and offers of scientific posts. Still, he was self-perceptive in reflecting on the universal implications of his early attempts to elicit attention from Nernst and Einstein and on his uphill struggles for recognition of his work in adsorption and X-ray diffraction studies. His later conviction of the value of apprenticeship to a master of the discipline may derive partly from a tinge of regret that he had not enjoyed the kind of apprenticeship in physical chemistry in Budapest that he might have experienced in Berlin with a mentor such as Nernst. Had Nernst not written him long ago that matters of disagreement could best be settled by face-to-face discussion? Rather, Polanyi rued the "complete ignorance of the professor of theoretical physics at the University of Budapest who accepted the substance of my theory as a Ph.D. thesis in 1917 (may his departed spirit forgive me these ungrateful remarks!)."[129]

We will return in the last two chapters of this book to the themes of rewards, recognition, and authority in science as used by Polanyi and other writers. These themes were fundamental to the historical and philosophical work of Thomas Kuhn and to the sociological theories of Robert Merton, for instance. They also can be found in the popular writings of the scientist J. D. Bernal, a father-figure in the social studies of science, as we will see in chapter 6. We turn in the next chapter to Polanyi's research in chemical dynamics where he achieved perhaps his greatest scientific recognition at the time of his original investigations. In this field he became one of the unrivalled leaders of physical chemistry.

# FOUR

# Chemical Dynamics and Social Dynamics in Berlin and Manchester

Polanyi is best known and revered among chemists and physicists for his theoretical and experimental work in chemical kinetics and dynamics, especially his development of the theory of the "transition state," or, what his colleague and American collaborator Henry Eyring (1901–1981) called the "activated complex." Polanyi, Eyring, and Eugene Wigner together are recognized by chemists today as the founders of the field of modern chemical dynamics.[1] Polanyi and Eyring made an enduring contribution in defining and defending use of the "semi-empirical method" in theoretical chemistry, which synthesized kinetic theory and thermodynamics with quantum mechanics and the theory of the electron-valence bond.[2] It is this collaborative work, its reception, and its implications in theoretical chemistry that is the main focus of this chapter, along with Polanyi's career transition from Berlin to Manchester and his earliest philosophical paper at Manchester in 1936 on the role of the inexact in science.

## Simple Atomic and Molecular Reactions

Polanyi pursued different research programs more or less simultaneously during the period from the 1910s through the 1930s in Berlin. One focus was the adsorption of gases on solid surfaces. Another was X-ray diffraction studies of natural fibers and metals, including research on the strength of crystals and investigations of the solid state. Yet another preoccupation was the activation and mechanism of chemical reactions, including Polanyi's short-lived use of a radiation hypothesis to explain some simple chemical reactions in the presence of light. He then went on to develop a method to calculate rates of reaction for gases, such as sodium vapor, with chlorine gas using chemical deposits along the sides of a reaction tube, leading to work on the transition-state theory in 1929.[3]

One of the most studied types of simple chemical reactions in the 1920s was the reaction of a hydrogen molecule with a chlorine or bromine molecule in the presence of light, sometimes referred to as a photochemical process.[4] In 1928, Cyril Hinshelwood of Oxford characterized this reaction as the "Mona Lisa of chemical reactions [which] still smiles its bewitching smile, leaving us in doubt whether even yet the secret has been completely fathomed."[5] Some chemists and physicists in the early 1920s favored a radiation hypothesis with use of a light quantum to explain chemical reactions that are sensitive to light and that occur more quickly than predicted by classical kinetic theory. The working hypothesis was that radiation induces formation of an intermediary complex or free radical which serves as a transition point in the reaction.[6]

An earlier version of this hypothesis came from the French physical chemist Jean Perrin and his student René Marcellin around 1913, using Svante Arrhenius's classical activity formula for chemical reaction. In his study of the influence of temperature on the rate of inversion of sucrose, Arrhenius suggested in 1889 that equilibrium exists between inert and active molecules of the reactant, with only active molecules taking part in the inversion process. The Arrhenius equation

$$\ln k = \ln A - E/RT$$

was rewritten in the equivalent form $k = Ae^{-E/RT}$ where $k$ is the rate constant, $E$ is the difference in heat content between the activated and inert molecules, and $A$, the proportionality factor, is the number of collisions at unit concentration of reacting molecules. Of the remaining parameters, $T$ is the temperature, and $R$ is the gas law constant.[7] Perrin and Marcellin took $E$ to be the energy which moves a molecule from a stable to a transition "critical" state, and Marcellin (who died in the First World War) wrote a thesis in 1914 demonstrating that, in addition to the Arrhenius activation energy, the rate constant had to contain an activation entropy term.[8] The source of the activation energy remained unclear, whether it be radiation energy, vibrational energy, or kinetic energy along the line of centers of colliding molecules.

Polanyi turned to the study of chemical reaction velocities and kinetics beginning with papers published in 1920. He noted that the rate at which bromine ($Br_2$) molecules dissociate is 300,000 times too slow for explaining the observed rate for reaction of a bromine atom with a hydrogen molecule in the formation of hydrogen bromide. By way of explaining energy of activation, Polanyi proposed an external energy source, perhaps, he said, radiation in the ether.[9] This notion was not as far-fetched, as it now sounds since reference to an 'ether' was often invoked in physics and chemistry around 1920. A letter from Reginald Herzog to Polanyi in October of 1920, for example, mentions the notion of an ether-like (*ätherartig*) bonding for secondary valences in some molecules.[10]

Radiation theory was the subject of discussions at a Faraday Society symposium in 1921 and at the first and second international Solvay chemistry conferences in 1922 and 1925.[11] A colleague encouraged Polanyi in this line of argument and suggested that he should look at the earlier paper by Marcellin on photochemistry energy values.[12] Max Born wrote Polanyi, however, of his skepticism about an ethereal radiation theory of reaction velocity, and he confided in his letter that James Franck shared Born's misgivings.[13] When Born wrote Einstein asking his opinion, Einstein, who already had told Perrin that he thought that the radiation theory was doubtful, replied "Polanyi's ideas make me shudder. But he has discovered difficulties for which I know no remedies."[14]

Invited to a special session of the American Chemical Society in Minneapolis in 1926, Polanyi found opinion hardening in favor of his Berlin colleague Max Bodenstein's theory of chain reactions and Hinshelwood's kinetic theory of collisions.[15] Polanyi was in Minnesota along with Bodenstein, Hinshelwood, F. A. Donnan, Irving Langmuir, Jean Perrin's son Francis Perrin, and the Oxford physicist Frederick A. Lindemann. Bodenstein had proposed the idea of chemical chain reactions in 1913, explaining rapid reactions as a result of the simultaneous formation of new unstable molecules with stable molecules so that rapid, even explosive, reactions can develop. Jens Anton Christiansen and Hendrik A. Kramers in 1923 argued that a chain reaction need not start with a molecule excited by light, but simply by two molecules colliding violently. G. N. Lewis of the University of California at Berkeley was one of those who initially defended the radiation hypothesis, but Hinshelwood presented convincing experimental evidence at the meeting for the falling off of the rate-coefficient for chemical reaction at low pressures, as predicted by the collision theory, which he favored over the radiation hypothesis along with Bodenstein and Lindemann. By the late 1920s, Hinshelwood's textbook *Kinetics of Chemical Change in Gaseous Systems* (1926), confidently ascribed reactions exclusively to collisions, independent of radiation. The Nobel Prize in Chemistry for 1956 would go to Hinshelwood and N. N. Semenoff for their work in kinetics, including chain processes and collisions.[16]

Other approaches by Polanyi to reaction mechanisms and reaction velocities were to prove more successful in the late 1920s and 1930s. Among these was his development of the experimental method called "highly dilute flames" for studying the course of reaction of two reactants—for example, sodium vapor and chlorine gas—in an evacuated vessel. This work connected with Institute director Fritz Haber's interest in the early 1920s in chemiluminescence, or, the emission of light as a result of a chemical reaction with little or no heat.[17] In Polanyi's technique, two gases were allowed to enter a tube from opposite ends of the tube in which the pressure was sufficiently low (on the order of 0.01–0.1 mm of mercury) that the length of the mean free path for the gas molecules was greater than the diameter of the tube. For the reaction of sodium vapor with chlorine gas, liquid sodium at one end of the tube produced a vapor at a temperature of 3,000°C, diffusing into the tube. Chlorine gas was introduced at the opposite end of the tube (which was 1 meter in length and 13 centimeters in diameter). Chemical reaction occurs in the middle of the tube, accompanied by gradual clouding of the tube wall, the deposit of a sodium chloride precipitate, and strong luminescence (the highly dilute flame) recorded as the D-line of sodium vapor. The length of the reaction zone in the middle of the tube is a measure of the reaction velocity for sodium vapor with chlorine gas.[18] Since the light was a "cold" light generated at relatively low temperature, it had popular appeal for science journalists, who called it "light out of the salt-cellar" and drew comparisons with Thomas Edison's invention of the incandescent lamp. Erika Cremer later wrote William T. Scott that Polanyi's colleagues recited a lighthearted doggerel when gathered for the annual Christmas get-together of Haber's institute.[19]

Meines Kindes schönster Traum
Ist kaltes Licht am Weihnachtsbaum.

[My child's prettiest dream would be
To see cold light on the Christmas tree.]

In collaborative work with Hans Beutler, who was skilled in assembling apparatus, and later with Stefan Bogdandy and Hans von Hartel, who also was an ingenious technician, calculation of the velocity constant $k$ for this and other reactions gave a value suggesting that the collision diameter for the reaction process was some seven times larger than the value deduced from the kinetic collision theory. Thus, Polanyi reasoned that the chemical

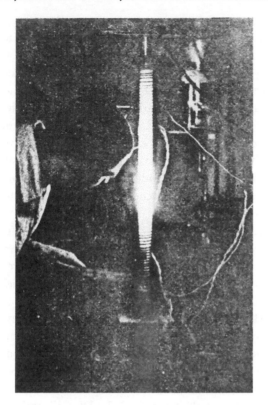

Figure 4.1. The "cold light" apparatus for chemical reaction of sodium vapor with chlorine gas in Polanyi's laboratory at Fritz Haber's Institute in Berlin. (In Michael Polanyi, *Atomic Reactions* [London: Williams and Norgate, 1932], 38, fig. 1. Courtesy of John C. Polanyi.)

reaction must take place in steps, a hypothesis given additional force by the fact that the light was distributed over a much greater length of the tube than the precipitate itself.[20]

Polanyi began turning his attention at this time toward the new quantum mechanics and its applications to chemical atoms and molecules, not only under the influence of his protégé Eugene Wigner, but also Fritz London. Polanyi sometimes found it frustrating to try to keep up with the younger men mathematically as he confided to his diary in 1929 about conversations with Wigner on quantum mechanics, chemiluminescence, and reaction velocities. About this time Henry Eyring arrived in Berlin on a National Research Council fellowship to work with Polanyi. Eyring had taken an engineering degree in mining and metallurgy at the University of Arizona and a PhD in chemistry at Berkeley. He moved to the University of Wisconsin in 1927 as an instructor and then became a postdoctoral assistant during 1927–1928 with the physical chemist Farrington Daniels, whose influence turned Eyring's research interests toward reaction kinetics. Daniels encouraged Eyring to embark for Berlin, where Daniels thought he should study with Max Bodenstein, now in his late fifties and widely recognized as one of the world's great experts on chemical reaction velocities. The Moscow physical chemist Alexander Frumkin was visiting Madison and, while walking along Lake Mendota together, Frumkin suggested to Eyring that he should get to know Michael Polanyi, "an outstanding chemist" who also was in Berlin.

Accompanied by his wife Mildred Bennion, Eyring set off for Germany in 1929 by way of stopovers in Norway, Sweden, and Denmark.[21] When Eyring arrived in Berlin, both Bodenstein and Polanyi were in Princeton as guests on the occasion of the dedication of the new Frick Laboratory. Bodenstein remained in the United States to lecture during the fall term at The Johns Hopkins University, with the result that Eyring ended up collaborating with Polanyi. Eyring's stay in Berlin was short, made possible by a National Research Council Fellowship for the academic year 1929–1930. By the time that Polanyi and Eyring's first coauthored paper appeared in 1930, Eyring was back across the Atlantic, working as a chemistry instructor at Berkeley. In 1931, he left California for an instructorship at Princeton University, where he joined Wigner on a faculty that included as well the distinguished British-born physical chemist Hugh S. Taylor.[22]

Princeton was an exciting place for physics and chemistry in the 1930s. Writing about his studies at Princeton, where he did a joint PhD program with Wigner in physics and Eyring in chemistry, the American chemist Joseph Hirschfelder recalled that his Princeton coursework included lectures by John von Neumann in theoretical physics, Wigner and Hermann Weyl in group theory, and a course in quantum mechanics from Paul Dirac, who was visiting Princeton (and would eventually marry Wigner's younger sister).[23] Some of these Princeton colleagues would reconvene in Polanyi's new academic home in Manchester in 1937, as we shall see.

At Eyring's arrival in Berlin, London was completing a paper on chemical combination in two-, three-, and four-atom systems, in which he derived an equation giving the potential energy of a system of three atoms X, Y, and Z as a function of variation of interatomic distances of the three atoms, each having one (uncoupled, s-level) electron. London described the energy in terms of both coulombic (or electrostatic) interaction of electrons and the exchange (or quantum mechanical resonance) interaction of electrons.[24] What was really significant in London's work for chemists was his extension of quantum mechanics from a static theory of electronic and nuclear forces, as in the treatment of the homopolar valence bond in a hydrogen molecule, to the dynamics of chemical reaction.[25] The classical Arrhenius equation for predicting rates of chemical reaction might now be approached using values for the activation energy $E$ that were formulated through fundamental principles of quantum mechanics and electron resonance theory. Polanyi encouraged Eyring to assist him in applying London's results to the theoretical problem of chemical activation energy, leaving to Hartel the work on chemiluminescence in which Eyring had been assisting.[26]

The simple chemical reactions that Eyring and Polanyi studied included sodium or potassium vapors with chlorine, bromine, or iodine, as well as the "Mona Lisa" reaction of hydrogen gas with chlorine or bromine gas. In addition, a new simple reaction was catching the limelight in Berlin in 1929: the conversion of para- to ortho-hydrogen. At Haber's Institute of Physical Chemistry, this conversion was the subject of research for two Hungarian-born brothers, Ladislaus Farkas and Adalbert Farkas. Working with Polanyi's friend Karl F. Bonhoeffer, Ladislaus Farkas finished his dissertation in 1928 on the mechanism of photochemical decomposition of hydrogen iodide. In 1929, Bonhoeffer and Paul Harteck isolated one of two different forms of molecular hydrogen, whose existence Heisenberg and Friedrich Hund had first predicted in 1927. The prediction had been that molecular hydrogen exists in two forms: an ortho form, with parallel nuclear spins on the two hydrogen nuclei, and a para form, with antiparallel nuclear spins. Heisenberg predicted that molecular hydrogen is a mixture of para- and ortho-hydrogen in the ratio of 1:3 para/ortho at room temperature.[27] In 1928, Bonhoeffer and Harteck obtained the rarer para-hydrogen at room temperature and showed that it could be converted into ortho-hydrogen (the higher energy form) at high temperature and pressure, by electric discharge, or by platinized asbestos.[28] The Farkas brothers turned to this area of investigation, and Adalbert completed his dissertation on ortho-para conversion. He hypothesized that the conversion proceeds through the dissociation of molecular hydrogen to hydrogen atoms which, if properly oriented, attack other hydrogen molecules to produce hydrogen molecules of different symmetries.[29]

## The Theory of the Transition State and
## Defense of the Semi-Empirical Method

After talking with Farkas, Eyring concluded that it was feasible to do a theoretical treatment on the energy requirements for the reaction. It seemed possible, too, to combine the older thermodynamics of activation energies, in the form of Arrhenius's rate equation, with the newer work on exchange energies for three-atom and four-atom systems. The Eyring-Polanyi collaboration resulted in two coauthored papers, with Polanyi completing the final versions after Eyring had left Berlin for Berkeley. One paper appeared in *Die Naturwissenschaften* in 1930, and the second, more detailed, paper in *Zeitschrift für physikalische Chemie* in 1931, with the title "Über einfache Gasreaktionen" (On simple gas reactions).[30] Eyring quickly published a follow-up paper in the *Journal of the American Chemical Society* after presenting it at a symposium on applications of quantum theory to chemistry during the annual meeting of the American Chemical Society in 1931.[31]

Eyring and Polanyi's starting point was London's approximate solution to the wave equation for the atomic transition complex $H_3$, resulting in an equation giving the variation in potential energy of a system of three hydrogen atoms as a function of changes in interatomic distance. This solution makes use of the Born-Oppenheimer approximation, published by Max Born and Robert Oppenheimer in 1927 during their collaboration while Oppenheimer studied with Born in Göttingen. In the Born-Oppenheimer approximation, they assumed that the electronic wave function depends upon the positions of atomic nuclei but not upon their motions, since nuclear motion is so much slower than electron motion. [32]

In his treatment of $H_3$, London suggested that many chemical reactions are electronically adiabatic (no change in electronic quantum numbers), in the sense that they do not involve any electronic transitions or electron jumps. Using approximations, London derived an equation for the binding energy holding three monovalent atoms together. The binding energy is the energy that must be supplied to break the bonds in a system in which the dissociated state is the state of zero energy. This binding energy is the negative of the potential energy of the system. London's equation gives the variation with interatomic distances of the potential energy $E$ of a system of three atoms, X, Y, and Z, each having one (uncoupled, s-level) electron. Coulombic interactions ($A$, $B$, $C$) of the pair of electrons on X and Y, Y and Z, and Z and X figure in the equation, as do the resonance or exchange energies ($\alpha$, $\beta$, $\gamma$), although London declined to try to sort out the relative contributions to the electronic energy system of the coulombic and exchange energies. He suggested that it should be possible to construct a potential-energy surface giving variation of potential energy for all possible interatomic distances in the three-atom system. This is the task that Polanyi proposed to Eyring, with three systems in mind for study.[33]

$$H + H_2 \text{ (para)} \rightarrow H_2 \text{ (ortho)} + H$$

$$H + HBr \rightarrow H_2 + Br$$

$$H + Br_2 \rightarrow HBr + Br$$

For the atoms $Y + XZ \rightarrow YX + Z$, as the reactant Y is brought up to XZ along a straight line, the potential energy of the system increases at first slowly and then more rapidly to a maximum, decreasing as the products $YX + Z$ are formed. The energy difference between the highest energy point in this path, corresponding to the formation of YXZ, an intermediary transition form, and the energy level of the initial reactants is approximately the activation energy. The most favorable reaction path can be distinguished from alternate paths.[34] As Michael Polanyi's son John Polanyi later expressed the problem, "What sort of forces does the atom feel, as it approaches [a diatom]? Is it immediately trying to climb over an energy barrier? When does it start to feel some attraction, as the new chemical bond begins to form? In answering these questions, one must understand the structure of the *transition state*."[35] The resulting potential energy diagram consists of what Eyring and Polanyi called two energy "valleys" stretching out parallel to each of the coordinate axes, separated by "passes" shaped somewhat like a "saddle," or "col." At the top, or the "saddle point," there is a shallow basin (later known as "Lake Eyring") with gaps leading to the valleys. On the graph, the upper right-hand portion represents the energy of the reactants, since X and Z (abscissa axis = $r_1$) are at their normal intramolecular distances apart at the beginning of the process, but X and Y (ordinate axis = $r_2$) are very far apart; $r_1$ is small and $r_2$ is large.

$$Y + XZ \rightarrow YX + Z$$

In the course of the reaction, the system passes from the upper left-hand corner to the lower right-hand corner of the potential-energy surface. In doing so, it passes up the bottom of the "vertical" valley, through the gap at the top of the pass, into the shallow basin, then out of the basin at the other gap and finally down the "horizontal" valley.

The originality of Eyring and Polanyi's visual and verbal approach is remarkable. The argument was explicit that you could "see" the energy process as a mass point or rolling ball (*Kugel*) moving up a valley until

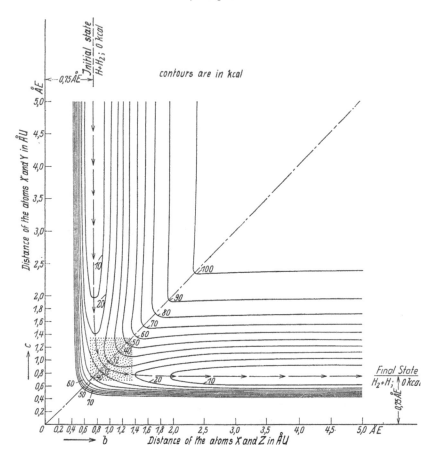

Figure 4.2. Diagram for resonance energy of three hydrogen atoms in a straight line as a function of their distances of separation (the "resonance mountain range"). The broken arrow shows the direction of reaction, and the shaded section is the saddle region of the mountain range. (In Michael Polanyi, *Atomic Reactions* [London: Williams and Norgate, 1932], 21, fig. 7. Courtesy of John C. Polanyi.)

reaching a divide with a neighboring valley and then either returning or passing over into the adjoining valley. As they wrote in their 1931 paper:

> The chemical initial and final state are two minima in the energy which are separated by a chain of energy mountains. The lowest pass, which leads over this energy mountain that causes the inertia of the reaction, yields the height of the activation energy . . . The heat of activation can thus in principle be found by checking all possible pathways from the energy valley of the initial

state to the energy valley of the final state, and determining to which height the energy goes up for the respective pathways . . .[36]

. . . What can we learn from the picture of a rolling ball? . . . [W]hen the ball is placed on the maximum of the saddle point, every small collision will suffice to let it roll into the valley; the kinetic energy with which it reaches the floor of the valley (neglecting the small kinetic energy originating in the collision) is the same as the height of the saddle . . . the proportion in which the energy of the ball arriving at the floor of the valley distributes itself in a rolling along the floor of the valley (translation) and a back and forth motion perpendicular to the axis of the valley (vibration), also yields the proportion in which the activation energy has to be distributed over translational and vibrational energy of the initial state in order to lead to the reaction.[37]

Polanyi's biographer, the physical chemist William T. Scott, credited this visual approach solely to Polanyi, while noting Eyring's flair for making useful approximations and his role in first dubbing the whole approach semi-empirical in the sense that they inserted empirical data in some of the calculations rather than using purely mathematical quantities.[38]

In order to calculate specific energy values, Eyring and Polanyi drew upon London's work and a theoretical paper by the Japanese physicist Yoshikatsu Sugiura, who studied with Born at Göttingen in the late 1920s. Eyring and Polanyi also took data from Adalbert Farkas's experimental study of ortho/para-hydrogen and from P. M. Morse's published experimental work on the $H_2$, HBr, $Br_2$ system.[39] For the para/ortho-hydrogen reaction, Eyring and Polanyi adopted what they called a half-empirical procedure (*halbempirisches Verfahren*) for determining energy of activation. Sugiura had proven that resonance energies are considerably higher than coulombic energies.[40] Eyring suggested that they use Morse's experimental spectroscopic values for the hydrogen molecule's binding energy, in relation to atomic distance, as a first approximation for the hydrogen molecule's total energy, given the small coulombic contribution to binding energy, which he estimated at only 10 percent. Morse's equation, which took into account both the molecule's heat of dissociation and its half quantum of vibrational energy at the lowest level, and Morse's data, published in *Physical Review* in 1929, gave energy values from spectra of molecules made up of atoms that react a pair at a time. The results could be plotted on a contour diagram with $r_1$, that is, the XZ distance, as abscissa and $r_2$, that is, the XY distance, as ordinate, and the various contour lines passing through points of equal energy.[41] In his earlier work with Hartel on chemiluminescence, Eyring's job had been graphing the potential-energy surfaces, a task at which he felt especially at home because of his earlier experiences in making contour maps while studying geology and mining.[42]

Eyring and Polanyi subtracted the small *theoretical* Sugiura coulombic energy from the *experimental* Morse energy (as if it was total energy) in order to get a value for resonance energy as a function of interatomic distance. They then constructed an energy plot for the total energy against distance by superimposing the estimated coulombic curve (a basin, or "coulomb hole") onto the semi-empirical resonance curve ("resonance mountain"), and read off the resulting values ("potential mountain.")[43]

After introducing several ad-hoc correction factors into the procedure, Eyring and Polanyi arrived at a value for energy of formation of the transition complex. In applying this approach to the ortho/para-hydrogen system, for example, the result is that in the system

$$H + H_2 \text{ (para)} \rightarrow H_2 \text{ (ortho)} + H \text{ or } Y + XZ \rightarrow YX + Z$$

the distance between Y and X initially is very large, whereas the distance between X and Z is closer, at approximately 0.75 Ångström; then as the hydrogen atom (Y) approaches the para-hydrogen molecule XZ, the potential energy increases adiabatically until the distance between X and Y equals the distance between X and Z, at approximately 0.95 Ångström, corresponding to an activation plateau of about 19 kilocalories. The reaction can now go to completion, or the system may return to its former configuration. The activation plateau corresponds to the potential energy, and since the zero point energy in the saddle region is considerably less than the zero point energy in the initial state, most of the zero point energy needs to be subtracted from the saddle-point potential energy. Using the value of 6.5 kilocalories for zero point energy results in an activation energy of approximately 13 kilocalories, which, hardly coincidentally, compared favorably to Farkas's experimental value of 4–11 kilocalories.[44] In his papers in *Journal of the American Chemical Society* in 1931, Eyring mentions another possible result at the saddle point: "The quantum mechanics provides still another alternative. The point may pass through the energy mountain without bothering to go over the divide," that is, tunneling.[45]

As a theoretical methodology, the philosopher of science Jeffry Ramsey emphasized that Eyring and Polanyi were doing something unique in the way that they mixed theory and experiment.[46] They were certainly conscious that their "half-empirical," or, semi-empirical method would be controversial, as it turned out to be. However, they called attention to the fact that they were demonstrating for the first time that chemists' understanding of chemical reaction has to take into account "a term not considered before," namely, the exchange or resonance energy between the two "outer" atoms, Y and Z, in the reaction Y + XZ → YX + Z. They thought they had shown clearly that the kinetic collision theory was inadequate.[47] Their approach was approximate and semi-empirical, but it achieved the aim, as they put it, of serving "to deepen our insight into the adiabatic reaction mechanism."[48]

In the next few years, Eyring and Polanyi pursued slightly different lines of development of their theory.[49] After Eyring returned to the United States in the fall of 1930, one of his first papers presented their joint results in English, while also offering new experimental data and extending the method to four-atom systems, such as $H_2 + I_2 = 2HI$. Eyring made the surprising prediction that hydrogen and fluorine should be unreactive at room temperature, contrary to current observations. He presented this paper at the Indianapolis meeting of the American Chemical Society at the end of March in 1931 with Hugh Taylor in the audience. Taylor knew that some recent results suggested that the explosive reaction of fluorine with hydrogen occurs only due to surface catalysis. He invited Eyring to visit at Princeton, where he soon became a member of the chemistry faculty.[50]

Polanyi continued to receive inquiries in Berlin from young men who wanted to work with him for a year or so on grants and fellowships. Meredith G. Evans intended to come from Manchester to Berlin in the spring of 1933 to work with Polanyi, but Evans's health problems and Polanyi's forced retirement from the KWG cut short that plan. Evans arranged to spend the 1933–1934 academic year in Princeton collaborating with Eyring and Taylor, and he then returned to Manchester where Polanyi now was directing the physical chemistry laboratory. Richard A. Ogg Jr., a Stanford PhD graduate who had a National Research Fellowship in 1932–1933 with George Kistiakowsky at Harvard, joined Polanyi in Manchester after learning that he was leaving Germany.[51]

Working independently of each other now, Polanyi and Eyring each set out to combine the theory of activation energy with predictions of reaction rates in order to derive the concentration and rate at which activated complexes pass through the critical configuration of the transition state. Wigner and Hans Pelzer first succeeded in this endeavor in 1932, using statistical methods and the London-Eyring-Polanyi (LEP) surface to calculate the rate of conversion of para-hydrogen to ortho-hydrogen, with emphasis on the saddle point as the point of control for the rate of reaction.[52] In a second paper, Wigner discussed the possibility of quantum-mechanical tunneling through the barrier of a potential-energy surface as an alternative

Figure 4.3. Polanyi's research group (*Abteilung*) in Berlin in 1933, including secretarial staff. Top row: Kurt Hauschild (technician and glassblower), Erika Cremer, Michael Polanyi, Frau Gehrte, Frau Weissenberg, and Juro Horiuchi. Front row: Hille, Irene Sackur, an unnamed Russian researcher, and Andreas Szabo. (Courtesy of Archiv der Max-Planck-Gesellschaft, Berlin-Dahlem.)

mechanism in chemical reaction.[53] Erika Cremer, who completed her PhD in physical chemistry in 1927 with a thesis under Bodenstein on chemical kinetics, helped Polanyi with calculations on the resonance mountain.[54]

At Princeton, Eyring's collaborators included the young theoretical physicist John A. Wheeler and the physical chemist Joseph Hirschfelder.[55] Eyring wrote Polanyi in October 1933 of his calculations with Wheeler on activation energies of $H_2$ with various halogens, taking account of directed valences and other factors.[56] In 1935, Eyring completed a theory of absolute reaction rates, choosing to publish his mathematically difficult paper in the *Journal of Chemical Physics*, rather than in the *Journal of the American Chemical Society*.[57] The theory is "absolute" because it predicts an absolute rather than a relative rate by combining the use of quantum mechanics (to calculate a potential energy surface) with the application of statistical mechanics (to calculate reaction velocity). Eyring arrived at a probability for the activated complex and then multiplied this probability by the rate of its decomposition in order to get the rate of reaction.

Eyring used the term "activated complex" rather than "transition state" in setting out his theory, designating the symbol $M^1$ for the activated complex. In fact, he meant to use the traditional transition-state symbol $M^*$ for the activated complex, but his secretary mistyped it, and thus a new symbol was born.[58] The new symbol first appeared in a paper coauthored by Eyring with W. F. W. Wynne Jones in 1930.[59] The absolute rate of reaction differs from the rate predicted by simple kinetic theory since the equilibrium constant for an activated complex is different from the ordinary equilibrium constant, because the activated complex is dissociating as it moves across the energy barrier (becoming products or reverting to reactants). Eyring distinguished his semi-empirical approach from what he called the intuitive approaches of the collision hypothesis, and he identified his method with the approach by Pelzer and Wigner in their 1932 paper.[60]

Polanyi and Meredith Evans published a paper in 1935 characterizing their approach as the "transition-state method."[61] They defined the probability of the transition state in terms of an infinitesimally thin layer of phase space. Systems with representative points within this layer were considered the transition state.[62] As in Eyring's work, Polanyi's fundamental equation expressed the instantaneous reaction velocity as a function of the equilibrium constant of the transition state and the thermal velocity of the representative point of the reacting system at the top of the energy barrier.[63] Perhaps mindful of Eyring's work and of criticisms of their whole approach, Polanyi showed some personal concern with continuity and priority in his 1935 paper, taking pains to relate it to his earlier publications of 1920 and 1928 (the latter coauthored with Wigner) which, Polanyi said, "foreshadowed" his method of the transition state. He also noted the significance of the transition-state approach for studies of reactions at high pressure, which was one of his longtime interests.[64]

From the time of their first joint publication in 1931, Eyring and Polanyi's work was met with considerable criticism. Eyring's 1935 paper very nearly was not published in his journal of choice. He sent the manuscript to Harold Urey, editor of the *Journal of Chemical Physics*, but it took personal pressure from both Wigner and Taylor before Urey agreed to take it; he complained that "the method of proof is unreliable and the result is spurious."[65] Hirschfelder recalled that Taylor suggested to Urey that Eyring add an appendix (which he did) explaining how his theory reduces to the simple kinetic theory and could be used to calculate the rate of collision between two rigid spherical molecules.[66] Some people came to refer disparagingly to Eyring's theory of absolute reaction rates as the "absolute theory of reaction rates."[67]

There were plenty of critics in addition to Urey. Among them were Harvard University's Albert Sprague Coolidge and Hubert M. James, who published in 1933 a groundbreaking *ab initio* calculation for the bond energy of hydrogen that predicted the ground energy state of the hydrogen bond with 98 percent accuracy. Starting from fundamentals required time. It was said that it took them more than a year to complete the calculation. They criticized the semi-empirical method as "a happy cancellation of errors" involving the "neglect of terms of considerable importance."[68] Frederick Lindemann, the Oxford physicist who had taken an active part in the radiation debates of the 1920s and currently was advising Winston Churchill on armaments policy, disparaged the semi-empirical transition-state theory and theory of absolute reaction rates at a conference in 1936.[69]

Hirschfelder, Keith Laidler, and Samuel Glasstone were among Eyring's Princeton students who went to work to apply the method. They became frustrated, however, that experimental work lagged so far behind their theoretical calculations.[70] In September 1937, Polanyi hosted a meeting of the Faraday Society in the Chemistry Theater at Manchester. Eyring, Wigner, and Taylor were among the participants, and Eyring at the time was making Manchester his headquarters for a four-month stay in England.[71] The Faraday meeting was contentious. E.A. Guggenheim, a researcher at the Ramsay Laboratories at University College in London, was outright hostile, characterizing Eyring's paper for the conference as unintelligible. Guggenheim objected to the use of experimental data to adjust the equation in order to successfully predict the outcome of the reaction, and he objected to the very hypothesis of an indefinite activated complex.[72] The physical chemist E. Alun Molewyn-Hughes and the physicist Ralph H. Fowler, both at Cambridge, as well as the physical organic chemist Louis Hammett of Columbia University, expressed some skepticism. Molewyn-Hughes spoke on behalf of those physical chemists who were loyal to classical collision theory by saying:

> To those who, like myself, are more at home with inorganic than with vector analysis, the kinetic method still appeals, and we cling to it unrepentantly, encouraged by the reflection that it is this particular method which made any statistical method at all possible . . . [the kinetic approach] adheres more closely than the others to the phenomenal aspects of the problem . . . and looks wrily at all hypothetical terms (such as the fictitious frequency $kT/h$) which are utilised with full confidence by the other two methods [of thermodynamics and statistical mechanics]. Which method appeals most to the

chemist is largely a matter of personal taste, depending on the times he allocates to the bench and the desk.[73]

If Molewyn-Hughes seems to have preferred the bench to the desk, Eyring made a strong statement in favor of the desk in his formal paper, sounding somewhat like a theoretical physicist, as he confidently claimed, "In so far as we can accept quantum mechanics as exact, every problem of chemistry can be answered from direct calculation by a sufficiently skillful mathematician."[74] Eyring's comment clearly echoed the self-confident statement of Paul Dirac, who famously had begun a 1929 paper on quantum mechanics by saying that applications of this new theory would reveal "a large part of physics and the whole of chemistry."[75]

Regarding the relative merits of thermodynamics, statistical mechanics, and kinetic or collision theory, Eyring, Wigner, and Polanyi were all in agreement. They would follow the most convenient procedure, which, as Eyring put it, was to regard thermodynamics as a branch of statistical mechanics.[76] Wigner noted, however, that there remained many sources of discrepancy between theory and experiment in the transition-state method, among them the necessity for quantum corrections on the nuclear motion and the possibility of Born-Oppenheimer breakdown. But the method would do for now.[77]

The problem, of course, was that quantum mechanics could not provide exact calculations for most molecular systems. As a consequence, Polanyi asked his colleagues at the Manchester Faraday Society meeting why chemists should not use a method, employing energy surfaces, that at least explains why the activation energy for the para/ortho-hydrogen reaction is far less than the bond energy of the disrupted links, and that shows better agreement with experiment than prediction based on preliminary dissociation of hydrogen into atoms. The approximative success of the semi-empirical method was just fine, claimed Polanyi: "Personally I attach no importance at the present stage to a precise numerical agreement between theory and experiments, but I believe that the theory can claim to give a reasonable picture of the mechanism of chemical reactions which would otherwise remain in the dark."[78] This attitude was fully consistent with Polanyi's attitude in his collaboration with Fritz London on the application of quantum mechanics to adsorption. London had said in his paper on dispersion and molecular forces that, in view of the complexity of the problem, "the goal of the theory is not so much to present numerically precise determinations, which can be compared to experimental data only with a large uncertainty, but rather to present the proof that the more phenomenologically oriented theoretical pictures already present . . . are indeed to a large extent justified."[79] As Hirschfelder wrote in 1941, semi-empirical methods were invaluable in placing "the intuition of the skilled organic chemists on a mathematical basis . . . The method . . . is sufficiently flexible so that it can be made consistent with any set of chemical facts and still agree with the basic principles of quantum mechanics." This statement seemed to concede that an ad hoc methodology is inherent in semi-empiricism.[80]

Chemists generally seemed to assume that transition intermediates in simple chemical reactions must remain hypothetical and could never be studied directly during the course of chemical reactions.[81] Some authors date a clear victory for the transition-state theory to a meeting of the London Chemical Society at Sheffield in 1962 when new experimental methods of flash photolysis were reported to attain single-collision conditions for atoms and product molecules that corroborated the soundness of the fundamental principles of Eyring and Polanyi.[82] Nobel laureates Dudley Herschbach, John Polanyi, and Ahmed Zewail all trace their investigations of reaction mechanisms to the pioneering work of Michael Polanyi and his colleagues. Herschbach's first research with molecular beams, when he was appointed to the chemistry department at Berkeley in 1959, focused on reactions of alkali atoms with alkyl iodides, the systems studied forty years earlier by Michael Polanyi.[83] John Polanyi, who had been a student of Meredith Evans and Ernest Warhurst at Manchester, obtained the first spectroscopic evidence in 1980 for generalized transition stages in the reaction of a fluorine atom with diatomic sodium. In 1987 Zewail used laser pulses that last only tens of femtoseconds (1 femtosecond = $10^{-15}$ second) to directly observe the transition state in the decomposition reaction of iodine cyanide: $ICN \rightarrow I + CN$.[84] All this was a great vindication of the earlier work of Eyring and Polanyi in the 1930s.

## Manchester Science in the 1930s and 1940s

When Polanyi arrived in Manchester in the fall of 1933, he joined a department that long had enjoyed a reputation as one of the foremost chemistry departments in Great Britain. Manchester's place in the history of chemistry dates back to one of chemistry's founding fathers, John Dalton. Other distinguished nineteenth-century Manchester chemists included Edward Frankland, Henry Roscoe, and Carl Schorlemmer. These pioneers were succeeded by the physical chemist Harold B. Dixon and the organic chemist

William Henry Perkin Jr. in the early decades of the 1900s when chemistry found its home in the buildings of the Roscoe, Schorlemmer, Perkin, and John Morley Laboratories and in the formerly private library and laboratory bequeathed to the university by Henry Edward Schunck. More recent faculty members were Jocelyn Thorpe, Arthur Lapworth, Robert Robinson, Walter Haworth, and Chaim Weizmann. Haworth received a Nobel Prize in 1937 for his researches on carbohydrates and especially the constitution of vitamin C, and Robinson received the Nobel Prize in 1947 for his investigations of plant products of biological significance.[85]

On the physics side at Manchester, the new physics laboratory of the early 1900s, directed by Arthur Schuster, was at the time the fourth largest in the world, after The Johns Hopkins University, Darmstadt, and Strassburg. Schuster wooed Ernest Rutherford to Manchester where his work from 1908 to 1919 became legendary in radioactivity and nuclear physics before Rutherford moved to Cambridge and the Cavendish Laboratory. Rutherford's successor, William Lawrence Bragg, built up studies in X-ray crystallography at Manchester, and Bragg's work was familiar to Polanyi from his own researches in Berlin in the 1920s. Physicists were on the university senate committee that hired new members of the Chemistry Department. Bragg helped hire Polanyi. P. M. S. Blackett, who took over Bragg's laboratory in 1938, and the mathematical physicist Douglas Hartree were on the committee that in 1938 hired the organic chemist Alexander Todd, who was a Nobel laureate in 1957 as were Rutherford (1908), Bragg (1915), and Blackett (1948).[86]

After Polanyi declined the position offered him at Manchester in January 1933, he was fortunate in the renewal of an offer in May 1933 from Manchester's vice-chancellor. The offer was made after an initiative from Frederick Donnan and consultation by Vice-Chancellor Walter Moberly with faculty, external advisers, and the university senate. Donnan wrote Polanyi informally what the terms would be: a salary of no more than £1,250 yearly and no possibility of embarking on construction of a new physical chemistry laboratory for two or three years. Polanyi's position would anticipate the upcoming retirement of Arthur Lapworth, and no one else would be hired in Lapworth's stead. The position originally offered to Polanyi in late 1932 was no longer available (the previous position had been a replacement for Robert Robinson after he moved to Oxford in 1928, and it was in negotiation with the organic chemist Ian Heilbron after Polanyi originally had turned it down).[87] Heilbron, who held a position at Liverpool, was thrilled that Polanyi and he might arrive at Manchester together, and he wrote Moberly that if Polanyi also came to Manchester, then there was every prospect of establishing Manchester as a chemistry center second to none in the country.[88] Not everyone was thrilled. The eighty-five year old Henry Armstrong, still speaking on behalf of his vision of British chemistry from South Kensington, wrote a protest in *Nature* that was quoted in major newspapers: "The introduction of a foreign outlook into Manchester is most undesirable at the present time." A British chemist should have been hired, reiterated Armstrong, and not a refugee from Germany.[89]

Setting up his research program at Manchester was going to be no easy matter. Some of Polanyi's Berlin apparatus, which he planned to ship, had been purchased with funds from the Rockefeller Foundation and the Studiengesellschaft. These included high-quality ground-glass joints and stopcocks, mercury-vapor vacuum pumps, and gas-circulating pumps; a machine lathe and accessories; and two hand bellows. Hartel had designed the dilute flame apparatus, but it was too delicate for shipment from Dahlem. Polanyi filed paperwork to allow his mechanic, Martin Schmalz, who had been with him since 1927, to move to Manchester, and permission was granted for a six-month stay for Schmalz, which eventually became two years. Two students who had been working in Polanyi's Dahlem laboratory, Andreas Szabó and Lotte Werner, moved to Manchester, as did the Japanese chemist Juro Horiuti (also spelled Horiuchi) who had begun studying the effect of substituting deuterium for hydrogen in hydrogen catalysis reactions. Horiuti later became head of physical chemistry at Hokkaido Imperial University in Sapporo. Meredith Evans, who had planned to work with Polanyi in Berlin during 1933–1934, made things easier for Polanyi in Manchester before embarking for a year of study in Princeton. He set up electrolytic cells that could generate heavy water for Polanyi. Upon his arrival in Manchester, Schmalz immediately went to work teaching instrumental techniques to the eighteen-year-old Ralph Gilson, Leslie Robertson, L. K. Parker, and other lab technicians.[90]

Polanyi set up his new laboratory with separate rooms for different experimental projects, making rounds daily to talk with his collaborators. Cecil Bawn and Meredith Evans's brother Alwyn G. Evans, a third-year graduate student, worked on the dilute flame experiments with Schmalz and Gilson, helping them master techniques for the apparatus. Fred Fairbrother had been teaching advanced chemistry lectures since receiving his doctorate at Manchester in 1926. Along with Ernest Warhurst, who was a third-year student, they built equipment to use sodium resonance in studying gas reactions. Bernard Cavanagh, a senior lecturer at the University of Melbourne, arrived in January 1934 to work with Horiuti on hydrogen catalysis and suggested the application in their research of Heinrich Wieland's discovery that enzymes in some bacteria, such as *Escherichia coli*, can act as a catalyst for hydrogen reactions.[91]

When Polanyi had accepted the second offer of a position at Manchester, he laid out a wish list to Lapworth, with whom he had discussed the earlier offer in 1932. Polanyi wrote to Lapworth in June 1933 that he normally collaborated in Dahlem with eight to ten fellow workers, along with four more from laboratories outside Haber's institute. This is about the right size for his needs, explained Polanyi, since he also could collaborate with junior people at the rank of lecturer. What is required for a first-rate laboratory, he told Lapworth, is about fifteen rooms of medium size, with total floor space of four hundred square meters, supplemented by another four hundred square meters for heating room, workshop, library, and so forth. A building and its equipment would cost around £20,000 to £25,000, supplemented by approximately £1,000 for Polanyi's eight to ten personal research assistants. He wanted a salary of £1,500 rather than the £1,250 previously mentioned, and he expressed optimism that perhaps 50 percent of the cost of the building and installation of his laboratory would be funded by the Rockefeller Foundation.[92]

The industrialist and philanthropist Sir Ernest Simon, previously a mayor and parliamentary representative of Manchester and now the university's treasurer, was on Polanyi's side, writing Vice-Chancellor Moberly in November 1934 about Polanyi's optimism that the Rockefeller Foundation might support capital expenditures for his Physical Chemistry Institute and reporting later in the month that Lawrence Bragg had the view that the Chemistry Department was not proving itself worthy of its two new professors.[93] Moberly was not at all optimistic, however, and for good reason. Lauder Jones had written Moberly the previous June, before Polanyi arrived, that the Rockefeller Foundation was in no position financially to award a grant of this magnitude, given the current economic situation in the United States.[94]

Polanyi was beginning to feel desperate about his research facilities at Manchester by November 1934, when he remarked in a memorandum that the current chemistry student laboratory benches and fittings were quite unchanged since Roscoe had designed them in the nineteenth century. In a letter to the mathematical physicist Geoffrey I. Taylor, who was a member of the national University Grants Committee, Polanyi noted that he hardly needed to explain to Taylor what is needed for a school of twenty research workers in modern physical chemistry. "Failing this, there is no hope."[95] Two more memos followed in the spring of 1935 on the occasion of Lapworth's retirement. Polanyi insisted on the need for new electrical, optical, low-temperature, and other physical apparatus in his physical chemistry laboratory, making the argument that he could never have developed his experimental methods had he been in Manchester rather in Berlin. Polanyi worried that progress would bypass his new laboratory just when application of the new physics to chemical problems was opening up a new chapter that was as much chemical physics as physical chemistry. In a telling comment, to be reiterated in his later philosophical writing, Polanyi wrote, "Supreme genius might overcome any difficulties. But the research laboratory of a university should be a place for humbler abilities to develop their faculties."[96]

Small grants of Rockefeller monies continued to come to Polanyi after his move to Manchester, just as they did to Heilbron, Todd, and other research programs in the university. In the late summer of 1935, Melvin Calvin arrived in Manchester after writing Polanyi to inquire about the possibility of a postdoctoral stipend. The university supported Calvin for his first year, and Polanyi's Rockefeller grant kicked in for a second year. Since Calvin's doctoral work had involved surface interactions of atoms, he was set to work with Horiuti who was scheduled to return to Japan soon.[97] The Rockefeller research stipends in physical chemistry, however, increasingly had to be oriented toward biochemical and medical research, as Lily Kay, Robert Kohler, and other historians have highlighted in their studies of the Rockefeller Foundation. A note of October 1933 among the papers of the Caltech quantum and theoretical chemist Linus Pauling makes the new general guidelines clear:

> Warren Weaver suggests sending in a request for extension of the fund . . . He says their policy is to support a biological program* (*especially *quantitative* work in biology). Atomic physics is definitely out. My work may be included because of its bearing on biological problems, although ordinarily work in organic chemistry would not be.[98]

Polanyi's interests in biochemistry legitimately went back to his very first published work as a medical student in Budapest with Ferenc Tangl.[99] Thinking along biological lines toward the end of 1935, Polanyi suggested to Calvin that they should note the chemical resemblance between platinum metal and a porphyrin molecule, such as heme, in the process of hydrogenation activation. If biological oxidation is a process of dehydrogenation and reduction, porphyrin crystals could be used instead of platinum to study hydrogen activation. Polanyi had heard that R. P. Linstead at Imperial College in London had synthesized a porphyrin analog that would be stable at the high temperatures used by the Manchester team, and so Polanyi sent Calvin off to London to learn how to make and purify phthalocyanine.[100]

Upon his return to Manchester, Calvin rejoined the hydrogen activation group which included E. C. Cockbain, who later worked at the Malaysian Rubber Research Institute in London, and Daniel D. Eley, who became professor of physical chemistry at the University of Nottingham. They worked first with zinc porphyrin compounds and then turned to hemoglobin, an iron-containing compound of great biological significance as the carrier of oxygen in the blood and a substance on which many biochemical researchers, including Pauling, were concentrating. Eventually, Eley concluded that in the absence of bacteria, which had confused some of their initial results, hemoglobin and related compounds could catalyze a para-hydrogen transformation but not the hydrogen-deuterium reaction.[101]

In the next few years Polanyi won renewed funding for his work from the Rockefeller Foundation: £300 in June 1936 for an assistant in biochemistry, based on the promise that Polanyi would seek university money for a permanent lecturership in this area, and £600 in 1937 for a two-year stipend for T. H. H. Quibell, who had come from Cambridge, to continue studies of porphyrin compounds and their derivatives because of their importance in blood and other tissue metabolism.[102] As Calvin was finishing his postdoctoral residence at Manchester in the summer of 1937, Polanyi recommended him to Joel Hildebrand for a job at Berkeley. Once there, Calvin extended his research on porphyrins to focus on chlorophyll. He received the Nobel Prize in 1961 for his discovery of the mechanism of photosynthesis. In his Nobel lecture, Calvin traced the origins of his discovery back to his studies with Polanyi on metalloporphyrins, such as hemoglobin and chlorophyll, and their electronic behavior. [103]

In the late 1930s and 1940s, more and more of Polanyi's research programs focused on hydrocarbons and their reaction mechanisms. Lapworth and Robinson had been pioneers at Manchester in the 1920s in the introduction into organic chemistry of reaction mechanism schemes that used the early Lewis-Langmuir electron valence theory, the hypothesis of organic ions, and a hypothesis of electron-induced "polarization" within hydrocarbon compounds. The British physical organic chemist Christopher Ingold made a huge impact in the field of organic reaction mechanisms in 1934 with an influential essay in *Chemical Reviews* titled "Principles of an Electronic Theory of Organic Reactions."[104] In addition to Polanyi's research program on hydrogen exchange and hydrogenation, pursued with Charles Horrex, R. K. Greenhalgh, J. H. Baxendale, A. R. Bennett and Warhurst, Polanyi focused on the application of transition-state theory to the study of free radicals and organic reaction mechanisms. [105]

Polanyi had already begun in Berlin to study the reaction velocity of organic halides with sodium atoms in the vapor phases as a function both of structure and the strength of the carbon-halogen bond. He argued that the reaction velocity depends, too, on the extent to which the carbon atom's substituents, other than halogen, shield the reacting bond, and he investigated how a reaction can lead to the inversion of configuration known as the Walden inversion. This inversion of structure sometimes occurs when a substituent directly attached to an asymmetric carbon atom is replaced by a new substituent with the result that the final configuration of the product is the opposite of that in the starting material. In Polanyi's interpretation, the product has to be the final stage of a transition state. [106]

When Ogg arrived in Manchester in 1933 on his two-year postdoctoral fellowship, he began studying the kind of reaction in which an electron departs from one atom, leaving it positive, and moves to another atom, making it negative. Coulombic forces then pull the two atoms together in a process that Polanyi and Ogg called "harpooning," which they tried to represent using contours for potential energy states and the intersections of the contours. This work was taken up, too, by Meredith Evans and by G. N. Burkhardt, among others in Polanyi's research groups. [107] John Polanyi later described harpooning in compelling imagery:

> "Harpooning" occurs when a charge-carrying "whaler" molecule comes within range of a charge-attracting "whale." The electron leaps the gap between the molecules, giving rise to a residual positive charge on the whaler and a negative one on the whale. Since plus and minus attract over large separations, the whaler pulls in the whale. A "reaction" has occurred binding one molecule to another, while the two remain at large separation. The hurled electron is the harpoon. [108]

The Polanyi group had strong competition in the field of organic reaction mechanisms, some of it close to home. Ingold, E. D. Hughes, and other coworkers at University College London were studying mechanisms of reactions of alkyl halides, trialkylsulfonium salts, and other substances with water or hydroxide ion. Using radioactive ions and changes in optical activity, they distinguished two primary mechanical paths for the substitution process. In one process, the new bond is formed at the same time that the old bond is breaking, and the incoming group and leaving group are both partly bonded in the transition state to the carbon atom under attack.

In a second type of mechanism, the bond to the leaving group is broken before a new bond is created, so that the reaction proceeds through an intermediate free carbonium ion to which substituents are attached. Hughes and Ingold wrote a series of papers during 1935–1937 on Walden inversion and the circumstances under which it occurs, relating it to their proposed reaction mechanisms. Ingold was in residence at Stanford University during 1932 when he worked out some of his ideas for the *Chemical Reviews* paper of 1934. Gerald E. K. Branch and Polanyi's former collaborator Melvin Calvin incorporated Ingold's approach into their coauthored and widely used *The Theory of Organic Chemistry* (1941).[109]

Ogg, who had taken his PhD at Stanford in 1932 before his Harvard and Manchester postdoctoral years, returned to teach at Stanford in 1934 just at the time that his papers coauthored with Polanyi began to appear with some differences in approach to organic reaction mechanisms from Ingold's.[110] Polanyi was keenly aware of competition with the Ingold research school. He wrote Ingold of the need to be sensitive to priority concerns especially among junior researchers after one of Polanyi's coworkers complained that his work had received inadequate credit from Ingold. As for their competition, Polanyi added, rivalries are all to the good for the field. Polanyi also corresponded with Hinshelwood about differences on the matter of reaction velocities in esterification.[111]

When Warren Weaver called on Polanyi during a visit to the Manchester laboratories in June 1936, he found Polanyi in a good mood, happy in his post and, as Weaver wrote in his diary, virtually promised a new building for his work. Polanyi was in a joking mood, telling Weaver that his laboratory floor was so weak and unstable that they had to make instrument readings by first standing on the right foot, then on the left foot, and then taking the average.[112] Things seemed to be looking up, and just two days later, on June 18, Polanyi wrote Friedrich Glum, the general director of the Kaiser Wilhelm Gesellschaft, that he wished to visit Berlin in July, accompanied by a Manchester architect, in order to get information that could be useful for building a new Institute for Physical Chemistry in Manchester. Polanyi wanted to see the new Kaiser Wilhelm Institute for Physics, directed by Peter Debye, along with other facilities. A favorable reply came back from Von Cranach, in Glum's absence, mentioning that the Physics Institute was still under construction and that Debye spent most of his time in Leipzig, but that Polanyi would be welcome in mid-July.[113]

By May 1937, Polanyi had word that Ernest Simon had given the go-ahead that Polanyi could proceed with building plans for a new installation for physical chemistry excluding only his request for a new laboratory for medical students. The vice-chancellor asked Polanyi to complete the plans by the end of July, but Polanyi made what turned out to be a very big mistake in requesting more time. He wrote the vice-chancellor that he had to think carefully about his needs on technical matters such as electrical services and protections against vibrations from Burlington Street. He needed more time. The vice-chancellor responded that Polanyi should let Simon know that he would not be in a position to submit final plans in July. The laboratory would not be built until several years after the Second World War.[114]

Later in the summer, in August 1937, Thorfin R. Hogness, a chemist at the University of Chicago, reported to the Rockefeller Foundation on his observations during a visit to Polanyi's laboratory. Polanyi was interested primarily in general problems of the mechanism of chemical reaction and energy transfer between molecules. He was devoting some attention to biological problems but had not gone very far in this direction. Polanyi was dissatisfied with the slow progress of physical chemistry research in the last few years, and for the last couple of years he had been devoting a great deal of attention to economics, influenced in part by his view that public ignorance in economics was making the populace susceptible to anti-Semitic propaganda. Calvin later said that toward the end of his fellowship, he found it increasingly hard to talk with Polanyi about physical chemistry, rather than other subjects of more interest to Polanyi.[115] When Todd arrived in Manchester in 1938, he, too, found Polanyi preoccupied with matters outside the laboratory and engaged in polemical discussions over political and economic questions with Manchester colleagues such as Blackett, the historian Lewis B. Namier, and Alfred P. Wadsworth, the editor of the *Manchester Guardian*.[116]

During the war years Polanyi was hard-pressed to find research assistants and money to fund them. Like Todd, Burkhardt, and other scientists in Manchester, and as he had done in Berlin, Polanyi consulted for industry and received both personal fees and stipends for research assistants in exchange for his services. He continued consulting for Tungsram in Hungary until September of 1937. He declined an initial offer from the Manchester firm Imperial Chemical Industries (ICI) for consulting, because ICI wanted him to do research in his Manchester laboratory and to clear any results with ICI before publishing them. A year later, in 1938, Polanyi worked out a contract with the vice-chancellor's approval for spending five days of the year in ICI's facility at Billingham and for engaging in talks with company representatives at his laboratory office in Manchester in return for a stipend of £150 per annum without any restriction on publication. This activity, he told the vice-chancellor, would give him new technical knowledge and open up prospects for his students.

Todd, Burkhardt, and F. S. Spring all carried out work for ICI in their Manchester laboratories, with Todd, for example, receiving a sum of £50 per year and his research student £25 in 1940. Polanyi requested permission from the university in the summer of 1941 for E. T. Butler, a conscientious objector, to work in his laboratory on iso-butylene polymers, which would be manufactured by ICI and would bring Butler a subsidy of £75 per year. Todd expressed the view shared by many of his colleagues that the industrial grant system was a valuable asset to university research as long as the projects were not necessarily dictated by industry and results could be published except when of essential value to industry in international competition.[117] Only one or two research students were able to work with Polanyi during the war, one of whom was Peter Plesch, the son of the Hungarian refugee Janos Plesch who had been Polanyi's family doctor in Berlin.[118]

Peter Plesch became part of a research group in Polanyi's laboratory that worked at the request of ICI on the production of synthetic rubber by means of polymerization of isobutene. Polanyi, Fairbrother, and Alwyn Evans began working on this project in 1942, and Polanyi brought in Plesch and H. A. (Hank) Skinner in 1944 from their wartime assignments elsewhere in industry. The isobutene polymerization process was erratic, the molecular weight and yield of the product were extremely irreproducible, and the reaction rate could become almost explosive before suddenly shutting down altogether. Plesch and Skinner brought a new perspective to the problem. They noticed that when the reactor—which contained hexane as the solvent, isobutene as the monomer, and titanium as the catalyst— remained opened, the reaction did not stop and that it could be restarted by blowing moist air through the reaction mixture. Finally, Plesch realized that water in the form of atmospheric moisture was serving the role of what he christened a "cocatalyst" (in analogy to a coenzyme) in successful polymerization. This was a major breakthrough for ICI and an important discovery for Plesch and Polanyi's laboratory.[119]

Just as in Berlin, Polanyi found himself reliant in the 1930s and 1940s on a network of funding that included government, industrial, philanthropic and university monies, and he coordinated research projects among a changing array of students, postdoctoral collaborators, and salaried assistants. A major difference from Berlin lay in his teaching responsibilities. During the war years Polanyi taught both beginning and advanced students, including a first-year course in physical chemistry for the honors degree and a third-year elective course on quantum mechanics. Fairbrother took over some of the lecture duties during 1943–1944 when Polanyi had a throat infection. Warhurst and Alwyn Evans also lectured on aspects of physical chemistry and Burkhardt on organic chemistry. What was unusual for many of the beginning students was Polanyi's use of the newest theories of chemical bonding, including quantum wave equations, hybridization of electron orbitals, and electron density maps taken from Linus Pauling's chemistry textbooks.[120]

Derry W. Jones and Brian G. Gowenlock were among a small number of undergraduate students who pursued a hybrid physics-with-chemistry program in the mid-1940s. In addition to chemistry courses, they attended lectures on properties of matter given by Blackett when he was not in London; George D. Rochester on heat, light, and sound; John G. Wilson on magnetism and electricity; and, in their third year, S. Keith Runcorn on topics including relativity and quantum physics. Bernard Lovell was starting up what became his famous radio telescope facility at Jodrell Bank. Jones found Polanyi to be a slow, quiet, and deliberate lecturer, notable for his slight accent. Polanyi discouraged the students from taking notes during lectures and provided after each lecture typed, duplicated notes of the lectures, complete with diagrams of X-ray powder photographs. Jones and the other students did not realize from Polanyi's lectures that he had pioneered some of the topics on which he spoke.[121]

By war's end, Polanyi was thoroughly immersed in a different kind of study and writing than his scientific work. When he moved from the Chemistry Department to the Social Studies Department in 1948, his successor was Meredith Evans, who had returned to Manchester from his position at Leeds. In November 1949, Evans wrote the new vice-chancellor, John Stopford, that the physical chemistry building, which had been slow in starting, was now going ahead rapidly, that he had some forty-six active research workers, and that he thought that his group was one of the best groups of young men in the country. Evans had a close and affectionate friendship with his mentor Michael Polanyi, but he had watched him become more and more drawn toward his philosophical and political thinking, a state of affairs that concerned Evans and others about the future of the department and its leadership. Evans thought, too, that Polanyi had made a mistake after his arrival in Manchester by retreating into his research in physical chemistry rather than interesting himself in department matters as a whole. In doing so, it could be inferred that Polanyi was following the course that he had learned in Berlin rather than adapting to the British university milieu.[122]

## The Semi-Empirical Method and the Philosophy of Chemistry

In the 1930s, objections to Polanyi and Eyring's work on the transition state had registered resistance, on the one hand, to the surrender of chemistry to the mathematical physics of quantum mechanics and reluctance, on the other hand, to substitute for the foundational calculations of *ab initio* theorists the give-and-take, or *ad hoc*, accommodations of the semi-empirical method. Disputes in the 1930s over the relative merits of the atomic-orbital-valence-bond theory versus the molecular-orbital theory similarly registered these concerns about the relative merits of useful approximations and quantitative rigor.[123] Albert Sherman, who was Eyring's student at Princeton, coauthored a paper with the mathematical physicist John Van Vleck in 1935 on this issue. They wrote:

> One must adopt the mental attitude and procedure of an optimist rather than a pessimist. The latter demands a rigorous postulational theory, and calculations devoid of any questionable approximations or of empirical appeals to known facts. The optimist, on the other hand, is satisfied with approximate solutions of the wave equation . . . that . . . give one an excellent "steer" and a very good idea of "how things go," permitting the systematization and understanding of what would otherwise be a maze of experimental data codified by purely empirical rules of valence.[124]

In Polanyi's efforts in the 1930s to build up his research program at Manchester, he stressed to colleagues in the university that chemistry was going through a huge transformation, because the application of the new quantum physics to chemical problems had opened a new frontier. He explained in memos that experimental and theoretical methods of physical chemistry were being adopted from the new physics, but that they were applied to subjects of traditional chemistry. More changes were in the offing, he predicted, as the physical outlook was brought to bear not just in physical chemistry, but in the whole domain of organic chemistry, as he was attempting to do in his own work at Manchester.[125]

In 1936, just before he hosted the Faraday Society meeting in Manchester, Polanyi wrote his first philosophical essay. It shows clearly his concern to defend the semi-empirical method and the use of the hypothesis of an unobservable transition-state intermediary. He was aware that invisible entities always have suffered tenuous status in science as objects that are in the process of evolution from transcendental to empirical status. Sometimes invisibles are confirmed as "real" entities, as in the demonstration in 1929 by Fritz Paneth of the physical existence of the alkyl radical using a lead mirror technique.[126] Paneth was a friend of Polanyi's in Berlin, and Polanyi recognized that his own hypothesis of an intermediary complex H-Br-Br or YXZ in a transition state was no less hypothetical than the alkyl radical had been before 1929.

Polanyi's transition-state theory was all the more difficult to advance because, like the atomic-orbital theory, it was an attempt to combine traditional chemical ideas with the new quantum physics. As noted by philosopher of science Jeffry Ramsey, explanations of the form and structure of transitory intermediates are inevitably suspect within quantum chemistry, since orientation and quasi-independent constituents are not part of strict quantum mechanical justification.[127]

This conundrum over the physical existence of hypothetical and short-lived chemical substances brought Polanyi directly to the question of the similarities and differences between chemistry and physics, a subject on which he reflected in a letter to *Philosophy of Science* in 1936. The journal was a new quarterly periodical, founded in the United States, with its first issue appearing in 1934. Its editor was William M. Malisoff at the University of Pennsylvania, and its editorial board included Rudolf Carnap and Herbert Feigl from the Vienna Circle philosophers. Polanyi was one of forty-two eminent scientists and philosophers who agreed that their names might be placed on the journal's advisory board. Others included the scientists Percy W. Bridgman, Irving Langmuir and Eugene Wigner, and the philosophers Hans Reichenbach and Moritz Schlick.

Articles in the first issues of the journal concentrated on physical science and quantum physics, the social sciences and applications of mathematics, but not chemistry. In his essay Polanyi wrote, "The subject of chemical concepts as opposed to physical ones has always been fascinating to me because it shows the great value of inexact ideas." Chemistry, Polanyi continued, is a world of ideas expressed by such terms as "relative stability," "affinity," "tendency," "inclination," and "general expectation" as descriptions of behavior. "There is not a single rule in chemistry which is not qualified by important exceptions." Chemists would have been ill-advised, he continued, perhaps with Coolidge and James in mind, to heed physicists' counsel to abandon vague methods and to restrict investigations to fields where only exact laws pertain. The development of chemistry, wrote Polanyi, "would at that moment have stopped dead."[128] Neither the characterization of substances nor the synthesis of compounds could be achieved by exact methods. Chemistry is an art, which depends on enlarging the investigator's field of awareness, a theme that Polanyi would reiterate in later essays and lectures, *Personal Knowledge*, and *The Tacit Dimension*. "Just link up two or three of the atoms of physics, and their behavior becomes so complex as to be beyond the range of exactitude."[129]

Of his father's approach to transition-state theory, John Polanyi has said that Michael Polanyi was most at ease with the domain of mechanical and molecular science where he could visualize molecules and make them tangible.[130] The unfamiliar novelty in the 1930s of Eyring and Polanyi's visual maps of potential energy systems and their language of rolling balls and moving points gradually became the usual custom of pedagogical and calculation techniques in modern physical chemistry by the 1960s, with computer-generated programs an essential part of undergraduate education in chemistry by the century's end.

The importance of the inexact was not the only theme in Polanyi's philosophy of science that can be traced explicitly to his experiences of research on chemical reaction mechanisms in Berlin and Manchester. In 1963, at about the same time that he reminisced in published essays on the scientific community's resistance to his work in gas adsorption and X-ray diffraction, Polanyi sent a letter to Peter Plesch in which he recalled his earlier personal resistance in the 1940s to information that had undercut his own pet hypothesis for explaining the seemingly random effects in the isobutene polymerization that they had studied together. Polanyi confided that he had delayed the discovery of water as a co-catalyst because he wanted to be able to interpret the erratic results as evidence of a novel phenomenon of rapid exchanges of vibrational energy between molecules undergoing a chain reaction. Such an interpretation would have been a greater achievement than an explanation in terms of known reaction mechanisms. "I tried not to think of water," he wrote, although Plesch had noted at the time that chemical literature mentioned the role played by water in some reactions involving catalysis by metal halides. Polanyi emphasized in his letter to Plesch not only the difficulty of overcoming a preconceived hypothesis, but also the scientist's drive to achieve recognition. "The main stumbling block was a fault of mine which has made me miss a number of discoveries," he wrote Plesch. "It was excessive ambition"; "aiming at a greater discovery than that which the problem in hand could offer and thus missing its solution."[131]

This experience, like Polanyi's many other professional experiences as a physical chemist, including the work in gas adsorption, X-ray crystallography, chemical dynamics, and reaction mechanisms, played a profound role in the character of his later philosophy of science. Not only did he draw upon his work in chemistry and physics in order to develop an interpretation of the psychological and personal aspects of science, but he also called upon his experiences in order to illustrate the social character of recognition and reward for scientific discoveries and scientific theories. Polanyi's departure from the scientific culture of Weimar Berlin and from Haber's institute, his disappointment in his inability to duplicate in Manchester the kind of resources that he had enjoyed in Berlin, and his efforts to observe and understand his new British milieu in the tumultuous 1930s all contributed about this time to Polanyi's loss of interest by the 1940s in undertaking new work in physical chemistry and to his increasing attention to economics and politics. His economic and political writings were rooted, as discussed in the next two chapters, in the economic and political liberalism of the late nineteenth century and in the anti-Communism and anti-Stalinism of the twentieth century. Polanyi's economic and political philosophies, no less than his experiences of a long scientific career, would serve eventually among the fundamental foundations of what he intended to be a new kind of philosophy of science.

# FIVE

# Liberalism and the Economic Foundations of the "Republic of Science"

When Michael Polanyi was a student at the University of Budapest, economics was taught in Austro-Hungarian universities principally in the Faculties of Law and Social Sciences (Rechts und staatswissenschaftliche Fakultäten). Economic organization and political economy were all the rage in intellectual circles at that time. Robert Musil wrote in his 1930 novel *Der Mann ohne Eigenschaften* (translated as *The Man without Qualities*) that physics and economics were the main topics of interest in Viennese society before the First World War, replacing theology as a contentious area of debate.[1]

The urgency in discussions about the ideal form of economic organization greatly intensified during the turbulent events of the Soviet Revolution, the breakup of the Austro-Hungarian Empire, the hyper-inflationary crisis of 1921 to 1923, and the worldwide Depression that began in the early 1930s.

Partisans of liberalism and the free market often drew loosely upon the laws of physical science in order to provide a rationale for the existence and character of natural laws in politics and economics. An example is Serafin Exner's 1908 inaugural lecture as rector at the University of Vienna. Referring to the physicist Ludwig Boltzmann and his probabilistic kinetic theory of gases, Exner told his audience that the doctrine of freedom and individual rights drew upon the "insight due to the latest Viennese physics: that an orderly system could arise out of an aggregate of strictly independent and unconstrained individual components."[2]

Thus, Michael Polanyi's interest in economics was hardly unusual, and a preoccupation with economics ran in Polanyi's family. His eldest brother, Adolf, who eventually made a career as an engineer, wrote a short monograph on the Japanese economy based on a trip to Japan in 1905. Michael's older brother Karl became widely recognized as a pioneer in economic history following publication of his book *The Great Transformation* in 1944. Michael's older son George wrote books and pamphlets in economics, and Karl's daughter Kari Polanyi-Levitt became an academic economist at McGill University.[3] Michael Polanyi's interest in economics accelerated in the late 1920s, and this absorption formed a bridge between his career in chemistry and his later immersion in philosophy of science.

The brothers Michael and Karl differed profoundly in their appraisals of capitalism and Socialism and in their prescriptions for economic reform (while George largely agreed with his father and Kari with hers). For both Michael and Karl, there were deep social and moral dimensions in their economic theories, and these concerns are repeated in Michael's philosophy of science.[4] In particular, the brothers disagreed on the relationship between economics and politics. Michael was convinced that economics, for the most part, must operate separately from the political sphere, and Karl was adamant that an economic system is embedded in its political system. For Michael, a modified laissez-faire capitalist economy is necessary to insure individual freedoms, whereas for Karl, the ideal system is a guild-like or functionalist socialist economy that addresses problems of social and material welfare. For the two brothers, the liberal value of individual freedom is a yardstick of economic goodness, with Karl willing to concede some individual freedom in favor of social welfare and Michael in favor of scientific expertise. The most wrenching source of disagreement between them was Karl's sympathy and support of Soviet economic policies during the 1920s and 1930s, which did not waver in the face of news of Stalinist purges or of Michael's attacks on Soviet policy and on central planning.

Economic and political matters are tightly bound up together, and it often is difficult to sort them out, as Karl argued. This chapter focuses principally on economic discussions from the 1910s through the 1940s, particularly as addressed by Michael and Karl Polanyi, and on how economic views played a crucial role in the way that Michael Polanyi came to describe scientific organization and scientific ethos in his free-market "republic of science." One of Polanyi's best-known articles is his widely read essay "The Republic of Science: Its Political and Economic Theory" (1962) published in the science-policy journal *Minerva*.[5] As we see in the concluding section of this chapter, Michael Polanyi's description of the role in science of institutional arrangements owes an unspoken debt to the functional conception of economic systems that Karl defended.

## Red Vienna and Austrian Economics

When their father died of pneumonia in 1905, Karl Polanyi was nineteen years old. His older sister Laura was married and his brother Adolf was working, while Sophie and Michael still were at home with their mother. Michael was fourteen. In a letter of 1957, Karl recalled, "Except for our father and my wife, I have never loved anyone as dearly as I loved you, and our first differences (?)some twenty years ago darkened my life as his death had done."[6] The differences to which Karl alludes likely had to do at least partly with the brothers' polarized attitudes toward economic questions and toward the Soviet Union.

By the time that Michael Polanyi left Budapest for Karlsruhe in December 1919, Karl had been in Vienna for a good six months. He joined the editorial office of the daily Hungarian émigré newspaper *Bécsi Magyar Ujság* (Viennese Hungarian news) and took on responsibilities for covering the world economy, international politics and questions of science and ideology. He also began to contribute to the *Oesterreichische Volkswirt* (Austrian economist), a magazine modeled on Great Britain's *Economist*. In 1924, he became the *Volkswirt*'s senior editorialist, with special responsibilities for international affairs. He was well known for his fluency in English and his expertise on Russian affairs.[7] He remained there until 1933.

Once in Vienna, Karl largely steered clear of the political activism and the political parties that had occupied his youth in Budapest. In the autumn of 1920 he met Ilona Duczynska, a Hungarian Communist Party member who had served in the Hungarian diplomatic corps in Switzerland during Kun's government. She had studied engineering in Zurich in the early years of the war and had been arrested in Budapest for distributing anti-war literature. After the fall of Kun's government, she fled to the Soviet Union and then joined Hungarian refugees who were living in a pension in the Viennese suburb of Hinterbrühl. The Hungarian Communist Party expelled Duczynska in 1922 because of her criticism of factionalism within the party, and she began working for the Austrian Social Democratic Workers Party, which had won a majority of votes in the Viennese city parliament elections of May 1919 (elections that were open to all adults both male and female). She and Karl married in 1923 and set up house in a flat that had belonged to her family overlooking the Danube in the city that became known for the next fifteen years as "Red Vienna." Karl Polanyi often remarked that Vienna's Socialist experiment was "one of the high points of western civilization."[8] Its achievements have been described by historian Anson Rabinbach, who notes that the Austrian Social Democratic Party was unique among European Socialist parties in basing its politics in the concept of *Bildungspolitik* and cultural transformation, with a strong focus on education, including adult education. By the end of 1933, the city had built more than sixty-one thousand new apartments, mostly in large housing blocks, with parks, swimming areas, schools, kindergartens, gymnasia, and health facilities.[9]

In Vienna, Karl's social circles included the Hungarian émigré community, the editorial group of the *Oesterreischische Volkswirt*, members of the Bund der religiösen Sozialisten of the Austrian Socialist Party, and scholars and students at the Volkshochschule, a socialist adult-education institute directed by Walter Schiff.[10] In addition, Karl held regular seminars on economic theory at his flat, and he was invited by Schiff to give seminars on guild socialism at the university.[11] Family members and old friends were living in Vienna. Karl had long been close to his uncle, Karli Pollacsek, a lawyer in Vienna, and to Karli's wife, Irma. Karl's sister Sophie and her husband Egon Szecsi, a lawyer whose health had been shattered during service in the war, moved from Budapest with their children to Hinterbrühl and then into the city.[12]

The Hungarian émigré community published several periodicals, including the *Bécsi Magyar Ujság*, an originally right-wing but later Communist-tinged newspaper that Oscar Jászi came to control in June 1921. It was a colorful paper with international, cultural, and literary articles that achieved a circulation of thirty-five thousand to forty thousand daily. Jászi aimed to make it "an organ of the left wing of the emigration, in a spirit of free, nondogmatic socialism." Initially some Communists remained with the newspaper, but they strongly objected to Jászi's Sunday edition articles of July 17 and 30, 1921, titled "Danton or Napoleon?" and "Tolstoy or Lenin?" The final straw for the Communist journalists was Karl Polanyi's article criticizing Lenin's New Economic Policy. One of the Communists expressed outrage at what he called a sufficient betrayal of Communist doctrines that its author would be welcomed back into the arms of the conservative Horthy regime.[13]

The Communists had already left the newspaper by the summer of 1922 when Karl argued in *Bécsi Magyar Ujság* in favor of a Christian Socialism based in a system resembling the guild Socialism of G. D. H. Cole described in Cole's *Self-Government in Industry* (1917) and *Guild Socialism Restated* (1920). Moving toward the English tradition of Fabian gradualist and constitutional Socialism, Karl called his position "socialist Christianity," distinct from conservative "Christian Socialism" and accepting a Tolstoyan creed that rejected the institutionalized church as well as the divinity and resurrection of Jesus Christ.[14]

As an expert on Soviet external and internal political and economic relations, Karl penned well over a hundred newspaper and magazine articles on Russian questions alone during the next decade. His child Kari, born in 1923, was put to work with a pair of oversized scissors to make clippings for her father from the *Times* of London, *Le Temps* of Paris, and the *Frankfurter Allgemeine Zeitung*. Thursdays were hectic, since that was the day that the *Volkswirt's* messenger arrived at the flat to collect the copy for Karl's next article.[15] His coverage of economic matters soon embroiled him in controversies within Austrian academic circles in economics, as did his teaching of courses at the Volkshochschule and at the university.

One of the most radical ideas around 1920 for reorganization of the economic system in Austria and Germany was Otto Neurath's proposal in 1919 for a centrally planned economy in which money would be abolished altogether and managers for the economy would rely on social statistics to guide production and distribution. Neurath considered this proposal a purely technical matter and not a political one.[16]

Neurath had been a member before the war of the economics seminar at the University of Vienna led by Eugen Böhm-Bawerk, a former minister of finance. Other members of that seminar included Joseph Schumpeter, Otto Bauer, Rudolf Hilferding, Emil Lederer, and Ludwig von Mises, all of whom developed influential and differing viewpoints that established them as leaders in twentieth-century economics. Böhm-Bawerk and Friedrich von Wieser were disciples of Carl Menger, the founder of the so-called Austrian School of Economics. The defining moment for the Austrian school was the *Methodenstreit* (debate over methods) in the 1880s between Menger and the German economist Gustav Schmoller. Menger argued for turning Viennese economics toward the development of generalizable theories based in empirical data and quantitative studies of consumer choices and the utility of goods, rather than the labor value of goods. In contrast, Schmoller eschewed deductive theories for his preference of comparative historical economics that detailed histories of industries and institutions with an eye to the role that economists can take practically in the making of public policy. Schmoller and his colleagues in the German School of Economics were instrumental in founding the Verein für Sozialpolitik in 1872 and advocating concessions to the labor classes under the Wilhelmine monarchy such as extension of public ownership and redistributive tax policies.[17]

As a member of Böhm-Bawerk's prewar seminar in Vienna, Neurath had begun investigating the question of how to run an economy under conditions of modern warfare. He gained practical experience when he became head of the General War and Economics Section of the Scientific Committee for War Economics in the War Ministry in Vienna and in 1919, president of the Central Planning Office of the short-lived Bavarian Soviet Republic. In 1919, Neurath published *Durch die Kriegswirtschaft zur Natural-wirtschaft*, where he suggested that a war economy provides a transitional state toward more extensive central planning in peacetime.[18] As mentioned in chapter 2, Walther Rathenau, a leading official within the Allgemeine Elektrizitätswerke Gesellschaft (AEG) electrical cartel in Germany, similarly led a wartime section in Germany's Ministry of War in which the procurement and disbursement of raw materials was centralized. This office (the Kriegsrohstoffabteilung, or Raw Materials Section) became for many German Social Democrats an exemplar for central planning as described in Rathenau's 1918 pamphlet *The New Economy*.[19]

Ludwig von Mises reconvened some of the original members of the prewar Böhm-Bawerk seminar in a group that met biweekly during the academic year from 1920 through 1934. As a monetary theorist and head of a government agency after the war, Mises followed on Menger's argument for a theoretical approach that studied the origins of money and exchange, the formation of prices, and the unintended consequences of human action, which Mises described as leading to "spontaneous orders" in economic life.[20] In this formulation there was consensus with the kind of thinking that Serafin Exner had expressed in his rector's inaugural address in 1908.

In 1920, Mises published his "Economic Calculation in the Socialist Commonwealth," and in 1922, *Die Gemeinwirtschaft: Untersuchungen über den Sozialismus*. In these books he made the objections that Neurath's moneyless economy with central planning cannot be a rational economic activity because no single manager can master all the possibilities of production in order to make judgments of value. Crucially, Mises argued that calculation is impossible because there are no prices without a free market and there can never be an efficient use of capital goods without private property.[21]

Karl Polanyi participated in this Socialist calculation debate. He was well versed in recent economic literature that contrasted the subjective theory of value, advocated by Menger's school, with the classical labor theory of value of David Ricardo and Karl Marx. Menger had developed an early theory of marginal utility, which was superseded by that of William Stanley Jevons, Léon Walras, and others. In this theory, the consumer is a crucial source of valuation of goods, and prices are determined by the final utility to the consumer of a good when the consumer becomes indifferent to acquiring an additional unit of the good. Under this circumstance, producers find production of more units to be of only marginal utility in meeting demand. This presumption of the marginal product undercut the Ricardian and Marxist emphasis on labor as the chief source of value, a critique with which Karl Polanyi agreed. Karl Polanyi discerned in both marginalist and neoclassical economics persistent links to former ideologies that "associated laws of commerce [with] the laws of nature and consequently the laws of God."[22]

Karl published articles on Socialist calculation in 1922 and in 1925 in the *Archiv für Sozialwissenschaft und Sozialpolitik*. He argued that the labor theory of value is erroneous and he endorsed the ascendant view that economic planning must make use of marginal analysis. He agreed with much of Mises's critique of Neurath, but he did not accept the assumption common to both the neoclassicist Mises and the then-Marxist Neurath that socialism *by definition* entails a centrally planned economy. Instead, Polanyi proposed that the ideal economy is one of functional Socialism (*Funktioneller Sozialismus*) in which the basic economic units are branches, or functions, of industry instead of private corporations (as in laissez-faire liberalism) or offices of state ownership (as in central planning). In his plan, there would be two kinds of prices: those that are regulated (*Festpreise*) and those that result from bargaining between economic units (*Vereinbarungsziffern*).[23]

This conception of functional Socialism was similar to Otto Bauer's ideas as a political leader of the Social Democratic Party in "Red" Vienna. Bauer suggested in 1919 that industrial sectors could be organized as economic units, with each sector governed by a tripartite body of employee, consumer, and state representatives who would settle conflicts among the various interests. Similarly, in Berlin, Rathenau and Emil Lederer suggested coordination and planning within some branches of industry, for example, mining and industries processing raw materials, as in the capitalist practice of monopolies and cartels. Lederer, although a student of Mises, was closer to Hilferding in taking a more Marxist point of view.[24] Schumpeter, too, came to advocate during the late 1920s a system of cooperative organization of industrialists and workers who would work out wage, price, and employment settlements. By the 1930s, however, with both Germany and Italy claiming to be following corporatist principles, the approach became associated with Fascism.[25]

After the war Friedrich von Hayek began working for Mises in the Office of Accounts in Vienna for settling international debt claims and, later, in Mises's Austrian Institute for Business Cycle Research. In addition to frequenting the Mises Circle, Hayek and the young lawyer Herbert Fürth started a discussion group in 1921 on economics and politics, which became known as the Geistkreis. The members of Hayek's Geistkreis, who met fortnightly for over a decade, included philosopher and jurist Felix Kaufmann along with the economists Oskar Morgenstern, Fritz Machlup, and Gottfried Haberler. There was considerable overlap in membership with the Mises Circle.[26]

Kaufmann provided a bridge to yet another intellectual circle, the Thursday evening discussion meetings that became known as the Vienna Circle, organized by Moritz Schlick in 1924 (see chapter 7). The group's programmatic pamphlet of 1929, titled "The Scientific World Conception: The Vienna Circle," associated the Vienna Circle with the goal of an empirically based scientific approach that could "shape economical and social life in accordance with rational principles."[27] Karl Popper was an occasional visitor in the Vienna Circle, and he met Karl Polanyi in 1924 at one of Karl's seminars on Socialism. Karl Polanyi participated in a study group on cooperation among the sciences, which was run by Rudolf Carnap, and he delivered a public lecture in 1930 on economic statistics as part of the Ernst Mach Society's public lecture program.[28]

In the 1920s, Popper and Hayek independently adopted a skeptical view toward the confidence in empiricism and calculation expressed in the Vienna Circle's manifesto and the implications of the manifesto for economics, as well as for philosophy. Popper and Hayek's skepticism was shared by both Michael and Karl Polanyi. They all strongly opposed programs for Socialist calculation and central planning. Hayek voiced his opposition in his 1929 PhD thesis in economics, which was translated into English in 1933 as Hayek's *Monetary Theory and the Trade Cycle*. Popper's rejection of the economics of Socialist calculation and central planning can be seen clearly in his more politically oriented book *The Open Society and Its Enemies*, written in the early 1940s. There he wrote that institutions never follow the specifications of an initial project, such as planning, because of unforeseen results from innumerable individual human decisions. As Menger had said in 1883, law, language, the state, the currency, and the market all are in large measure the spontaneous result of social evolution, as are the price of goods, the interest rate, land income, salaries, and so on.[29]

By the early 1930s, the National Socialists had made huge gains in the German election in the summer of 1930. The bankruptcy in May 1931 of the Austrian bank Credit Anstalt worsened the banking and currency crisis that had begun with the New York stock market crash in 1929. Hayek left Vienna in 1932 for a position at the London School of Economics, and

Schumpeter moved from the University of Bonn to Harvard University in that same year. Adolf Loewe, then teaching in Frankfurt, later reminisced that in 1932 he began keeping a few suitcases packed in case of emergency.[30] In the next decade the foremost members of the German and Austrian schools of economics emigrated, with some of the German left-wing economists, including Lederer, finding jobs at the New School of Social Research in New York City. As discussed by Claus-Dieter Krohn, many of the Austrian neoclassical economists obtained appointments at elite private East Coast universities like Harvard and Princeton.[31]

In Karl Polanyi's view, an incompatibility between capitalism and democracy and the divorce between the market economy and political democracy was at the heart of liberal Europe's crisis.[32] His views were regarded as "Red," and he began to fear that his politics would not be tolerated at the *Oesterreichische Volkswirt* under a new political regime. Following the suspension of the Austrian constitution and the free press in March 1933, along with a huge demonstration by Austrian National Socialists, he began to think of leaving central Europe, finally making up his mind to do so after Michael Polanyi and his family moved to Manchester in the fall of 1933. Karl arranged to move into a spare room in the London house of his friends Irene and Donald Grant. The *Oesterreichische Volkswirt* gave him three months' leave with pay so that he officially remained a journal correspondent until Hitler's takeover of Austria in 1938. Ilona remained in Austria for a while, organizing press and radio publicity for the Schutzbund workers, a wing of the Social Democratic Workers Party that opposed the Christian Socialist regime of Engelbert Dollfuss. After the repression in February 1934 of the Schutzbund uprising, which Ilona called "the first armed resistance to fascism" in her later book on its history, Kari Polanyi joined her father in England, as did Ilona in 1936.[33] Another casualty of the Dollfuss regime was the informal Vienna Circle and its formal Ernst Mach Society, which was dissolved by the government in February 1934 on the grounds of disseminating Social Democratic propaganda.[34]

## Michael Polanyi and Economic Theory: Negotiating Keynes and the Austrians

Michael's work in physical chemistry thoroughly absorbed him in Berlin in the early 1920s. He visited the Soviet Union for the first time in 1928, on the invitation of the physical chemist Abrahm Joffe. In his diary Polanyi noted down facts about prices, wages, and overcrowding in Russian urban apartments. In a letter to his sister Mausi, he wrote, "The economic system

functions so badly that one cannot judge from the result what its fundamental and dubious principles are. Everything is permeated by brutal and
stupid fanaticism considering all other opinions as devilish nonsense. The
tone of voice in public is a distasteful, monotonous cursing."[35] His diary
records for December 1928 list three economics books for reading: Carl
Landauer's *Das Wesen der Wirtschaft* (Nature of the economy) (1928), Hubert D. Henderson's *Supply and Demand* (1922), and one of Dennis Holme
Robertson's books.[36] The economists Henderson and Robertson were pro-
tégés at Cambridge of John Maynard Keynes.[37]

Sometime in 1928, along with Leo Szilard and John von Neumann, Polanyi began following a seminar in economics led by Jacob Marschak. Polanyi later told Thomas Kuhn in an interview that the seminar members all
were trying to understand the Russian phenomenon. Polanyi contributed
to the Marschak seminar with a paper on the Soviet Union.[38] Marschak,
who was born in Kiev, had fled to Berlin in 1919 and studied with Lederer,
with whom he would write a book in 1937 on the economic descent of
white-collar workers into the working class. Marschak was one of the contributors to the Socialist calculation debate, writing in 1924 that the Socialist system could be more efficient in pricing than the free market because of
the monopolist practices of capitalism. He would flee to Paris in 1933 and
then to Oxford, eventually taking a position at the New School in 1938
and at the University of Chicago in 1943. He directed the Cowles Commission for Economic Research from 1943 to 1948, which developed a general
equilibrium theory and econometrics.[39]

By early 1930, Polanyi organized his own study group that would meet
occasionally for Sunday evening dinner and discussion at the Kaiser Wilhelm Society's Harnack Haus. In a draft of a letter to Erwin Schrödinger,
Polanyi wrote that he hoped to get together his scientifically trained colleagues with economists. Those who began to come to the dinners fairly
regularly included Szilard, Von Neumann, Fritz London, Eugene Wigner,
Paul Harteck, Marschak, and the economists Gustav and Toni Stolper (but
not Schrödinger). At the group's second meeting, Fritz Haber attended,
along with Polanyi's host during his 1928 visit to Russia, Abraham Joffe.
Polanyi also made efforts to get Wichard von Moellendorff, a protégé
of Rathenau's in economic planning during the previous war, to attend.
Moellendorff was now Director of the State Material Testing Office and of
the Kaiser-Wilhelm Institute for Metals Research in Berlin-Dahlem.[40]

Gustav Stolper was an important figure in Berlin as a German Democratic Party member of the Reichstag and editor of *Der deutsche Volkswirt*,
which he founded in 1926. The *Deutsche Volkswirt* initiated theory-based

debates on current economic issues and entertained little input from the German historical school of economics. It was said by contemporaries to have influenced the laissez-faire approach and deflationary policy of Chancellor Heinrich Brüning's government during 1930–1932, which included the cutting of wages and prices and the refusal to fight unemployment by a program of public works. This was a policy that later was held by some historians to be partly responsible for the downfall of the Weimar republic.[41] The permanent staff of the paper included Landauer, along with Stolper's wife Toni. Contributors included Joseph Schumpeter, who was a personal friend of Stolper's, Hayek, Haberler, Oskar Morgenstern, Wilhelm Röpke, and from the so-called Kiel school, Marschak and Lederer.[42]

Polanyi wrote an article for the May 1930 issue of *Der deutsche Volkswirt*. This was his first economic publication. In this short essay, Polanyi argued for government support of pure, or fundamental, science even when practical benefits might not be immediately obvious as in medicine or engineering. Science, he wrote, is not simple scholarly pleasure or self indulgence, but intellectual *"work"* (*Gedankenarbeit*). There needs to be a requirement that government support will not be withheld from research on the grounds that it is pure or *"genuine"* (*echten*) research, he wrote.[43]

Many writers for the *Deutsche Volkswirt* opposed deficit-financed public works programs and inflationary programs, while National Socialist proposals were dismissed for their antiliberal and anticapitalistic views. Prominent in the *Volkswirt* discussions were the theories of Schumpeter and Hayek, who agreed that the only way to prevent a depression is to cut short a boom, such as the boom in the United States in the 1920s, and that if a depression seems underway, to recognize that attempts to cure it by monetary or fiscal policy will likely worsen the situation. In Hayek's 1929 thesis, "Monetary Theory and the Trade Cycle," and in his 1931 book, *Prices and Production*, he modeled the economy as a closed system in equilibrium so that that a slump in a trade cycle is a sign that the system will head back toward equilibrium and should be left alone. During a boom, growth has become unbalanced, with investment in the expansion of industrial capacity outstripping the supply of savings. Recessions restore the balance between investment and savings. Thus, patience must reign during inevitable periods of unemployment, and an elastic supply of currency makes the situation worse—not better.[44]

In contrast to this view, which won out in Stolper's periodical, was the argument from the Cambridge school of economists as expressed by John Maynard Keynes in the *Treatise on Money*, a book that Röpke reviewed favorably in the *Deutsche Volkswirt* in 1931. Keynes focused on the demand side of the economy, rather than on supply, advocating cutting taxes and increasing public spending when there is a slump. Depressions result from a lack of investment, requiring the priming of the pump by an expansionary monetary policy. Keynes's approach was by no means an abandonment of the private enterprise system. As he wrote in his pamphlet *The End of Laissez-Faire* in 1926, unemployment and other economic evils should not be ignored but "can be cured partly by the deliberate control of the currency and of credit by a central institution" and partly by the collection and dissemination of economic data.[45] This was an argument that Stolper did not accept, fearing that government intervention and government programs paved the way to a planned economy that was the aim of both right- and left-wing opponents of the market economy.[46]

The first meeting of Michael Polanyi's Sunday evening Harnack Haus group focused on a document on philanthropy published by the United States National Bureau of Economic Research. During his visit to the United States in the summer and early fall of 1929, Polanyi had talked with Jászi, who now was teaching at Oberlin College, and Polanyi himself was thinking about taking an academic position in America. He had made notes during his trip on prices, wages, and labor relations in the United States, just as he had done the previous year in the Soviet Union. He himself was a recipient of American philanthropy via the Rockefeller Foundation's provision of funds for his and other laboratories at the Kaiser Wilhelm Gesellschaft.

The next Sunday meeting focused on the Soviet Union. Polanyi asked Toni Stolper to talk at the meeting, following a 7 p.m. dinner, about the Five-Year Plan that had been initiated by Stalin in 1928 for the development of industry and collectivization of agriculture. There would be no smoking, he promised her.[47] There were at least nine or ten meetings through the middle of 1931, and Polanyi took some pride in the group's discussions and accomplishments. His laboratory colleague Karl Söllner told William T. Scott that the Sunday discussion group marked "the exact time when Polanyi had begun to be less interested in physical chemistry."[48] When later recalling Marschak's seminar and the Sunday meeting group, Polanyi told Kuhn in 1962 that he always had intended to give up science at one stage or another and do something else and that he never had looked forward to being a scientist in old age and to competing with young people.[49]

Polanyi made more trips to the Soviet Union in the early 1930s. He was invited by Nikolai Semenoff to a chemical kinetics conference in Leningrad in September 1930, where he was offered a position by Semenoff at Joffe's institute. Polanyi declined the offer, but agreed to visit and give

Figure 5.1. Michael Polanyi during an institute excursion in June 1931.
(Courtesy of Archiv der Max Planck-Gesellschaft, Berlin-Dahlem.)

lectures occasionally. He returned in 1932 to lecture at Joffe's institute in Leningrad, and again in 1935, by which time he had moved from Berlin to Manchester.[50] Polanyi and Joffe were good friends who independently were doing fundamental research in X-ray crystallography and solid-state physics with industrial applications and running physical chemistry laboratories. Joffe's responsibilities were considerably greater, since the Leningrad Physico-Technical Institute that he had established in 1918 provided some of the oversight for fifteen other research centers and over a hundred factory physics laboratories. Polanyi knew that Joffe had tried to protect fundamental research from both ideological and bureaucratic interference since the 1920s, urging Moscow administrators at a conference in Moscow in 1931 to recognize the danger of project-specific, long-range planning that might constrain research and harm long-term industrial productivity.[51]

Soon after settling in Manchester in late 1933, Polanyi visited the Department of Economics and Social Studies, where he made a friend in John Jewkes, who later recalled that Polanyi often lunched with chemists, economists, and other faculty members at a large table in the campus restaurant. Polanyi wrote Toni Stolper in September shortly after his arrival in Manchester that he had "already had fights with my progressive friends here."[52] In the spring of 1935, when he traveled to Moscow for a conference, he met with Nikolai I. Bukharin, who had helped devise Bolshevik policy during 1925–1927 and currently was editor of *Izvestia*, where he wrote of the dangers of Fascism in Europe.[53] Polanyi also visited with his niece Eva Striker and her husband, both of whom were arrested later, like Bukharin, on charges of plotting against Stalin and the state.[54]

Upon his return from Moscow, Polanyi began to put together a long paper critical of Soviet economics in which he closely examined claims and data from the first two Soviet Five-Year Plans. The flavor of Polanyi's essay can be seen from the following: in 1933, Polanyi, reported, the Soviet Union's Red Dawn Knitted Goods Mills received nineteen different sets of instructions. The plans were altered as follows: the output plans, seven times; the productivity of labor plan, four times; the cost of production, eight times. The plan for 1933 was endorsed on January 4, 1934. *This* was central planning.[55]

The paper appeared in the November 1935 issue of the *Manchester School of Economic and Social Studies* and was republished by Manchester University Press in 1936 as the twenty-three-page pamphlet *USSR Economics—Fundamental Data, System and Spirit*. Drawing both upon data and his personal experiences in Russia, Polanyi described his observations of a basic diet of bread and potatoes and crowded housing while conceding improvement in education and industry. The government mollified its citizens, he said, with the promises of full employment and a new Socialist state that would eliminate the old inequities of the czarist regime.[56]

Polanyi sent copies of his pamphlet to economic and political commentators, and he was pleased with the journalist Walter Lippmann's compliment in *The Good Society* in 1937 that Polanyi was "an exceptionally gifted observer." Jászi cautioned Polanyi, however, that he might be too trusting of Russian statistics and too optimistic about so-called improvements. Sir Ernest Simon congratulated him on his groundbreaking attempt to estimate quantitatively the achievements of Russian industry and agriculture.[57]

Polanyi seems to have entertained the idea of making a film on economics for several years.[58] After completing his article on the USSR, he made some notes outlining a film and sent them to his Manchester friend Hugh O'Neill, who was a senior lecturer in metallurgy, a Roman Catholic, and someone with whom Polanyi felt at ease discussing chemistry or economics or religion.[59] In the fall he sent Toni and Gustav Stolper, who had emigrated to New York City, a reprint of his USSR article and told them of his plans to make a film, saying he had discussed the idea with a filmmaker in Vienna but had decided to pursue the project in England. He was thinking of a film an hour and a half long divided into four sections.[60] He also wrote Jászi about the plan.[61]

Within a few months, as his laboratory colleagues later recalled, Polanyi enlisted his glassblower and his technician to construct a glass apparatus using water and a vacuum line to illustrate the law of supply and demand with water flowing in and out of flasks and beakers. Wilfried Heller and Cecil Bawn later said that these were the only experiments that the Manchester staff ever saw Polanyi perform entirely alone.[62] As described in Scott and Moleski's biography of Polanyi:

> Gilson and Syar built a second machine with a conveyor belt and a series of tubes through which colored balls dropped into colored containers on the belt. The belt represented Polanyi's key conception, the circulation of money, and the balls the payments for such things as wages and interest, consumer goods, bank savings and capital expenditures.[63]

A third machine was built, too, with "lots of wheels," which functioned like an analog computer in making possible the study of effects of increasing the money supply and making other changes in the economy. This machine appears to have been influenced in its design by the technicians' familiarity with Douglas Hartree's work at Manchester on a clockwork mechanical analyzer for solving differential equations. Melvin Calvin and other personnel in Polanyi's physical chemistry laboratory were not happy with the economic project's interference with the amount of time that technicians could devote to help with student apparatuses.[64]

Polanyi's use of fluid and kinetic models was not new in economics, at least conceptually. The French physiocrat François Quesnay argued in the mid-eighteenth century that the advance of capital, like the initial surge of blood as it leaves the left ventricle of the heart, keeps the monetary system going. Gaspard Grollier's design of a rolling-ball clock and hydraulic machine may have influenced the visual content of Quesnay's *Tableau Economique*, which included images of a zigzagging circular flow independent of Quesnay's textual analysis.[65] There is good reason to think that David Hume's study of hydrostatics played a role in his emphasis on the flow and circulation of money, with Hume making explicit analogies between money and water and writing of the tendency of water to reach its own level. Later hydraulic machine models included one designed in the 1950s by A. W. H. Phillips, while he was a student at the London School of Economics.[66]

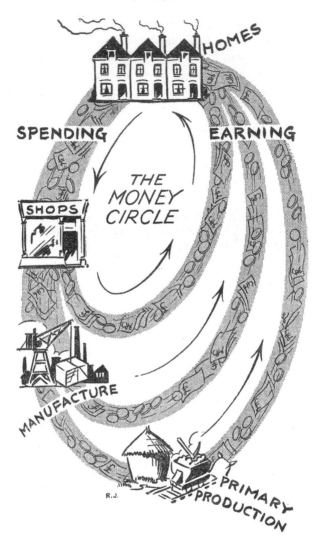

Figure 5.2. "The Money Circle." (In Michael Polanyi, *Full Employment
and Free Trade*, 2nd ed. [Cambridge: Cambridge University Press,
1948], 4, fig. 1. Reprinted with the permission of Cambridge
University Press and John C. Polanyi.)

For the film, Polanyi leaned toward the conveyor belt analogy or model,
describing it in a talk in June 1936 to the Association for Education in Citizenship, an organization in Manchester chaired by Simon.[67] The film was
to be a series of diagrams, with its central image the sketch of what Polanyi
called the money circle, or belt, that flows through a specific number of

businesses and homes, linking production on the one side and consumption on the other, with spending (prices) and earning (wages) as part of the process. The criticism by Toni Stolper that his tableau was oversimplified did not dim his enthusiasm.[68]

Polanyi's initial plan for the film was based in his economic readings and discussions up through 1935 and especially on John Maynard Keynes's *Treatise on Money*, published in 1930. At Christmastime in 1936, Polanyi read Keynes's new book *The General Theory of Employment, Interest, and Money*, which had been published the previous February. Polanyi gave a pseudo-Keynsian lecture in February 1937 to the Manchester Political Society on faults in the laissez-faire system: unfair distribution of income, the principle of complete government separation from the market, and the lack of a sense of communal responsibility. In March he gave a talk at the Manchester Historical Society in which he warned that economic crisis and unemployment could bring demands to England for a Communist or Fascist economic solution, which would be disastrous. By this time he was finishing a script for his film, and he persuaded Sir Samuel Turner of Rochdale, a philanthropic friend of the university, to provide funds for completion of the first part of the film. In the meantime, he prepared a course of six lectures on economics to give in university extension classes.

Gaumont Instructional Films made the soundless film. The first screening of *An Outline of the Working of Money* debuted to an audience of 150 people at the Manchester Statistical Society on March 9, 1938.[69] The *Evening News* reported the next day that "Money Is Star of This Film" made by a "new Walt Disney":

> The inventor, producer, and director is Professor Polanyi, a brilliant Hungarian on the staff of the Chemistry department of Manchester University . . . [A]nimated figures of workmen are seen leaving for factory, farm, and mine . . . Womenfolk are shown going forth day by day buying commodities . . . and the part played by banks is shown.[70]

Polanyi arranged screenings of the film at the Association for the Education in Citizenship, the Manchester Historical Association, the technical employees of Imperial Chemical Industries in Norwich, the Hope Street Church Social Study Group in Liverpool, the Film Society of the Bank of England in London, and other venues. Nor did Polanyi ignore professional economists, arranging for viewing of the film in Paris at a conference he attended on Lippman's book *The Good Society*, where Hayek, Mises, Röpke, and other prominent scholars were present.[71] J. D. Bernal arranged a viewing for the left-wing Association of Scientific Workers in London.[72]

Encouraged by interest in the film, Polanyi decided to revise it, working with Jewkes in what turned out to be a successful request to the Rockefeller Foundation for £1,000 in funds to be awarded to the Economics Research Section at the University of Manchester to further develop its experiments in the production of diagrammatic films illustrating the functioning of economic processes. Polanyi's application reported that £500 had been raised by private subscription in support of the first film. Rockefeller funds now were requested in support of making a sound recording of commentary for the film and for making different versions for different audiences. The film also needed to be made faster and less jerky. Money would be used, too, for exploring the feasibility of making a second film called *Population and Economic Life*. A letter of support came from Henry Clay, economic advisor to the Bank of England, applauding Polanyi's brilliance of mind and suggesting that the fact that he was not a professional economist enabled Polanyi to imagine ways of presentation that would be almost impossible for a professional.[73]

By the time that the revised film, titled *Unemployment and Money*, was completed in the spring of 1940, war had broken out in Europe. The new film version was about forty minutes long with only about 20 percent retained from the first version. There was a showing in London in April, with several hundred invitations sent out and Karl's help enlisted for turning out an audience.[74] Polanyi shipped the negative of the film and a sound track to the Rockefeller Foundation. He feared that he would be unable to work any further on the film in England because of the war. The Rockefeller Foundation allocated some funds, which it designated as the last monies for the purpose of assisting with travel expenses to the United States in connection with the film.[75] Polanyi was unable to come himself for the first showing of the film in the United States at the Film Library of the Museum of Modern Art on the afternoon of November 7, 1940. About fifty people attended, including Joseph Willitts and Tracy Kittredge from the Rockefeller Foundation, J. B. Condliffe of the University of California at Berkeley, and Loewe and Marschak, who were teaching at the New School for Social Research in New York. Oskar Morgenstern and some of his students were there from Princeton. Toni Stolper and Polanyi's sister Laura also were present, but not Karl, who had begun teaching at Bennington College in Vermont that fall.[76]

The film made stops over the next few years at college campuses across the country, with the Museum of Modern Art providing copies at the cost of $75 for the six reels of the film. Condliffe wrote Kittredge of his impression that the film was not useful in teaching, because parts of it were too simplified to interest the serious student and too complex for a truly popular presentation. Morgenstern's reaction was similar, but delivered with the caveat that he was not in great sympathy with Polanyi's presentation of Keynes's theories, which Morgenstern regarded as a superficial attempt to understand the modern economy. Kenneth Spang in the Economics Department at Yale thought the mechanism of credit creation was poorly explained and that the quantitative relationship of old to new investment was not made clear. Unfortunately Willitts himself had not been impressed at the MOMA showing.[77]

Polanyi also gave three silent copies of the film and one sound copy to the Workers Educational Association, the organization for which his brother Karl had been teaching since 1933. The soundless version was shown most frequently, because it was more easily adapted by individual teachers to their classes. The reactions were not much different from those at the elite American universities. J. I. Roper reported that elementary students drew the wrong conclusions from the film, for example, that it is wrong to save, while advanced students wanted to know about central bank policy, the influence of the interest rate, international trade, and other matters. H. Dawes criticized aspects of the film, including the assumption that the major part of a plant's value will be lost by ageing before renewals are undertaken. Mrs. S. G. Raybould objected to the film's explanation that a boom comes to an end because all employable labor is employed, noting that there was over 10 percent unemployment in the boom years of 1929 and 1937.[78]

Although Polanyi continued to request further funds during the next couple of years from the Rockefeller Foundation in order to revise his film further, and his sister Laura attempted to see Willitts personally, Polanyi shifted the emphasis in letters to Willitts by the end of 1944 to descriptions of other projects. These projects included a long article on patent reform; a book eventually published by Cambridge titled *Full Employment and Free Trade* (1945), on what he called his logical study of economics and social planning; and a book on scientific life as adumbrated in his essay, "The Autonomy of Science."[79] Polanyi originally had explained his determination to make a film by saying that he believed "that verbal explanations are incapable of conveying a popular understanding of these matters," and it is not surprising that he incorporated some of the illustrations from his film into *Full Employment and Free Trade*.[80]

The 1945 book drew upon the film along with Polanyi's 1935 pamphlet on economics in the USSR, his 1944 article on patent reform, and analysis

Figure 5.3. Adolf, Karl, and Michael Polanyi with their sister Laura
Striker in the 1940s. (Courtesy of John C. Polanyi.)

of discussions of British postwar economic policy. A particular target for
Polanyi was the 1942 British White Paper, known as the Beveridge Report,
and William Beveridge's 1944 book *Full Employment in a Free Society*. Beveridge, who had been director of the London School of Economics from
1919 to 1937, was an influential architect of programs for social welfare in
Great Britain.

Among economists who read Polanyi's manuscript before the book's
publication was Lionel Robbins, who held the chair of Political Economy
at the London School of Economics from 1929 to 1961 and was director of
the Economic Section of the Offices of the War Cabinet during the war. He
initially opposed Keynsian theory and the Cambridge school of economics,
as well as the Fabian orientation of Beveridge and other Socialist-oriented
members of the London School of Economics. Robbins had pushed for
hiring Hayek in 1933, an appointment that Beveridge, always in favor of
hiring the best people, supported.[81]

A main goal in Polanyi's *Full Employment*, as in his economic films, was
to make an accessible presentation of Keynes's main ideas, augmented by

an explanation of his differences with Keynes and by his own views on economic organization in postwar Britain. First and foremost, Polanyi identified Keynes as an economic and political individualist and defender of capitalism and democracy like himself. "A correct Keynsian policy should regenerate free competition and re-establish capitalism on renewed foundations," wrote Polanyi in his book's preface, echoing Keynes's own conclusion in *The General Theory*:

> It is not the ownership of the instruments of production which it is important for the State to assume . . . The advantage to efficiency of the decentralization of decisions and of individual responsibility is even greater, perhaps, than the nineteenth century supposed . . . But, above all, individualism, if it can be purged of its defects and its abuses, is the best safeguard of personal liberty.[82]

Like Keynes, Polanyi had some concern for the defects in capitalism that led to huge economic disparities and social injustices.[83] Keynes, who was educated in mathematics and had written a *Treatise on Probability* in 1921, stressed the role of subjective judgments and inevitable uncertainty in economics, rather than just the role of rational calculation. This was an argument congenial to Polanyi's views. Keynes further argued that economists needed to take into account not so much cold calculation in economic production, but the individual or corporation's temptation to take a chance and the satisfaction that is found in inventing something or constructing a factory, a railway or a mine.[84] For this reason, Keynes wrote, there is social and psychological justification for "significant inequalities of incomes and wealth, but not for such large disparities as exist today."[85] Keynes thought that considerable progress had been made to address unjust inequities through instruments of direct taxation, such as income taxes and surtaxes and death taxes. Polanyi, somewhat more conservatively, suggested that inheritance tax laws should be revised rather than using tax measures directly against disproportionate earnings or distributive injustice.[86]

Like Keynes, Polanyi advocated deficit spending, although he differed from Keynes in opposing public works projects. Concentrating on the circulation of money, or the money belt, Polanyi wrote that a state near full employment can be achieved when a cyclical depression appears to be approaching by filling the gap between savings and commercial investment, not by adjusting the rate of interest or granting government subsidies or taxing the rich, but by a "principle of neutrality" that allows the government to continue to meet normal public expenditure by decreasing taxes and running a deficit until the economy recovers from the impending slump. Necessary cash to meet normal public expenditures can be issued by the Central Bank in an amount equivalent to the proposed deficit until the width of the money belt expands back to normal.[87] This kind of state governance should address only the problem of the money supply and not matters, such as social welfare, that are political rather than economic in nature, according to Polanyi.[88] In short, the economy was not to be used for social engineering, a view consistent with the main thrust of the Austrian school of economics against the German historical and reformist schools.

In sections of the book on the USSR and on the wartime economy, Polanyi argued that deficit spending was precisely what provided the Soviet Union and Germany with economic recovery during 1933–1937, not compulsory labor regulations and large-scale public works projects.[89] Beveridge's approach in Great Britain, advocating public investment and government supervision of employment policy, "is precisely the connection against which the present book is protesting," wrote Polanyi, reiterating objections to the Roosevelt administration's public works projects (Works Progress Administration; later renamed the Work Projects Administration) and Hitler's jobs programs during the 1930s.[90]

If Keynes and Polanyi agreed on the values of individualism and democracy, however, they fundamentally disagreed on the naturalness of the economic order, even though they both spoke of cycles and equilibrating mechanisms. Polanyi took the side of Hayek, Mises, and Menger on this matter. In contrast, in Keynes's view, the postulates of Malthusian and Ricardian classical economic theory were based fundamentally in political preferences and not in natural laws. Keynes wrote of Ricardo's theory: "That it could explain much social injustice and apparent cruelty as an inevitable incident in the scheme of progress, and the attempt to change such things as likely on the whole to do more harm than good, commended it to authority."[91] The classical economists, Keynes continued, were wrong to consider full employment as a natural state, since we oscillate around an intermediary position well below full employment and considerably above a minimum employment that would be unsupportable. Keynes wrote:

> But we must not conclude that the mean position thus determined by "natural" tendencies, namely, by those tendencies which are likely to persist, failing measures expressly designed to correct them, is therefore, established by laws of necessity. The unimpeded rule of the above conditions is a fact of observation concerning the world as it is or has been, and not a necessary principle which cannot be changed.[92]

In contrast, Polanyi wrote near the conclusion of *Full Employment and Free Trade* that "I wholeheartedly accept the guidance of the 'invisible hand' for the mutual adjustments of productive units."[93] He argued for leaving alone not only the capitalists and consumers, but also workers, even in times of advancing full employment, when they might be tempted to demand higher wages that would lead to an inflationary condition widening the money belt beyond its ideal state. No government interference should be allowed in response to the workers' demands, but rather individual responsibility must be allowed to reign in this situation, too:

> It will require a great sense of responsibility on their part to refrain from such an action. The temptation can be avoided only by the clearest intellectual realization of the whole position on the part of employees and the public in general, *coupled with the assurance that the Government can be trusted to compensate the employees fairly by effective measures of redistributive taxation.*[94]

The "obvious" source of this taxation would be "the swelling profits of industry," according to Polanyi.[95] The government, in enacting such policy, should be guided by experts in the skill of economic administration and, in the case of Great Britain, Parliament should decide annually on the level of monetary circulation and corresponding level of unemployment.[96]

This somewhat surprising concession to government guidance excited scrutiny, as Polanyi expected. His optimism about the rational and self-denying behavior of the higher-wage-seeking worker likely elicited the sort of response from readers that Keynes made to one of Hayek's similarly optimistic suppositions about self-denial in the capitalist: "I doubt if such an individual exists."[97] Polanyi noted that some readers might identify his suggestion of any role whatsoever for government as a kind of "planning," but he denied that this was the case.[98] Rather, in his view, there are legitimate roles for government in many areas of economic and political life, where expert opinion without coercion may be relevant, which he portrayed as a weakened version of Keynes. Guidance is needed from "government economists" working within the parliamentary framework in consultation with the professional opinion of fellow economists.[99] As sociologist Charles Thorpe notes, Polanyi here demonstrates a political instinct for public deference toward expert authority, a view that reappears in his philosophy of science, as we will see later.[100]

Historian of science Adrian Johns notes Polanyi's recommendation in a 1944 article on patent reform that current patent law should be replaced by the establishment of a government license (as had already been proposed by economist Arnold Plant at the London School of Economics) and a one-time payment to inventors, based on the opinion of a committee of experts. Users of the invention would pay license fees, instead of a company (or individual) holding a monopoly on rights and profits.[101] Johns found comments about patent law in many of Polanyi's draft manuscripts during the early 1940s and argues that this is one of the links between the different projects on which Polanyi was working at the time.

At the conclusion of *Full Employment and Free Trade*, a book written against socialism and in favor of capitalism, Polanyi writes that private property rights are justifiable only on the grounds of their usefulness to the community.[102] In the patent reform article, Polanyi argued, as Johns notes, that patents on intellectual property smack of economic monopoly that prevents the free circulation of ideas and interfere with creativity and free trade. Polanyi had long served as a consultant to industry, and his recent experience in Manchester at Imperial Chemistry Industries (ICI) had convinced him that the pooling of patents by commercial laboratories was turning into a kind of planned enterprise that stifled individual creativity in the laboratories and ultimately would put a brake upon new products.[103] Thus for Polanyi, in the economic sphere no less than in the scientific sphere, there is a tension between community interests and individual freedoms that serves as an engine for motivating discovery and innovation. This is an insight that he would bring into his writings in the philosophy of science.

Jewkes thought Polanyi's patent scheme was unworkable, but he encouraged Polanyi in writing what became the 1945 book and liked the result. So did Gottfried Haberler, who in 1936 had joined Schumpeter in Harvard's Economics Department. Haberler, who was in touch with the Stolpers in New York, arranged re-publication by Macmillan of Polanyi's book and wrote a laudatory introduction, which nonetheless expressed skepticism that Polanyi's prescription could relieve a slump.[104] Polanyi had misgivings about many of his liberal friends' reaction to his book, writing Karl Mannheim that he thought that among his friends, his own view was "the most 'radical' Keynesian attitude which—incidentally—involved the least 'planning.'"[105] Karl Polanyi's opinion must have mattered to Michael. Karl offered congratulations for his brother's successfully showing the general social and political significance of Keynes's theory, but deplored Michael's preference for a market economy despite a "damning indictment" of its purest forms.[106] By this time, Karl had developed a full blown historical interpretation of capitalism that undermined its defenders' claims for its natural foundation in immutable laws. Karl simultaneously offered an argument for the social embeddedness of economic life that had significant resonance in the longer run in Michael's views on the character of science and scientific life.

## Karl Polanyi and *The Great Transformation*

After arriving in London in the fall of 1933, Karl Polanyi had set out to make contact with G. D. H. Cole and R. H. Tawney, as well as with Keynes and other academic economists.[107] At Tawney's invitation, Karl began teaching adult education courses, as he had done in Vienna. The courses were organized by the Workers' Educational Association (WEA) and by the Extramural Delegacies of the Universities of Oxford and London. He held classes once a week in small towns and villages in Kent and Sussex, preparing notes for lectures on economic history and international relations, including the social and economic history of England, which would serve as the basis for *The Great Transformation*.[108] Barbara Wootton, like Cole and Tawney an advocate of guild Socialism, was the director of tutorial studies at the University of London and favored Karl Polanyi's approach to economic history, which addressed the conditions of the working class under capitalism and the problem of the rise of Fascism.[109] Karl Mannheim, who had moved to London, recommended Karl Polanyi for a part-time teaching position in Liverpool through the Academic Assistance Council, and in the winter of 1935 Karl Polanyi made a two-month visit to the United States for lecturing.[110] Karl loved teaching and reflected in a letter to Jászi that he was fifty years old when he was led by his circumstances in England to research in economic history. "I earned my living that way as a teacher. For I was born to be one."[111]

Donald and Irene Grant introduced Karl to a new circle of friends and colleagues who were Socialist Christians and members of the Christian Left Group. They included Donald K. Kitchin, John Lewis, and John Mac-Murray. Together they launched a book project called *Christianity and the Social Revolution*, which was published by Gollancz, a left-wing publisher for books as diverse as the Webbs' *Soviet Communism: A New Civilisation* and George Orwell's *Road to Wigan Pier* and *Coming Up for Air*. Joseph Needham and Charles Raven were among the contributors to the *Christianity* volume, which sold well in 1935 and 1936. In the collaborating authors' view, Christianity's current mission was to foster a transformation of society rather than to preach outmoded doctrines such as the virgin birth and the resurrection. Marxism and Christianity must not be seen as irreconcilable doctrines because of the Soviet Union's state policy of atheism, but as inherently ethical doctrines aimed at social justice and the common good.[112]

Karl's contribution to the volume was "The Essence of Fascism," a revision of an article that had appeared in 1930 in a journal of the Bund der Religiösen Sozialisten. Here he laid out some of the arguments that would reappear in *The Great Transformation*, and he specifically criticized the "liberals of the Mises school." He wrote that the strained relationship between political democracy and laissez-faire capitalism in the nineteenth century had given way in the twentieth century to the subjection of the political to the economic order under Fascism and to the subordination of the economic order by the political order under communism. In Fascism, only economic life remains and "capitalism . . . becomes the whole of society." Fascism, he argued, aims to undermine both the labor movement and Christianity, while Marxism, although atheistic, is more promising in its communal and egalitarian aims.[113]

As he continued the argument, in a Worker's Educational Trade Union Committee pamphlet of 1937, "Europe Today," Karl argued that democracy cannot be maintained unless democratic principles are extended to all of society including the economic system: "This is commonly called socialism." The first Fascist groups under Mussolini had attacked trade union centers, workers' clubs and local labor headquarters. In contrast, the democratizing Bolshevik revolution prevented the return of czarism and the White Terror in Russia and brought education and industry to a population that had been illiterate and agricultural. As a consequence, "those who wish to strike at democracy direct their attacks against Socialism."[114]

By 1939, Karl was even firmer in his commitment to the Soviet Union, questioning in his essay "Russia and the Crisis" Western reports on the 1937 purge trials: "To stand for socialism and not for Russia is the betrayal of socialism in its sole existing embodiment. To stand for Russia without mentioning socialism would also be the betrayal of socialism."[115] In taking this position, Karl was one among many anti-Fascist intellectuals who saw the Soviet Union as a bulwark against Mussolini and Hitler and who were willing to overlook what they hoped were only temporary birth pangs in the emergence of a revolutionary new Socialism. A break between the two Polanyi brothers nearly occurred when Karl suggested that their niece Eva must have been treated by fair and judicial methods when arrested in the Soviet Union in 1935.[116] On these questions Karl and Michael Polanyi were futilely divided.

By the time the Germans broke the nonaggression pact of August 1939 and invaded the Soviet Union on June 22, 1941, Karl was in Vermont, where Bennington College's president had invited him to be a resident scholar. During his stay in Vermont, from 1941 to 1943, he finished *The Great Transformation*, a book, he said in his acknowledgments, whose main thesis was developed during the academic year 1939–1940 in conjunction with WEA classes in London, Canterbury, and Bexhill. He was supported for two academic years in Bennington by a Rockefeller Foundation Fellowship.[117] He had help with American publication of his book from three Bennington scholars: John A. Kouwenhoven, who had taught literature at Bennington College and was an editor at *Harper's* in the early 1940s; Horst Mendershausen, a German émigré economist at the National Bureau of Economic Research who was associated with Bennington College; and Peter Drucker, the Austrian émigré economist whom Karl had known in Vienna and who was teaching at Bennington College.[118]

Before publication of *The Great Transformation*, Karl and Ilona, who had been invited to teach mathematics at Bennington, went back to England in 1943, where Karl resumed teaching for the WEA. In 1947, he returned to the United States as a visiting professor of economics at Columbia University, where he lectured annually until retiring in 1953. To their dismay, by 1947, Ilona could no longer get a visa from the U.S. State Department on the grounds of her earlier membership and activity with Communist organizations in Hungary and Austria. In 1950, the Polanyis made their home in Pickering, Canada, near Toronto, and Karl Polanyi regularly commuted to New York City until his retirement from Columbia in 1953 at age sixty-seven. After retirement he was able to continue research, with a grant from the Ford Foundation, collaborating with colleagues at Columbia and elsewhere.[119]

In August 1941, Karl wrote Michael from Vermont that he would likely call his new book *Anatomy of the 19th Century: Political and Economic Origins of the Cataclysm.* "The main thesis is that the Cataclysm was due to matters of an economic order, the last 150 years having been eminently an age of economic determination." This disaster was due, he wrote Michael, to the attempt to separate the economic and political spheres, a theme that Karl reiterated in *The Great Transformation* and a point of view that Michel rejected in *Full Employment and Free Trade.*[120]

One of the most original arguments of *The Great Transformation* is the description of the unusual and short-lived character of the liberal free-market system. Keynes had argued in *The General Theory* that the postulates of classical economics are applicable only to a special case "that is not the economic society in which we actually live." Karl argued that the self-regulating market which came into being under special historical circumstances cannot endure.[121] Like all economic systems, it is embedded in a certain time and place. Once machines and plants were used for production in a commercial society, he wrote, the idea of implementing a self-regulating market was bound to take shape. The industrialist bought raw materials and labor, that is, commodified nature and man. All transactions were seen as money transactions, with the "natural" expectation that humans behave only in order to achieve maximum money gains.[122]

In England, the home of the self-regulating market, particular historical circumstances made all this possible by the early 1830s. Following the relinquishment of the Speenhamland law, which had guaranteed minimum income to the poor, the Reform Bill of 1832 and the Poor Law Amendment of 1834, a competitive labor market was established in which labor could move freely, resulting in the full establishment of the social system of industrial capitalism. Nineteenth-century economists gradually relinquished the humanistic assumptions of Adam Smith's political economy which, as historian and philosopher Margaret Schabas notes, recognized the role of government in providing public goods and guarding against the deterioration of the minds and bodies of their subjects.[123] Instead, now it was being said that economic society is distinct from the political state, even though, Karl argued, the actions of the political state had made the market economy possible in the first place. "From this time onward naturalism haunted the science of man."[124]

As historical and anthropological research made plain, liberal utopian ideas of market capitalism ignored that fact that economies as a rule are submerged in social relationships: "man . . . acts so as to safeguard his social standing, his social claims, his social assets."[125] We see this clearly in primitive economies, as well as in civilized societies, as Max Weber realized, Karl wrote.[126] In tribal societies, such as the Trobriand Islanders of Western Melanesia studied by Bronislaw Malinowski, the individual's economic interest is rarely paramount, as was the case, too, in the ancient civilizations of Babylonia and Egypt where reciprocity, redistribution, and the community were all important. The fact of social life is that the motives of human individuals are less determined by material want-satisfaction than by interests of standing and rank, status and security.[127]

The differences between what Karl called "formal" and "substantive" economic systems, as revealed by the comparative study of historical and contemporary "primitive" societies, became Karl's preoccupation in later years. In *The Great Transformation*, he traced this distinction back to Aristotle's *Politics* in which Aristotle distinguished between house-holding (the substantive or non-market economy) and moneymaking (the formal or money-market economy).[128] Ferdinand Tönnies had updated these distinctions in his 1887 classic *Gemeinschaft und Gesellschaft* by contrasting the close-knit communities of pre-industrialized life (*Gemeinschaft*) with the de-personalized market societies of modern life (*Gesellschaft*).[129]

For one hundred years, argued Karl, the stability of the laissez-faire market system was ensured by the balance-of-power system of the Holy Alliance and the Concert of Europe, along with the international gold standard, and the politics of the liberal state. At the core was the institution of banking, or *haute finance*, identified with the Rothschilds, which was subject to no one government. By 1904, the Concert of Europe was faltering, the catastrophic First World War ensued, and the economic system collapsed in the 1930s as liberals fought to prevent regulation or compulsion in the postwar market system. "The emerging regimes of fascism, socialism, and the New Deal were similar in discarding laissez-faire principles," he concluded, but with very different consequences.[130]

Liberals base their attacks against governmental participation in the economy on the alleged denial of freedom, which is said to be fully realized in free enterprise and private ownership independent from state regulation. Some measures of planning and integration of economic institutions are reconcilable with freedom, however, argued Karl, although implementation must safeguard the rights of the individual in society, including the "right of the individual to a job under approved conditions, irrespective of his or her political or religious views, or of color and race." Some degree of power and compulsion is necessary in any society, but not the Fascist solution, which is the rejection of freedom. In contrast, he insisted, the Socialist program upholds the claim of freedom.[131]

At the very end of *The Great Transformation*, Karl Polanyi returned to the themes of *Christianity and the Social Revolution*, remarking on the Christian doctrine of the uniqueness of the individual and the oneness of mankind which, he argued, are at the heart of Socialism. For Karl, Marxism was at its best in Marx's more humanist manuscripts written in 1844 and first released in the Soviet Union in 1932. Marxism is an incomplete form of Socialism because of its atheism, just as Soviet Communism is incomplete because of its limitations on freedom. But Socialism, and in particular what Karl Polanyi called a functional, or institutional, Socialism is humanity's best hope.[132]

*The Great Transformation* had a favorable, but small, reception in the United States, attracting the attention of American sociologists and institutional economists working in the tradition of Thorstein Veblen, whose *Theory of the Leisure Class* had appeared in 1899. Wesley C. Mitchell, who was Veblen and John Dewey's student, taught at Columbia University and helped found the New School for Social Research. John R. Commons was an economist at the University of Wisconsin–Madison and the author of *Legal Foundation of Capitalism* (1924) and *Institutional Economics* (1934).[133] It was their enthusiasm for *The Great Transformation* that led the economist Carter Goodrich and the sociologist Robert M. MacIver to lobby for a position for Karl at Columbia University.[134]

As Charles Kindleberger put it in a 1973 article on twentieth-century classics in intellectual thought, *The Great Transformation* was "slow in arriving and it has kept on coming." The book was rediscovered by economic historians in the 1950s and adopted by the youth culture of the 1960s as "gospel."[135] It caught on with American college and university students because of its attack on the myth of homo sapiens as economic man, and it attracted economic historians because of its argument that free trade is not a natural state of affairs but has to be planned, coming into existence only with the aid of protective tariffs and export bounties among other government measures.[136]

The book excited little attention when it first appeared in England in 1945 under the title that Karl preferred, *The Origin of Our Times: The Great Transformation*. Michael Polanyi's unfavorable reaction was not unexpected. As mentioned above, Karl had written Michael in 1941 of his plan to focus in the book on the disastrous consequences of believing that the economic and political spheres are separate from each other. Michael was then working on drafts for his various projects, including what would become *Full Employment and Free Trade*. In that book, Michael wrote explicitly:

> The principle of neutrality which I have advocated . . . is but a variant of the principle of separation of economics from politics. That latter maxim has recently fallen into discredit: partly on account of its abuse by those who upheld it to bar the State from fulfilling its humanitarian obligations; and partly through the influence of Marxist Socialism which has weakened the sense for the ordered division of powers which alone can preserve society from arbitrariness, corruption and oppression. We must restore respect for this maxim once more . . . [I]t is imperative that the new ground for which the State shall be responsible should be strictly delimited . . . [A] policy for Full Employment should be primarily concerned with Full Employment— and with nothing else. It should leave to other sections of governmental machinery . . . the issues of social security, equality, efficiency, and all the rest of them.[137]

Karl took Michael's argument to be aimed at him and replied:

> The separation of politics and economics is not the charge leveled by "Marxian" socialism against a market economy, but it is mainly *my*—non-Marxian—formulation of the characteristic of the 19th century society. I call this the *institutional* separation of the political and economic sphere. In this sense the position is much more that of Owen or Sismondi than that of Marx, who regarded, on the contrary, capitalist *property* as a *political* institution on which capitalist 'dictatorship' rested.[138]

Nor did Michael agree with Karl's distinction between formal and substantive economies, later writing that he disagreed with "your fundamental distinction between formal and substantial economics," preferring a conception where "a balance is struck between large numbers of particulars which mutually supplement each other. Economy is . . . the most general characteristic of life."[139] In this, Michael seemed to reiterate his view shared with the Austrian neoclassical economists that the actions of innumerable economic actors result in a dynamically balanced economic system that can be described in a general theory. In criticizing Karl's arguments about differences to be found historically and presently between formal and substantive economics, or between what Karl also called self-regulating market systems and reciprocative systems, Michael voiced the same kind of skepticism that was to be expressed strongly in the 1960s by some anthropologists who insisted that non-modern and non-western peoples were no less self-regulated and rationally calculating than Western Europeans. Karl and Michael's differing views, now so clearly in print, caused tension between the brothers. Michael remarked in a letter to Karl, "There is a good deal to clarify in this world, and some of it is given to you and me to elucidate at the expense of our peace of mind . . . we cannot wonder or complain if our mutual relations are strained as well."[140]

Karl's old friend Oscar Jászi disagreed with the main thrust of *The Great Transformation*, although, as with Michael Polanyi, the sharpest point of division fundamentally lay in Karl's continued sympathy with Marxism and the Soviet Union. In his diary Jászi wrote that he had not expected the success of Karl's book and, with some venom, noted, "Karli became a famous man through his semi-Marxist Kauderwelsch [i.e., Double Dutch]."[141] Jászi soon wrote frankly to Karl, "You stand on the line of historicism, whereas I cannot imagine a political life that is either fruitful or decent without the rehabilitation of natural law . . . I don't believe that a fragment of Assyria's economic history could give guidance of any kind for the solution of the economic and moral problems of today."[142] In a similar vein, Jászi wrote Karl Mannheim that Mannheim's philosophy "reduces individuality to zero, making it strictly determined, thus into a mere effect of social development."[143] Jászi wrote Karl that he was in Michael Polanyi's camp on basic issues.[144] Jászi missed or ignored the novelty of Karl Polanyi's attempt to argue for a humanist and functional socialism coordinated within a system of political institutions that are not dominated by any one power. Karl aimed to demonstrate the practical impossibility of the idealized models of both laissez-faire market capitalism and Socialist planning.[145]

In the United States and Great Britain, Hayek's economics won the day in the late 1940s. Economists in the 1940s and 1950s privileged those aspects of economic theory that focused on "measurable motives" in human behavior and the individual's rational pursuit of utility maximization. The individual—not institutions—became the primary parameter for study.[146] Milton Friedman's 1956 paper "The Quantity Theory of Money" marked a triumph of laissez-faire economics that emphasized a watchful eye on the demand for money and on the money supply rather than on aggregate demand, much less on economic inequities or social problems. Aggressive government intervention was judged useless because individual economic actors would rationally act so as to counter any effects of government policy initiatives. In the next half century the genre of liberal economics accepted by Michael Polanyi eclipsed the reformist visions of Karl Polanyi and of the German historical school among economists of Great Britain and the United States.[147]

## The Economic Foundations of the Republic of Science: Individualism and Institutionalism

Among Michael Polanyi's writings, one of his best known, widely read, and most often cited publications is "The Republic of Science." It appeared in the very first issue of the journal *Minerva* in 1962 as an inaugural piece for the journal. The economic and free-market language of the essay has often been remarked upon, but it is not obvious to a general reader how thoroughly and how long Polanyi had engaged with economic theory, and how Polanyi's economic preoccupations form a bridge to his sociologically inflected philosophy of science.[148]

Like Menger, Mises, and Hayek, and like his brother Karl, Michael Polanyi rejected the empirical and rational-calculation approach to economics that was characteristic of economic planning. He opted for a more theoretical economics that emphasized process and change. In *Full Employment and Free Trade*, while arguing for Keynes's priming of the pump, Polanyi clearly stated his agreement with the notion of the "guidance of the 'invisible hand' for the mutual adjustments of productive units."[149] He began drawing upon the economic language of spontaneity and unintended consequences as originally used by Menger and taken up by Serafin Exner at the beginning of the twentieth century and later by Hayek, for example, in Hayek's inaugural lecture at the London School of Economics in March 1933. Hayek argued that the greatest accomplishment of classical economics was the recognition of the mechanism that coordinates economic activity, arising "spontaneously as the unintended consequence of the actions of many individuals."[150] Hayek extended the notion of spontaneity from economic activity to general knowledge in his presidential address in 1936 before the London Economic Club, which he published in the London School of Economics journal *Economica*. Hayek later said that one of his greatest discoveries was the "utilization of dispersed knowledge," which is echoed in Polanyi's article in *Economica* in 1941 in which Polanyi compared the growth of scientific knowledge to two other "spontaneously arising orders," the market order and common law. Thus Hayek, like Polanyi, moved freely from economic theory to the theory of knowledge.[151]

If Polanyi drew upon Hayek, so did Hayek upon Polanyi, as they corresponded and read each others' work.[152] Hayek wrote in *The Constitution of Liberty* (1960) of mutual adjustments of individuals and the emergence of spontaneous orders in political, legal and market systems: "It is what M. Polanyi has called the spontaneous formation of a 'polycentric order.'"[153] This was a term that Polanyi employed in essays of 1946 and 1948 on "Profits and Polycentricity" and "Profits and Private Enterprise" in which he took a new tack in a revival of debate about central planning. Socialist calculation and planning are impossible, argued Michael Polanyi, because the running of an economic system involves problems of production, distribution, and allocation of capital, which are "polycentric" problems. The fact that a number of centers are involved in production or distribution, for example, makes it impossible to solve the problem centrally. Here scientific experience provides insight: as a scientist, "You . . . solve this problem for one center and proceed to the next center . . . This is the . . . method of successive approximation by proceeding from one center to the next which is quite universally used in all polycentric problems." An orderly dynamic system is not the result of central planning, but of adaptive change.[154]

The application of the economic theory of mutual adjustment, spontaneous emergence, and polycentricity to the problem of a theory of scientific knowledge soon was to become Polanyi's new preoccupation, but one still formulated within the commitment to individualism and freedom that defined his economic (and political) philosophy. In their biography of Polanyi, the authors Scott and Moleski, like Johns in his article on intellectual property, describe some of the drafts of projects on which Polanyi was working in the 1940s. In August 1941, he gave his typist Olive Davies twenty-five pages entitled "Beginning of a Book on the Scientific Life." The title and details changed in the next months, but he aimed at a book that would integrate his studies of the market with his views on the nature of science. His older son, George, encouraged Polanyi in the spring of 1942 to write a book that would focus only on the themes of his economic film, which led Polanyi into the writing of *Full Employment and Free Trade*, while he continued to draft essays for lectures, newspapers, and journals.[155] An unpublished essay of September 1942 called "The City of Science" lists characteristics of scientific truth: it upholds standards, inspires the devotion of geniuses, commands students' admiration, has universal and permanent convincing power, gains public confidence and interest, and produces useful knowledge and reasoning techniques. This essay reflects not so much Polanyi's economic views as it does his scientific career experiences in Berlin.[156]

The issue of organization of science and scientific planning would become a focus of Polanyi's writing in wartime and postwar Great Britain, along with his project for a study of scientific life rooted in scientific values and beliefs that he had experienced during his long scientific career. His economic publications ceased by 1949, but he had found in economics an important way of thinking about the institutional framework for science. Paradoxically, Polanyi's key conceptual formulation was rooted in some of the functionalist views of his brother Karl, while nonetheless tied to free-market economic theory.

The key lay in focusing on functions, institutions, and social relationships as well as on individuals acting in isolation. This was a shared point of emphasis with Karl's *Great Transformation*, namely, the objection to positivist confidence that empiricism and reason alone can arrive through calculation at ideal economic arrangements. As we have seen, the principal point of Karl's economics had been the historical analysis of how the laissez-faire system was embedded historically in political institutions that made the free-market system possible. Without those institutions, with their different functions such as the liberal state and the banking system of *haute finance*, the nineteenth-century system called laissez-faire could not have existed.

In turn, Michael Polanyi's philosophy of science would argue that genuine science is embedded historically in necessary scientific institutions and in social arrangements that alone make possible the development of stable scientific knowledge. In his 1943 essay, "The Autonomy of Science," Polanyi wrote that he would show in his analysis that within science "the individual scientist, the body of scientists and the general public, each play their part and that this distribution of functions is inherent in the process of scientific development, so that none of these functions can by delegated to a superior authority."[157] Thus, Michael used in his philosophy of science the very emphasis on tradition, function, and community that was characteristic of Karl's historical economics.

In *Full Employment and Free Trade*, as we have seen, Michael Polanyi described an economic system in which workers could be counted on to recognize their responsibility to keep the system in equilibrium because of assurances that the government would exercise effective means for their receiving proper rewards. A panel of economic experts would advise Parliament on the proper levels of money supply and employment. In his writing on patent policy, too, as we have seen, Polanyi warned against inventors' seeking monopoly on their discoveries and he suggested the establishment of a governmental panel of experts who would assign licenses and guarantee rewards. Almost simultaneously, in "The Autonomy of Science," Polanyi wrote of the influence over scientific research of "a small number of senior scientists" and "unofficial governors of the scientific community"— the "Influentials" who expertly make decisions on matters of academic appointments, journal publications, honors and awards, and research funds for all the institutions of science.[158] As for the fairness and reliability of the results, Polanyi wrote in his lecture "Dedication or Servitude," delivered at Durham University in 1946, "Inasmuch as each scientist is following the ideals of science according to his own conscience, the resultant decisions of scientific opinion will be rightful."[159] The system is one of community, in Tönnies's meaning of *Gemeinschaft* or of Karl Polanyi's description of preindustrial societies, where reciprocity, redistribution, and community were all important. The system also is one of interacting institutions, as described in the economics of Karl Polanyi, the American institutionalists of Veblen's tradition, and, to some extent, Keynes and his school.

The results of Michael Polanyi's long immersion in economics can be seen in many of his subsequent writings on the nature of science, but nowhere as clearly as in "The Republic of Science." It is the "republican" values and methods of liberalism, not those of social democracy, that rule in the city of science. Polanyi began the essay thus:

My title is intended to suggest that the community of scientists is organized in a way which resembles certain features of a body politic and works according to economic principles similar to those by which the production of material goods is regulated. Much of what I will have to say will be common knowledge among scientists . . . scientists, freely making their own choice of problems and pursuing them in the light of their own personal judgment are in fact cooperating as members of a closely knit organization . . . the activities of scientists are in fact coordinated . . . by mutual adjustment of independent initiatives . . . [which] . . . leads to a joint result that is unpremeditated . . . Their coordination is guided as by "an invisible hand" towards the joint discovery of a hidden system of things.[160]

He continues with a market analogy in which scientists take note of other scientists' published results, just as producers and consumers take note of prices and make mutual adjustments. As in the capitalist market, scientists aim to produce the best result available with their limited stock of intellectual and material resources. The authority exercised within science is established "between scientists, not above them."[161] Science is an organized community with institutions that function to guarantee just recognition and rewards, as well as the progress of the whole. The physicist John Ziman writes of Polanyi's emphasis on community in "The Republic of Science": "It is devastating, for it puts out of business most efforts to celebrate the production of scientific knowledge as if it were a mechanical process driven by logical clockwork."[162]

Science is not simply an economic community, however, and in much of "The Republic of Science," as we will see in the next chapters, Polanyi discusses the particular characteristics of scientific standards, values, beliefs, and methods that distinguish the sciences from other kinds of activities and systems, ending with his redefinition of the republic of science as a "Society of Explorers." These are themes that he developed in noneconomic writings in the 1940s and 1950s, including discussions on scientific organization in wartime and postwar Great Britain. In fact, Polanyi's arguments in this vein undercut his market analogy for science because he accepted the general principle that the members of a scientific community are working for a common task rather than acting exclusively from self-interest.[163]

Toward the end of the "Republic of Science," Polanyi returns to a subject dear to his heart, namely the theme of the tension between originality and tradition. Jewkes encapsulated the core of Polanyi's intellectual concerns in economics, politics, and science by saying, "There seems to be one central strand: how best to reconcile the safeguarding of individual liberty with the controls upon the individual inseparable from a complex and organized society, or as Polanyi succinctly put it, the relationship between spontaneous and social order."[164] We saw this theme, too, in this chapter, in Polanyi's discussion of patent reform. At the end of "The Republic of Science," as he comes back to the theme of freedom, Polanyi writes that he accepts Edmund Burke's thesis that freedom must be rooted in tradition, but with the caveat that freedom embedded in tradition must be capable of cultivating innovation and progress in modern industrial society.[165]

Debates in Great Britain over scientific planning in particular and not just economic planning in general occupied a great deal of Polanyi's attention from the mid-1930s until the late 1940s when it became apparent, as he conceded in "The Republic of Science," that his fears of government control of scientific research had been allayed.[166] Like Hayek's and Popper's publications, Polanyi's writings on economics, politics, and science during the 1940s constituted his own personal contribution to the war effort and to the struggle against totalitarianism which, for Polanyi, meant both Fascism and Marxism. As we will see, he debated matters of the organization and functions of science with British scientific colleagues who included the Socialist physicist Patrick Blackett and the Marxist X-ray crystallographer J. D. Bernal. His political allies among scientific colleagues included the ecologist Alfred Tansley and the zoologist John Baker. Discussions of scientific freedom and scientific institutions brought Polanyi more closely into sociological circles and into American scientists' postwar arguments about the future organization of the sciences in the United States. These activities in the politics of science and their results are the focus of the following chapter.

# SIX

# Scientific Freedom and
# the Social Functions of Science

When Polanyi arrived in Manchester in the fall of 1933, Labour Party leader Ramsay MacDonald had become prime minister in a national government formed in the summer of 1931. It was a national front: an uneasy coalition of the Labour, Liberal, and Conservative parties. The Political and Economic Planning (PEP) Group, which had opened an office in London in 1931, was circulating publications on planning. Polanyi, as a critic of Soviet planning, found these developments in Great Britain disturbing. As a liberal with conservative tendencies, he feared that his new home country would adopt central planning models not only for business and industry but for scientific research in the universities.

Equally alarming to Polanyi were favorable assessments of the Soviet Union that he heard reported directly from British scientists who were visiting Russia. Among British scientists, the X-ray crystallographer J. D. Bernal was among the most outspoken in his praise of Soviet support for science. Another proponent was the experimental physicist Patrick Blackett with whom Polanyi became a close friend despite their political differences. Bernal and Blackett were among British scientists who were enthusiastic about Boris Hessen's historical paper on the "Social and Economic Roots of Newton's *Principia*," which Hessen presented in London in 1931. While maintaining busy research schedules, Bernal and Blackett each began lecturing on the history of science and, to Polanyi's dismay, criticizing what they called an artificial distinction between pure and applied science. Along with the biologist Julian Huxley, the mathematician Hyman Levy and other scientists, they spoke on these matters to the public on BBC radio with the aim to influence British scientific and technological policy. In 1939 Bernal's book *The Social Function of Science* appeared, offering a

Marxist and social history of science to his British audience, as well as documenting an account of the current scientific scene in Great Britain with recommendations for changes in scientific organization.

As discussed in this chapter, Polanyi became outraged at Bernal's book and joined other scientists, as well as historians and philosophers, in establishing the Society for Freedom in Science (SFS). Polanyi's commitment to the work of the SFS was rooted in the experiences of his scientific career and in his economic and political views, but also in his emotionally charged reactions to the arrests and persecutions of family members and scientific colleagues in the Soviet Union. In trying to get support for activities and publications aimed against the pro-Soviet "Bernalists," Polanyi appealed successfully to the Rockefeller Foundation and worked in the 1950s with pro-democracy and anti-totalitarian intellectuals in the international Congress for Cultural Freedom (CCF).

Defeating the social relations of science movement in Great Britain became one of the essential aims in Polanyi's intellectual and political life around 1940. Bernal's views played a crucial role in Polanyi's decision to devote increasing amounts of his time to writing about what he and some others, for example his conservative friend John Baker, called "scientific life." For radically different reasons, Polanyi and Bernal (and Blackett) decided to shift talk about science from scientific method and scientific heroes to scientific communities and scientific practice—from the logic of science to the life of science. Polanyi and Bernal each argued for a social turn in studying the history and philosophy of science. Polanyi's arguments would find their full philosophical development in his explicitly philosophical work *Personal Knowledge* (1958) and in the essays of *The Tacit Dimension* (1966).

## British Science and Liberal versus Left Politics in the 1930s

Shortly after Polanyi and his family arrived in Manchester in September 1933, the university recommended that he rent Kenmore, a spacious house at Didsbury Park, south of the university. There were three large reception rooms, five bedrooms, two rooms for servants, and a guest room. The Polanyis employed three maids (although the university had recommended four), and Magda had an au pair. It was decided that George, who turned eleven years old in October 1933, would enter Bootham, a Quaker boarding school in York, while four-year-old John was at home or in day care.[1] Hugh O'Neill became a good friend, as did Lewis B. Namier. In March 1934, Polanyi was elected to the Manchester Literary and Philosophical

Society, giving him a foothold in the Manchester intellectual community outside the university.[2]

Manchester in the 1930s had a considerable reputation for both liberal and red politics. Its most famous "red" citizen had been Friedrich Engels, who worked in his father's textile firm, Ermen and Engels, during 1842–1844 and 1849–1870, becoming a partner in the firm in 1864. *The Condition of the Working Class in England in 1844* was a result of Engels's initial stay in Manchester, as was his 1848 *Manifesto of the Communist Party*, coauthored with Karl Marx in Paris in 1848. The *Manchester Guardian*, founded in 1821, was one of the great liberal British newspapers, editorializing in favor of laissez-faire and free trade but moving further toward the left by the early 1900s and supporting the women's suffrage movement, while deploring some suffragette tactics. On the side of the arts, Manchester was the home of the Hallé Orchestra, founded in 1858 by Charles Hallé and today noted as Britain's longest established symphony. Arthur Lapworth, who spearheaded the hiring of Polanyi in the Manchester Chemistry Department, was a skilled violinist who in earlier years had organized occasional musical evenings with the organic chemist William Henry Perkin, accompanied by professional players from the Hallé. It was said that Perkin had given up the violin when he could no longer practice on his uncle's Guarnerius. He then turned to the piano.[3]

In England Polanyi supported the Liberal Party. In 1940, he would speak as a Liberal Party representative alongside members of the Conservative and Labour parties at a conference of the Student Christian Movement. As he discussed definitions of "conservative" and "liberal" in his talk, "The Liberal Conception of Freedom," he said that conservative means traditionalist and that he had no quarrel with this kind of conservative philosophy, because "in England, tradition is Liberal."[4] For Polanyi, according to his own statement in a letter to Blackett in 1946, political concepts of liberalism were at the center of his thinking: "I see our problem not in the light of an economic dialectic between 'Socialism' and 'Capitalism' but in terms of civic ideas embodied in certain nations . . . To me in Russia it is the theory of the State which matters while the economic system is secondary, and also largely eye-wash. Nationalization in Europe is done everywhere admittedly for political and not for economic reasons. Such measures are largely the outcome of internal struggles in which the two leading political philosophies, the Liberal and the Marxist, are struggling for supremacy within European nations."[5]

When Polanyi arrived in England, the Liberal Party was in a period of decline after many of its supporters began voting Labour or Conservative in the 1920s. The party, traditionally identified with laissez-faire and free-trade economic policies, favored social reform and personal liberties in the late nineteenth century, and it adopted a Keynesian economic platform in the 1929 election. The platform was based in the Liberal Yellow Book *Britain's Industrial Future*, a result of David Lloyd George's Liberal Industrial Enquiry project. Lloyd George's illness during the elections of the summer 1931 led to Sir Herbert Samuel's leadership of the Liberal Party and his proposal of a coalition national government to be formed by the Liberal, Conservative, and Labour parties. The proposal met with favor from Labour leader Ramsay MacDonald and King George V.

The Labour Party had devoted considerable energy to dissociating itself from Marxism and to purging Communists from its ranks under Ramsay MacDonald when it formed governments in 1924 and 1929. It expelled MacDonald and some other Labour members when MacDonald entered the national government along with the Liberal and Conservative parties. In late 1933, shortly after Polanyi arrived in Manchester, the Liberal Party crossed over into opposition to the government on the issue of tariffs and protectionism. MacDonald continued to serve as prime minister until 1935, when he was succeeded by Conservative Party leaders Stanley Baldwin (1935–1937) and Neville Chamberlain (1937–1940).[6]

As noted by historian Martin Pugh in his study of British politics in this period, Labour circles frequently used the fashionable idea of "planning" during the 1930s, but Labour economic policy owed as much to Keynesian and Fabian thinking as to Socialism. The Labour Party's thirty-one-page program of action, *For Socialism and Peace* (1934), proposed only limited sectors under public ownership and concentrated during 1935 to 1939 on a small selection of public utilities and ailing industries.[7] Influential independent planning groups of the 1930s included the PEP Group on the Left and the Next Five Years Group toward the Right. The PEP Group was a private nonprofit that opened an office in London in 1931 and published many studies beginning in 1933. Its first director was Max Nicholson, whose *National Plan for Britain* appeared in 1931. Nicholson was a scientist—a noted ornithologist and civil servant in the British government who became director of the Nature Conservancy.[8] Julian Huxley, who had just completed coauthoring the multivolume *Science of Life* with H. G. Wells and his son G.P. Wells, was among other prominent members of the PEP Group think tank.

Huxley supported the Next Five Years Group, which included radical Tories like Harold Macmillan and Liberals like A. D. Lindsay, who was vice-chancellor of Oxford University and a frequent contributor to the *Man-*

chester Guardian. The Five Years Group originally was established at Oxford in 1934 as the Liberty and Democratic Leadership Group, but it lost most of its Labour members soon afterward. In 1935, the Five Years Group published *The Next Five Years: An Essay in Political Agreement*, followed by *A Program of Priorities* in 1937. Their arguments encompassed broad ideas for national planning and for dealing with international disorder, with specific calls for advocating the abolition of the means test, reduced tariffs, and extension of public control of some industries. Scientists who supported the group included Sir Richard Gregory, who had been the de facto or official editor of *Nature* since 1912, along with Lord Rutherford, Sir Oliver Lodge, and H. G. Wells.[9]

Polanyi, like many of his colleagues, was an avid reader of the novels of Wells, whose latest book in 1933 was a historical fiction titled *The Shape of Things to Come*. It played upon the Marxist notion of the dictatorship of the proletariat and envisioned a benevolent, but sometimes necessarily oppressive, technocratic and worldwide "Dictatorship of the Air" that would arise after the devastations of future wars and then wither away into a utopia.[10] Wells's early novels appealed to the young Polanyi because of Wells's faith in science and humanity, but, as he had confided to Kasimir Fajans late in 1914, after reading *The World Set Free*, it seemed hard to believe that any future society could be nonpolitical and run on purely rational principles, as Wells envisioned. In the late 1920s, Polanyi discussed Wells with Eugene Wigner during evenings in Berlin, and he characterized Wells in his diary as a positive spirit of the present. Increasingly, however, Polanyi found Wells too ready to excuse the Soviet government's abuses of freedom and brutal exercise of power in pursuit of a new kind of state.[11]

Wells had visited Russia soon after the Bolshevik Revolution, and he paid another visit in 1934. He was a gradualist in his approach to political change rather than a revolutionary, and he was a believer in the potential of a humanistic technology. James G. Crowther, the scientific correspondent for the *Manchester Guardian*, was another writer whom Polanyi read in Manchester and whom he had entertained in Berlin in 1930.[12] Crowther had first visited Russia in the summer of 1929 and met with Nikolai Vavilov, Abraham Joffe, and Peter Kapitsa, who spent his summers in Russia. Crowther again visited Russia in November 1930 and published *Science in Soviet Russia* upon his return.[13] He was instrumental in working with the Society for Cultural Relations with the Soviet Union (SCRSU) in order to arrange the visit of the Russian delegation, which included Vavilov and Joffe, to the International Congress for the History of Science in London in late June and early July of 1931. Crowther acted as guide for the delegation in London, taking Nikolai Bukharin to visit Kapitsa in Cambridge. Crowther organized trips to the Soviet Union in July and in August of 1931 under the auspices of the SCRSU. The July visitors included Julian Huxley and his brother Aldous Huxley; the August group included Bernal, the physicist John Cockcroft, and the biochemists Bill and Tony Pirie.[14] Upon their return, Julian Huxley published an upbeat account, *A Scientist among the Soviets* (1932), and Aldous Huxley published the considerably more pessimistic *Brave New World*.[15] Bernal enthusiastically reported "It was grim but great."[16] Most striking to all of them was Soviet economic support of science and technology. Nowhere in Europe nor in the United States was a higher proportion of national income being spent on research and development than in the USSR.[17] According to Nikolai Bukharin in a speech of April 1931, only the Socialist state properly valued science for its own sake, because the new state needed everything that science could provide.[18]

In 1931, Hyman Levy, a Marxist mathematician who taught at Imperial College in London, was invited to organize a series of British Broadcasting Corporation radio programs called "Science and the Changing World." The idea came from Mary Adams, a producer with the BBC, who knew Levy through his involvement with workers' education in the 1920s when she was a biology tutor. The results were sufficiently positive that Levy was asked to do further programs, and in 1934 he collaborated with Julian Huxley on the broadcast series "Scientific Research and Social Needs" which became a book edited by Huxley with a foreword by Levy.[19] By this time Huxley was famous in Britain. As Daniel J. Kevles notes, readers of the *Spectator* ranked Julian Huxley ahead of James Jeans, Ernest Rutherford, and Bertrand Russell as one of Britain's "five best brains."[20]

Levy was among many scientists who worried that science, in association with technology, was being blamed for a role in the social dislocations of the Great Depression. Julian Huxley, in turn, reminded listeners and readers that science is an important social activity and that this is a lesson for scientists to learn as well as statesmen and the public. In one of the broadcasts, Patrick Blackett, who then was head of the physics department at Birkbeck College in London, spoke on the topic of pure science with Huxley serving as moderator. Blackett agreed in the broadcast with Huxley's argument that science is a social activity, and he agreed with Huxley's view that scientists should not delude themselves into thinking that they could be aloof from politics. Scientists' interests were dependent on material and moral support from government and industry and the larger public. Scientists need to step up and make clear to the public the virtues and advantages of science at a time when one could see so clearly in Hitler's Germany the development of anti-scientific and anti-intellectual movements.[21] The notion, however, that there could be scientific consensus in matters of politics was pie-in-the-sky, as far as Blackett was concerned. Optimists about the disappearance of political differences (including H.G. Wells, for example) were wrong:

> No, there I disagree. As a matter of scientific observation, I find that my scientific colleagues, between them, represent all the possible outlooks you have mentioned . . . Don't be too optimistic. I am afraid that if society thinks that the scientist is going to be its saviour, it will find him a broken reed.[22]

On the matter of what was being called "planned capitalism" in British economic and political circles in the 1930s, Blackett was again skeptical, as he said in another BBC program, which would gain him renown as a committed "Red." Of the policies of the national government, he said, "You are being told there is a third way," a planned capitalism or planned economy:

> I believe that there are only two ways to go, and the way we now seem to be starting leads to Fascism; with it comes restriction of output, a lowering of the standard of life of the working classes, and a renunciation of scientific progress. I believe that the only other way is complete Socialism. Socialism will want all the science it can get to produce the greatest possible wealth. Scientists have not perhaps very long to make up their minds on which side they stand.[23]

Blackett was by this time a quite famous scientist for his work in nuclear physics and cosmic rays. Blackett and Polanyi had met in Berlin in 1930, and Polanyi had been excited to read British newspaper reports in the winter of 1933 that Blackett had discovered the positive electron (a discovery soon assigned by the scientific community to Carl Anderson at Caltech).[24] If he heard the BBC programs on science, however, Polanyi would not have been happy. He later offered the question, "Ought Science to be Planned?" in a postwar BBC broadcast program in the fall of 1948, giving answers well rehearsed from earlier debates in the lunch room or over the dinner table with Blackett.[25]

At stake for Polanyi were the fate of Liberalism in Great Britain, the defense and validation of "pure" science independent from social and economic needs, the maintenance of scientists' individual autonomy in choosing their scientific research against the claims of centralized planning, and

Figure 6.1. P. M. S. Blackett with his daughter Giovanna and his wife Costanza in Manchester in November 1948, after learning of his Nobel Prize in Physics. (Blackett Family Papers. Courtesy of Giovanna Blackett Bloor.)

the protection of stable scientific traditions that he saw dating back to the scientific revolution. More broadly Polanyi sought to do what he could to ensure freedom from oppression of individual rights and duties by a centralizing regime. These concerns became his increasing preoccupation by the late 1930s.

## Pure versus Applied Science

The language of distinctions between pure and applied science is a very old one, going back at least to Francis Bacon's designation of the categories of natural knowledge as "luciferous," or enlightening, and "lucriferous," or money-making. Bacon judged alchemy with suspicion, for example, because of its lucriferous practitioners and possibilities. Over the course of the development of the sciences after the seventeenth century, as the social origins of scientists drifted from an elite in the upper bourgeoisie to a larger cadre of middle-class teachers in colleges and the universities, the rhetoric of pure science separable from useful science increased. To be sure, university scientists worked in principle for long-term public good and the benefit of humanity, but they maintained the value of their work independent from any immediate useful payoff to their patrons in ministries of education, in legislatures, and, increasingly by the end of the nineteenth century, in local industries.[26] Prestige within the sciences corresponded to the reverse order of Auguste Comte's chronological history of scientific development running from mathematics through natural philosophy to chemistry and the biological and social sciences, a hierarchy that attributed the greatest reverence to the most theoretical sciences, with abstract mathematics at the apex, precisely because pure sciences display devotion to the natural divine without immediate use or monetary advantage.[27] As Steven Shapin notes in *Scientific Life: A Moral History of a Late Modern Vocation*, the physicist and editor of *Nature* Richard Gregory still articulated in 1916 a view of the divine in science when he wrote, "The conviction that devotion to the study of Nature exalts the Creator gives courage and power to those who possess it; it is the Divine afflatus [inspiration] which inspires and enables the highest work in science."[28]

Polanyi would come to disagree with many of Gregory's positions as editor of *Nature*, but not with the exalted and privileged value placed upon pure science. By 1933, when Polanyi arrived in Manchester, different opinions about the relationship between pure, or basic, and applied science were widely heard, as in the various BBC broadcast series. Levy, Huxley, Blackett, and others had been strongly influenced by some of the arguments about this relationship that were presented by the Soviet delegation that Crowther had helped organize in 1931 at the International Congress of the History of Science. Blackett later reminisced that he had been bored with people talking about scientists as if they were living in something like a social vacuum. Boris Hessen's paper, which argued that practical economic needs had driven Newton's revolutionary science, just as they drive modern science, drew Blackett's attention for the first time to the history of science.[29]

Most people, like Blackett, read Hessen's paper in the volume *Science at the Crossroads* rather than hearing it at the London meeting at the Science Museum. A special final Saturday session of ten-minute papers for the Russians had been arranged by Charles Singer, the president of the congress, but most of the scientists and historians attending the meeting had already scheduled a sightseeing trip for the day. At the suggestion of Lancelot Hogben, the Marxist statistician who held the chair for social biology at the London School of Economics, the Russians launched what the *Manchester Guardian*, with tongue in cheek, described as "The Five Days Plan" to turn the Soviet Embassy in London into a translating and publishing center. The result was the distribution of all the Russian papers at the Saturday, July 4, morning congress session. Three days later a bound volume of the essays appeared as the book *Science at the Crossroads*.[30] All these events were well reported in newspapers and periodicals, including the *Daily Mail*, the *Morning Post*, the *Manchester Guardian*, and the Communist Party's *Daily Worker*.[31] Bernal got a review of the book into the conservative weekly the *Spectator* because of his friendship with Celia Simpson, who managed its reviews, and the *New Statesman* published a short article, "Science and Politics in the Soviet Union," written by Bukharin.[32] Charles Singer's wife, Dorothy, was in the audience on Saturday morning and she was not impressed, writing Joseph Needham, "I listened hard and with open mind, and the impression gained was that they adopted toward Marx exactly the medieval attitude towards Aristotle—that nothing could be right unless it could be traced to his words."[33]

In contrast, Crowther was inspired. "Hessen's paper revealed to me a method of prosecuting the history of science which was more profound than the conventional one."[34] Crowther conceived the plan of applying the notion of science as a social product to more recent periods in British science, resulting in his publication of *British Scientists of the Nineteenth Century* in 1935, followed by *Famous American Men of Science* in 1937, *The Social Relations of Science* in 1941, and many subsequent volumes.[35] In Crowther's view, Hessen's essay "transformed the history of science from a minor into a major subject," and it did so by transcending the previous "antiquarian" nature of history of science in favor of showing that knowledge of the scientific past was essential for the solution of contemporary social problems.[36] Crowther recognized that Hessen's ideas were unwelcome among many scientists, as well as among historians, and during his

July 1931 stay in Russia, he witnessed Hessen's scientific colleagues at a banquet in Moscow arguing that a differential equation could not be a reflection of social and economic conditions. According to Crowther, when Lev Landau had an opportunity to use the first service for telegraphing pictures between Leningrad and Moscow, Landau transmitted a mathematical diagram caricaturing Hessen's ideas.[37]

Blackett, too, began writing and talking about scientific development using Hessen's perspective. In the 1933 essay "The Experimental Craft of Physics," he wrote of the experimental physicist as a jack-of-all-trades who must be a mechanic, glass-blower, carpenter, and electrician, as well as "enough of a theorist to know what experiments are worth doing and enough of a craftsman to be able to do them." Blackett wrote that there is a reciprocal relationship between technical innovations in the laboratory and industry, and he noted the many years that it may take the physicist to ready an apparatus for the experiments he has in mind and that the physicist gradually develops a hands-on and intuitive understanding in his research. The scientific knowledge that results, Blackett continued, is the result of the skill of the hands as well as the mind.[38]

In lectures that he began giving on the history of science in 1936, including a lecture series for undergraduates at Manchester during 1938–1939, Blackett told students and public audiences that science is practical in its origin and that it began flourishing during the Renaissance, in part because of new needs for navigation. Drawing from Hessen, he mentioned astronomy, mechanics, mining, hydraulics, ballistics, and other areas of scientific investigation in the seventeenth century, concluding with references to the principle of conservation of energy, the industrial revolution, and the current reaction against science due to the capitalist crisis, a recurring theme in BBC radio programs and the public media in the early 1930s. A simple problem in physics can be divided into two extreme types, Blackett suggested: the abstract mathematical method and the intuitive methods of everyday life, with the work of the experimental physicist lying between these extremes.[39] His remarks combined a close reading of Hessen with reading and references from Crowther, Bernal, and R. H. Tawney, and also from more traditionalist historians and philosophers, such as Norman R. Campbell, Richard Braithwaite, George Sarton, Frank Sherwood Taylor, Charles Singer, and Abraham Wolf.[40]

In the 1940s, Blackett's historical lectures incorporated views from Robert Merton's 1938 *Osiris* monograph, "Science, Society and Technology in Seventeenth-Century England." Blackett likely learned of the monograph from Crowther, who gave a series of lectures in 1937 at Harvard University. Upon his return to Manchester, Crowther pronounced Merton "the most able of all the coming men whom I met at Harvard."[41] Like Hessen, Crowther, Huxley, Bernal, and other left-wing intellectuals, Blackett self-consciously offered a radically different conception for the origins of Newtonian science than clichés that the scientific revolution was due to divine providence and personal genius.[42] The best way to understand science, as Blackett put it to Manchester Arts students, is to know science's history rather than trying to define a system of abstract definitions of its scope, methods, and conditions for which there is little agreement in any case.[43]

As we saw earlier, Polanyi had written while still in Berlin at the Kaiser Wilhelm Institute a public plea for the support of pure science, arguing in *Der deutsche Volkswirt* for government support of fundamental science even when practical benefits might not be immediately obvious.[44] Faced in England with fellow scientists publicly undermining the distinction between pure science and applied science, Polanyi expressed his alarm by bringing to bear his conversation with Bukharin during Polanyi's visit to the Soviet Union in 1935:

> [Bukharin] explained that the distinction between pure and applied science made in capitalist countries was due only to the inner conflict of a type of society which deprived scientists of the consciousness of their *social functions*, thus creating in them the illusion of pure science. Accordingly Bucharin [*sic*] said, the distinction between pure and applied science was inapplicable in USSR. In his view this implied no limitation on the *freedom* of research; scientists could follow their interests freely in USSR, but owing to the complete internal harmony of Socialist society they would, in actual fact, inevitably be led to lines of research which would benefit the current Five Years' Plan.[45]

Polanyi, like others in Britain in the 1930s and 1940s, tied the erosion of the distinction between pure and applied science to what he viewed as dangerous arguments for the social responsibility of scientists, central planning of scientific research, and relinquishment of individual freedom. Very specifically, he argued that defense of the existence of pure science and the right to pursue it was a Liberal value, with a capital L:

> To the Liberal, science represents in the first place a body of valid ideas . . . Each new addition to [branches of science] is the product of a continued application of certain methods of thought and observation which are characteristic of the branch in question . . . it is essential for the Liberal distinction between pure and applied science to keep steadily in mind that . . . [a]lthough

every moment of man's life depends on his handling of practical knowledge, yet none of this is science . . . so long as knowledge is merely viewed in its practical context, it can have no scientific interest . . . The discoveries made by the empirical crafts often prove later to be interesting objects of scientific investigation ("stimulation of science by industry"); and, on the other hand, knowledge gained and stored up by science is widely used by the modern inventor . . . ("application of science to industry"). But it should also be clear from our description of science as an organism of ideas, that scientific research, which is the growth of the organism, cannot be deflected from its internal necessities by the prospects of useful application.[46]

In these remarks Polanyi sounds a great deal like the historians of ideas Sarton and Taylor, who were writing in a semi-positivist mode on the internal logic of science in which "each scientific question suggests irresistibly new questions connected with it by no bounds but the bounds of logic."[47] Polanyi's main point, however, was the danger of the erosion of the distinction between pure and applied science.

## The Social Responsibility of Science and Scientists

In 1917, scientists, including H. G. Wells and the Nobel laureate chemist Frederick Soddy, proposed the formation of a National Union of Scientific Workers, a trade union organization that would enable scientific workers to "exercise in the political and industrial world an influence commensurate with their importance."[48] The union languished until 1927 when it was revived by scientists who included Bernal, Haldane, Julian Huxley, Sir William Bragg, and Richard Gregory. They renamed the group the Association of Scientific Workers and stated its mission as one of defining professional ethics, setting up a central register, and cooperating with the British Association for the Advancement of Science and other organizations to promote the interests of science in Parliament. A parliamentary Science Committee, with Lord Rayleigh as chairman, was established in 1929 during the second Labour government.[49] In an editorial in *Nature* at the end of 1933, Gregory argued that new government aid to science should be established in the form of block grants or outright endowments, and in a BBC program, Bernal called for a tenfold increase in grants to industrial research and for a new British organization for science.[50]

In 1933, Bernal published an article in *Cambridge Left* arguing that intellectuals must be engaged politically, joining with Gregory, Crowther, Ritchie Calder (who was the science correspondent for the *Daily Herald*), and others to call for the British Association to address scientists' responsibilities for their work and for the uses to which it is put by politicians and industrialists. Polanyi was alarmed at these ideas. So was A. V. Hill, who had considerably more influence than Polanyi in the British scientific elite as a 1922 Nobel laureate in physiology, a brother-in-law of John Maynard Keynes, and the secretary of the Royal Society from 1935 to 1945. Hill warned in a letter to *Nature* in 1933 that the Royal Society frowned on electing members who meddled in politics. Still, in 1934, the British Association added the social impact of science to its areas of study, and in 1938, the officers established a Division for the Social and International Relations of Science with Blackett, Bernal, Hogben and Levy serving on the division's governing committee alongside scientists of more moderate political perspectives.[51]

As discussed in chapter 1, the British scientific community was appalled at expulsions of scientists from Germany beginning in April 1933 and organized the Academic Assistance Council, later renamed the Society for the Protection of Science and Learning. Blackett, Bernal, C. P. Snow, and Aldous Huxley were among British intellectuals who established another group in 1935 called For Intellectual Liberty. It was a parallel organization to French scholars' Comité de Vigilance des Intellectuels Anti-fascistes, founded in 1934 by French left-wing scientists, including Paul Langevin, Jean Perrin, and Frédéric Joliot. Explaining French scientists' abandonment of laboratory workbenches for marches in the street and public speeches, Langevin expressed clearly the ideology of scientists' political responsibility, saying, "The scientific work that I do can be done by others, possibly soon, possibly not for some years; but unless the political work is done there will be no science at all."[52] For Intellectual Liberty had some six hundred members, with Aldous Huxley serving as its first president. The executive committee on which Blackett, Bernal, and Snow served with Kingsley Martin, Henry Moore, R. H. Tawney, Leonard Woolf, and E. M. Forster, met twenty-nine times from 1936 to 1939. They promoted meetings and seminars, sent delegates abroad, for example, to Paris, and arranged asylum and employment for refugees in cooperation with activities of the first Academic Assistance Council.[53]

If most British scientists agreed on the need to aid German refugee intellectuals, they disagreed in the mid-1930s on the proper response to German oppression at home and increasing militancy abroad. In 1933, some Cambridge scientists formed the Cambridge Scientists' Anti-War Group (CSAWG), an anti-Fascist but fundamentally pacifist group that took part in local demonstrations when the Royal Air Force had air displays at neighboring airfields at Mildenhall or Duxford.[54] Many of the CSAWG scientists were working in the laboratories of Rutherford and of the biochemist Frederick Gowland Hopkins.[55] For most of these scientists, anti-Fascism was synonymous with support for the Soviet Union. Recruitment of Communist Party membership swelled in Cambridge. Some of the new recruits were the later notorious spies Guy Burgess, Harold ("Kim") Philby, and Donald Maclean.[56] On the more moderate side, Nevill Mott was among scientists who joined the Society for Cultural Relations with the Soviet Union. "Most of us were politically committed in the Nazi period," he later wrote, "and were friendly to the Soviet Union because they were against the Nazis." Mott parted with the left, however, as did Blackett, on leftist pacifist policy against rearmament.[57]

The Spanish Civil War, Italian bombardment of Ethiopian civilians, and the German occupation of the Rhineland changed many minds. The CSAWG became actively concerned with British defense, and a group that included Needham, Pirie, Waddington, and R. L. M. Synge, along with volunteers in Manchester, began testing claims made by the Home Office for its gas protection measures in the event of air raids. They demonstrated that civilian gas masks were ineffective and published a Left Book Club book in 1938 titled *The Protection of the Public from Aerial Attack*.[58] Following discussion with Solly Zuckerman in the autumn, Bernal sent a memorandum to the secretary of war, as well as to Liddell Hart, the military correspondent for the *Times*. It called for scientific workers to be placed in national service. Sir John Anderson, who was in charge of national defense, met Bernal at a luncheon in January 1939 and enlisted him in the Civil Defense Research Committee. As Ritchie Calder recalled, someone reminded Anderson that Bernal was known as a "Red," eliciting Anderson's reply, "Even if he is as red as the flames of hell, I want him." Bernal served in operational defense during the war, and he masterminded research on the beaches of Normandy that was pivotal in the D day landings in June 1944. He received the Medal of Freedom in the United States in 1947.[59]

Bernal's *Social Function of Science* appeared in January 1939 and elicited one of Michael Polanyi's first strongly political publications in England. An earlier piece was Polanyi's review of Fabian Socialists Beatrice and Sydney Webb's 1936 book, *Soviet Communism: A New Civilization?*[60] The Webbs had visited the Soviet Union in 1932, the year of Polanyi's third visit there when he had lectured in Leningrad at Joffe's institute. Polanyi again visited Russia in 1935, where he met with Bukharin in Moscow and visited his niece Eva Striker, who took him on a tour of the city. They visited a new high-rise apartment complex built for workers and Polanyi asked if there was running water. Their guide took Polanyi and his niece to a single outdoor tap, out of sight in a courtyard, which serviced the entire block of apartments.[61]

The Webbs' book enraged Polanyi, and he published his review in the *Manchester School of Economic and Social Studies* in 1936. He wrote that *Soviet Communism* focused more on the political system than on the economic system and described in great detail the provisions of the Soviet constitution for a pyramidal structure based in a huge number of small meetings of associated citizens who discussed matters and elected representatives to the next level, with the whole rising "tier after tier," to a supreme group. What the authors failed to discover, Polanyi argued, was the "secret terror" that controls the whole process, belying the idealized picture of freedom and democracy. Puzzles arise in the Webbs' version of contemporary Russia, he noted, such as how collectivized agriculture was established in a country where the peasants fiercely opposed collective farms or how the result came about that a small number of Communist Party members secured two-thirds of all putatively elected positions. "We are told," Polanyi concludes, "that the enterprise of the Soviets is like the undertaking of a great engineering work of uncertain success," but "I cannot remember any engineering project from the Suez Canal to the flight of the Graf Zeppelin or the draining of the Zuider Zee during which the public expression of doubt, or even of fear that the plan will not be successful, is an act of disloyalty, and even of treachery."[62]

The "secret terror" of the Soviet Union became a terrible reality for Polanyi when his niece Eva Striker was arrested in Moscow in May 1936. She had moved to Kharkov in order to marry Alex Weissberg, a Viennese mathematical physicist who accepted a position in 1931 at the Ukrainian Physical Technical Institute (now the Kharkov Physical Technical Institute), one of the largest and best-equipped experimental laboratories in Europe. Its scientists included I. V. Obreimov, L. V. Shubnikov, A. Leipunskii, L. D. Landau, and B. Podolskii, most of whom had studied in Western Europe. Visitors included Niels Bohr, Paul Dirac, Paul Ehrenfest, and Fritz Houtermans. Physicists in Kharkov built the first Soviet linear accelerator and confirmed in 1932 the experiments of John D. Cockcroft and Ernest T. S. Walton on splitting the atomic nucleus.[63]

Arthur Koestler had known Eva Striker since they were children and had attended a Budapest kindergarten run by Eva's mother, Mausi. Koestler and Eva had a short romantic affair in 1930.[64] Koestler visited Eva and Alex Weissberg twice during his yearlong visit in the Soviet Union of 1932–1933, staying at their flat, which Koestler thought luxurious by Russian standards.[65] By 1934, Eva and Alex had separated, and she moved to Moscow where she became the arts director of the Dulevo Porcelain Factory. When arrested, she was taken from the home of her brother, who was a specialist in charge of the Patent Department of the Invention Office headed by Gyula Hevesi, a former official of the 1919 Hungarian Council Republic, in whose apartment she had sublet a room before moving to her brother's flat. The charge against Eva was that she had inserted swastikas into the pattern of her ceramic teacups and that she belonged to a subversive group that aimed to kill Stalin. The alleged evidence was a revolver found in her former room at Hevesi's flat. Much later, she learned from Hevesi, who spent the years between 1938 and 1946 in Soviet prisons and Siberian gulags, that officers of the People's Commissariat of Internal Affairs of the USSR (NKWD) had found Hevesi's revolver, which he had neglected to declare, and had hidden in Eva's room.[66]

Eva's estranged husband, Alex, whom she later divorced, went to Moscow and Leningrad to secure testimonies to help her. He was arrested in Kharkov in March 1937 on the charge of having hired twenty bandits to ambush Stalin while he was on a hunting trip in Caucasia. A letter from Einstein and a cable written to Stalin by Koestler with signatures from Joliot, Langevin, Perrin, and Michael Polanyi had no effect. Apparently, because he had just signed a protest on behalf of his former assistant Fritz Houtermans, Blackett, who visited Moscow for conferences in September 1935 and September 1937, declined to sign the cable for Weissberg. Following the Stalin-Hitler pact of August 1939, Weissberg was handed over to the Gestapo along with a hundred-odd other Austrian, German, and Hungarian Communists, including Houtermans. Through the intervention of Max von Laue, Houtermans was released by the Gestapo in Berlin. Weissberg suffered a great deal more, surviving prisons and camps in Poland throughout the war, eventually escaping from a concentration camp in Kavencin and joining the Polish underground, with whom he was hiding in Warsaw at war's end.[67]

Eva spent eighteen months in the Lubianka prison. After solitary confinement and signing an untrue statement, she slashed her wrists but was saved from death by a female warden. Eva's mother, who spent time in Russia in the mid-1930s, had been collecting affidavits from famous scientists, including Kapitsa, as well as appealing to the Austrian consul. Eva was released from prison in September 1937 and put on a train to Vienna, where she was welcomed back from her horrors by her brother George and his wife, Barbara.[68] Only a month earlier, Polanyi received a warning from a colleague, who recently had visited Moscow, that Polanyi's Russian friend

Alexander Frumkin, who was director of the electrochemistry department at Moscow University, worried that Polanyi's recent anti-Soviet publications on economics might endanger his relatives in prison.[69]

All these experiences, along with his own experiences during imprisonment in Spain, would become part of Koestler's chilling novel *Darkness at Noon* (1941). In an essay for the 1950 volume *The God That Failed*, Koestler criticized British scientists who remained friends of the Soviet Union. "The moral of this story is that Joliot-Curie, Blackett, and the rest of our nuclear Marxists cannot claim starry-eyed ignorance of the goings-on in Russia. They know in detail the case-history of at least these two of their colleagues . . . reliable, second-hand reports of hundreds of similar cases in Russian academic circles are available to them . . . Each of us carries a skeleton in the cupboard of his conscience."[70]

Michael's brother Karl had tried to convince Michael that Eva must have received fair judicial treatment in the Soviet Union after her arrest. Michael replied in a letter of 1944 that Eva had told him that she was pressured to confess "just a little" or else she would be shot without trial. She broke down, she told her Uncle Michael, and made false confessions implicating innocent persons. For this, she tried to kill herself.[71] Critics praised the fictionalized account in *Darkness at Noon* for "clearing up the mystery of what has happened in Soviet Russia," and it became the third most popular fiction book of the year in the United States.[72] In 1945, Koestler dedicated his collection of essays *The Yogi and the Commissar* to Michael Polanyi.[73] After Weissberg wrote an autobiographical book *Conspiracy of Silence* in 1952, with a preface by Koestler, Polanyi wrote a scathing review. Why, asked Polanyi, did Weissberg support the destruction of the political opposition and the collectivization of the peasantry for so many years before the Great Purge? Weissberg's book, Polanyi concluded, is the "standard biography of Modern Destructive Man at his most naïve and amiable best."[74] It was not until after Stalin's death that the public came to recognize the scale of Stalin's purges in which most likely eight million died and ten to fifteen million were incarcerated at some point in the labor camps.[75]

## Freedom of Science

The year 1939 marked Polanyi's increasing engagement in political writing and activity following the publication in January of Bernal's *Social Function of Science*, a revolutionary book in many respects. We will have more to say in detail about the book's contents in the next section of this chapter. The book quickly became a political target for liberal and conservative thinkers who were concerned about science policy. Bernal laid out a dazzling presentation of information and statistics about scientific research in Great Britain and proposed a major reorganization of research, including the formation of a central endowment based in government and industrial funds.[76] Worse, from Polanyi's perspective, Bernal praised Marx, Engels, and Lenin for their understanding of science and lauded Soviet achievements since 1917, detailing ten key problems identified by the Soviet Academy of Sciences in the third Five Year Plan that began in 1938. Bernal described dialectical materialism as a useful tool for Soviet science, coexisting with traditional methods of induction and proof. Among his examples of leading Soviet science, Bernal offered the research of Nikolai Vavilov and his institute for plant research as a "beautiful" instance of the integration of fundamental discoveries in "genetic principles" with useful purposes in agriculture.[77]

The example was ill chosen. Many Western scientists knew that Vavilov's work in genetics was under attack by the peasant turned experimenter Trofim Lysenko, who was claiming to be implementing a program of Lamarckian biology in agriculture. Already by 1935, Lysenko had accumulated sufficient power in the Communist Party to be able to force Vavilov out of his position as president of the Lenin Academy of Agricultural Sciences, although Vavilov remained head of his institute. A brief comment appeared in *Nature* in early 1937 on the recent Soviet conference of December 1936 in which modern genetic theory had been attacked for ignoring "the Marxian principle of the unity of theory and practice." Editorial comment in *Nature* later in 1937 lamented the fact that the conference illustrated "the atmosphere in which scientific investigators in totalitarian countries have to live and work."[78] Not all British scientists agreed with this view, and Bernal followed the line of argument of Joseph Needham and J. B. S. Haldane that genetics research was continuing apace in the USSR and that Lysenko's plant physiology might have some merit. It was some time before British scientists learned that Vavilov was arrested in August 1940, sentenced to death by a firing squad for espionage in July 1941, and imprisoned in Saratov, where he died of malnutrition in 1943.[79]

For Polanyi and others in 1939, it was infuriating that Bernal failed in any way to criticize the Soviet Union when it was well-known that trials and arrests had been directed at many Soviet scientists and that no Soviet scientist was free to cross the Soviet border. Polanyi immediately wrote a review in the journal *Manchester School of Economics and Social Studies* in February 1939. One of Polanyi's telling critiques aimed at Bernal's account of Soviet scientists' adherence to dialectical materialism. Likely drawing upon his own conversations with Russian colleagues such as Joffe, Semenov, and Frumkin, Polanyi wrote, "Not one in a hundred physicists believed this nonsense, but no one could dare to contradict the statement publicly."[80]

Six months later in August 1939 came the great shock to British anti-Fascist and Left intellectuals of the German-Soviet Nonaggression Pact. The agreement, signed in Moscow, renounced warfare between the two countries and pledged neutrality in the event that the other nation was attacked by a third party. Blackett was vacationing in Wales at the time, and he immediately received a request from Polanyi, who was vacationing in Ireland and asked Blackett to help him get defense work. Blackett's letter, dated August 26, 1939, assured Polanyi that he would do what he could, but that he lacked connections on the chemical side of things. At this point, Blackett wrote Polanyi that he was not sure himself what he would do in the war. As for the nonaggression pact, Blackett wrote Polanyi that he was taken aback and imagined that Polanyi felt a certain degree of *Schadenfreude* over what had happened.[81] Blackett soon joined the instrument section of the Royal Aircraft Establishment at Farnsborough, where he worked on the design of the Mark 14 bomb-sight. Within the next year he became a leading scientific adviser to the Anti-Aircraft Command and then to Coastal Command as a leader of operational research groups to coordinate data for military strategy.

In addition to writing Blackett from Ireland about war work, Polanyi wrote John Stopford, the vice-chancellor at Manchester, that, in view of the situation which seemed to be heading for war, he wanted to offer once more, as he had done previously, his service in defense of the country.[82] Polanyi became a naturalized British citizen in September 1939, but he still was unable to get war work.[83] The Foreign Office issued him an identity card for travel with the alias name of Michael Pollard, a surname also used by his sons during the war years.[84] In early June 1940, following the evacuation of British troops from Dunkirk, he volunteered to no avail to join O'Neill with materials testing in Derby. Some of the research in Polanyi's laboratory, particularly the experiments on the production of synthetic rubber for ICI, was potentially useful for the war effort, however.

Thwarted in his efforts to devote himself to full-time war work, Polanyi began concentrating on political and economic writing, publishing a series of essays in the summer of 1940 on planning and the Soviet Union under the title *The Contempt of Freedom: The Russian Experiment and After*. His essays on the Webbs and on Bernal, along with his 1935 article on economic statistics in the Soviet Union, constituted three of the four chapters of the book. The fourth chapter was a lecture, "Collectivist Planning," that

Polanyi gave in April 1940 to the South Place Ethical Society in central London.[85]

The theme of the chapter was the imperative need for the defense of freedom and liberty in the face of the coercive ideologies of collectivist powers such as Germany, Italy, and especially the Soviet Union. Polanyi noted the great appeal of what he called the "engineers' approach to society" for many intelligent, energetic, and progressive minds and then examined in his essay some examples of collectivism in military planning, economics, science, and cultural affairs in the West and the Soviet Union. He identified what he called "Supervision" of individual interests for the protection of the public good as the proper liberal alternative to collectivist "Planning," and he offered a remarkably fierce attack upon what he called the doctrine of extreme laissez-faire. Sounding almost like his brother Karl, Michael Polanyi wrote:

> The protection given to barbarous anarchy in the illusion of vindicating freedom, as demanded by the doctrine of *laissez-faire*, has been most effective in bringing contempt on the name of freedom; it sought to deprive it of all public conscience, and thereby supported the claim of Collectivism to be the sole guardian of social interests.[86]

Then, picking up themes from his economic film, Polanyi continued:

> Liberalism was misled to extremism mainly by failure to understand unemployment . . . It was thus held that all measures reducing the income of the rich and increasing that of the poor must produce unemployment; and most of the other proverbially dismal and inhuman conclusions of economic science arose from this central error . . . I believe that the adoption by Brüning in 1932 of a policy of retrenchment and deflation . . . was one of the most potent immediate reasons of the Nazi revolution, which might have been avoided by a policy of financial expansion, as inaugurated by Roosevelt a few months later.[87]

As for the collectivist approach to planning in science, Polanyi criticized and ridiculed planning procedures and results in the Soviet Union and offered a brief sketch of characteristics of the traditional "mental situation" of the scientist in order to show the impossibility of central planning for science. It was a sketch which he would elaborate using the same crucial elements in subsequent writings of the next few decades: the education of the scientist under a master's guidance, collaboration in a research school, the young scientist's devotion to patient research, and the process of his gradual feeling his way to independence. "We recognized here that a large number of independent activities can form a system of close co-operation . . . In the Liberal State the cultivation of science is public concern, in the performance of which the community is guided by *scientific* public opinion." Scientists—and not an external authority—should guide scientific development. Polanyi registered his own personal experiences as a researcher by arguing the necessity for a leader, or planner, of scientific research to be able to see the experiments performed: "Few scientists can do good work with more than a dozen personal collaborators."[88]

Early in 1941, Polanyi took issue with a December 28 article written by *Nature*'s editor Richard Gregory in which Gregory, using a theme in Bernal's *Social Function of Science*, wrote that people should refrain from ascribing "peculiar holiness to scientific truth" and should discard the idea of scientific detachment. Although we might think that Polanyi would have sympathized with the second suggestion, he abhorred the first one, responding that he could recognize nothing more holy than scientific truth.[89] Cyril Hinshelwood wrote Polanyi a note agreeing with Polanyi's riposte published in *Nature*.[90] By this time Polanyi had a new group of political allies, the founding members of the Society for Freedom in Science, whose leaders were the Oxford zoologist John Baker and the Cambridge plant ecologist Arthur G. Tansley.

In the brouhaha that followed the publication of Bernal's *Social Function of Science*, a response even more outraged than Polanyi's came from Baker, who in his "Counter-blast to Bernalism," published in the *New Statesman*, aggressively decried "Bernalism," which "is the doctrine of those who profess that the only proper objects of scientific research are to feed people and protect them from the elements, that research workers should be organized in gangs and told what to discover, and that the pursuit of knowledge for its own sake has the same value as the solution of crossword puzzles."[91]

In the fall of 1940, Baker proposed the formation of the SFS in a letter addressed to forty-nine prominent British scientists. In early 1941, Polanyi met with Baker and Tansley. As described by Polanyi's biographers, William Scott and Martin Moleski, Polanyi later identified that day as a decisive moment.

I see myself arriving to our first meeting in Oxford, shaking hands with Tansley, being received by you and your lovely Liena. I had written before a few essays on subjects of the kind, but it was your response which launched me finally in the direction of our mutual interest and brought me eventually to the entrance of my career as a philosopher. I had not realized until this moment, that this was the turning point of my life. At any rate the last of its turning points.[92]

The three of them drafted a circular to define the society's aims and recruit members. The physicist George P. Thomson agreed to serve as president; Polanyi, Tansley, and Henry Dale served as co-vice presidents; and Baker took on the duties of secretary and treasurer.[93] Many people objected to the circular, the German physicist Max Born among them. Born, who was teaching in Edinburgh, wrote Blackett that he would not join a society that coupled freedom of science with an attack on planning and Socialist scientists and labeled them "gangsters."[94] A. V. Hill, who had some sympathy with the SFS, cautioned Tansley in June 1941 to moderate the group's adversarial language:

> Remember, that Haldane and Blackett, for all their queer political notions, are useful and cooperative members of the [Royal Society] Council: I am sure that Bernal and Hogben will be the same when their turn comes to serve . . . We can keep them in order better by cooperating with them in scientific affairs than by formally setting up to oppose their political ideas in the name of science.[95]

The society's peak membership reached 430 members in 1946. Many of them were ecologists because of the influence of Tansley, who had been the founding president of the British Ecological Society in 1913 and editor of the *Journal of Ecology* from 1917 to 1938. Among British historians and philosophers of science, members included William Dampier, Gavin de Beer, Herbert Dingle, Alexander Findlay, Michael Postan, Charles Singer, Frank Sherwood Taylor, H. Hamshaw Thomas, E. T. Whittaker and Joseph Needham (who shared humanist Christian views with some of his political opponents).[96] The historians, for example, Taylor, who was curator of the History of Science Museum at Oxford, explicitly linked their objections against government planning for science to their rejection of economic interpretations in the history of science linking the growth of scientific knowledge to material needs. Among American members was the Harvard physicist Percy Bridgman, who made the themes of SFS the subject of a speech at the Boston meeting of the American Association for the Advancement of Science (AAAS) just after he received the 1946 Nobel Prize in Physics. In his lecture "Scientists and Social Responsibility," Bridgman expressed opposition to making science a servant of the state and called upon the legacies of the American philosophers Charles Sanders Peirce and William James for the notion that the community of science is a model for all democratic societies.[97]

Members of the SFS were concerned that the wartime need for enlistment of scientists in service of the nation would become a license for postwar planning and ongoing restrictions or directives for fundamental scientific research. The publication *Science in War* in mid-June 1940, an anonymously edited book written cooperatively by twenty-five scientists, argued that full use must be made of science in the war. Thousands of copies were sold.[98] More ominously for SFS members, the book *Science and World Order* appeared in early 1942, making papers available that were given in September 1942 at an international conference sponsored by the British Association for the Advancement of Science (BAAS) and the Association of Scientific Workers (AScW) and timed to coincide with Tanks for Russia week, which highlighted "tanks and thanks" to Britain's now-ally.[99] Early in 1942, the Labour Party issued a pamphlet, *The Old World and the New Society*, with principles for reconstruction after the war. It included the points that there must be no return to the unplanned competitive world of the interwar years; a planned society must replace the old competitive system; and the main wartime controls in industry and agriculture should be maintained to avoid the scramble for profits which followed the last war.[100]

Leftist scientists wrote a barrage of books in the early 1940s favoring greater governmental organization of science. The books were published for a broad audience, including Haldane's *Science and Everyday Life* (1941), Waddington's *The Scientific Attitude* (1941), and an abridged version of Crowther's 1936 *Soviet Science* (1942). Crowther's *The Social Relations of Science* also appeared in 1941.[101] In 1942, Tansley countered with *The Values of Science to Humanity* and Baker with *The Scientific Life*, in which he charged that the organizers of the three-day BAAS/AScW conference had barred all dissenting voices.[102] Friedrich Hayek, another member of SFS, wrote Polanyi in the summer of 1941, asking him to join in the critique against the Left. Like Polanyi, Hayek spent the war years teaching classes and writing. Although some London School of Economics faculty were called to duty in various government departments when the war began, Hayek was not offered a post, although he had been naturalized as a British subject in 1938.[103]

I attach very great importance to these pseudo-scientific arguments on social organization being met and I am getting more and more alarmed by the effect of the Propaganda of the Haldanes, Hogbens, Needhams etc. I don't know whether you've seen the latest instance, C. H. Waddington's Pelican on the Scientific Attitude. I think this last specimen is really contemptible.[104]

Hayek wrote his own diatribe against Waddington in his book *The Road to Serfdom*, which appeared in 1944.[105] In a chapter titled "The Totalitarians in our Midst," Hayek offered a savage critique of two "authors whose sincerity and disinterestedness are above suspicion . . . I hope in this way to show how the views from which totalitarianism springs are now rapidly spreading here." Hayek's targets were Waddington and the historian E. H. Carr, with Hayek also referring to Crowther as a more extreme case than the other two. In Waddington's *Scientific Attitude*, wrote Hayek, he treats freedom as "a very troublesome concept for the scientist to discuss, partly because he is not convinced that, in the last analysis, there is such a thing." Quoting Waddington, Hayek charged that he anticipates an economic system which "will be centralized and totalitarian in the sense that all aspects of the economic developments of large regions are consciously planned as an integrated whole" while maintaining a "facile optimism that in this totalitarian system freedom of thought will be preserved."[106]

In response to Hayek's challenge, Polanyi published "The Growth of Thought in Society" in *Economica* in 1941, focusing on Crowther's book on the social relations of science and warning that science, and specifically British science, must combat influential scientists who were using their links with the government to adopt a hierarchical "corporate" organization for science and undermining its traditional "dynamic" form of organization. Using examples from physics, for instance, crystal formation during cooling of a molten material, Polanyi explained dynamic order as "an ordered arrangement resulting by spontaneous mutual adjustment of the elements" independently of any single directing force or constraint applied to individual particles. Hayek similarly had used the language of spontaneity and unintended consequences in his inaugural lecture at the London School of Economics in March 1933.[107] Polanyi, too, was now comparing the growth of scientific knowledge to other "spontaneously arising orders" in the natural world, the economic world, and common law.[108] In his 1941 essay, he also cited Gestalt psychology as a source of the idea of dynamic order.[109]

During the war years, Polanyi wrote articles on the autonomy and freedom of science for various journals and magazines, and he was a guest on BBC radio broadcasts in June 1944 to talk about science and the decline of freedom and, in 1948, on the planning of science.[110] His talks and lectures included the Riddell Lectures at Durham, published in 1946 as *Science, Faith and Society*. In 1945 the SFS began publishing "occasional pamphlets"; the series eventually included fifteen essays, the last of which was by Baker in 1962. In addition to Polanyi, Baker, and Tansley, authors of these pamphlets included Frank Sherwood Taylor, Warren Weaver, N. S. Hubbard, Jean Pelsener, and George Thomson. Several of the anonymously edited pamphlets focused on Soviet science and especially on what became known as the Lysenko Affair.

In the early 1940s, inquiries began to be made officially about Vavilov from the Royal Society to the Soviet Academy of Sciences but with no result. In June 1945 British scientists were among foreign scientists invited to Moscow for a celebration of the 220th anniversary of the Academy of Sciences. Among those who attended were Needham and Bernal. Blackett wanted to go but was denied a visa by the British government on the grounds of his continued involvement in confidential work. British delegates to the meeting heard two accounts of the disappearance of Vavilov, whose brother Sergei was president of the Soviet Academy. One story was meant, they were told, for internal dissemination. It was claimed that Vavilov had been shot during the war while trying to escape from Russia. The second story, for external consumption, was that Vavilov had died during the war "while breeding frost-resistant plants." It appeared that he had died sometime between 1941 and 1943 along with some of his coworkers in genetics. Sir Henry Dale resigned from the Soviet Academy when its secretary failed to reply to the request from the Royal Society for Vavilov's time and place of death, and the Royal Society severed its relations with the Soviet Academy of Sciences after their decision to support Lysenkoism in 1948.[111] Until 1948, many British scientists, among them Needham and Haldane, treated the Soviet genetics debate as a normal scientific controversy between two research schools, a point of view explained in Nils Roll-Hansen's historical study *The Lysenko Effect: The Politics of Science*. In the late 1940s and the 1950s, the gradual recognition of the scale of the extermination of Soviet geneticists destroyed explanations of the episode as a normal scientific debate. Among the first exposés were C. D. Darlington's article in *Nineteenth Century* in 1947, "Retreat from Science in Soviet Russia," and Conrad Zirkle's book *Evolution, Marxian Biology and the Social Scene* in 1949.[112] Eric Ashby spoke on a BBC Third Programme broadcast in January 1948, breaking from his long support of the Soviet Union, with the statement that "formerly when biologists were liquidated, it was always given out that their liquidation has nothing to do with their biology; whereas now, the pretense is dropped."[113]

An early contribution to this discussion about Vavilov was Polanyi's essay "The Autonomy of Science," which dealt with information and rumors about Vavilov and Lysenko as he had heard them by 1942. Polanyi gave this talk, later retitled "Self-Government of Science," as a lecture at the Manchester Literary and Philosophical Society in February 1942, and he published it in the society's journal. It was this talk which Polanyi came to see as a kind of prospectus for a book on scientific life.

At the time his political concerns mainly were central planning, the distinction between pure and applied science, and individual freedom in science and in the state more broadly. He began by discussing the scientist's position within a tightly bound scientific community and some of the mechanisms by which this community works, such as scientific journals, refereeing and decision-making by what he called leaders of scientific opinion, the development of standard knowledge in textbooks, and the transmission of scientific tradition from one locale to another. Using an implicitly center-and-periphery model, Polanyi argued that science had developed naturally in Europe and could take root elsewhere only if

> the government of a country succeeded in inducing a few scientists from some traditional centre to settle down in their territory and to develop there a new home for scientific life, moulded on their own traditional standards . . . science as a whole is based—in the same way as the practice of any single research school—on a local tradition, consisting of a fund of intuitive approaches and emotional values, which can be transmitted from one generation to the other only through the medium of personal collaboration. Scientific research—in short—is an art . . . There can be therefore no question of another authority replacing scientific opinion for the purposes of this function; and any attempt to do so can result . . . in the more or less complete destruction of the tradition of science.[114]

Polanyi then turned to recent developments in Germany and the Soviet Union, first quoting from Heinrich Himmler, head of the Nazi SS and the Gestapo: "Science proceeds from hypotheses that change every year or two. So there's no earthly reason why the party should not lay down a particular hypothesis as the starting-point, even if it runs counter to current scientific opinion."[115] Just as the Nazis favored some fields of science and destroyed others, so, too, in the Soviet Union, Polanyi explained, it was well known that the state had been intervening in genetics and plant breeding since around 1930. By 1939 it appeared that Lysenko had acquired sufficient power within the party and the state that he persuaded the Commissariat of Agriculture to prohibit the methods of genetics in plant-breeding stations and accomplished the elimination of Mendelism entirely from education and practice. When Vavilov protested these actions at a conference sponsored by the journal *Under the Banner of Marxism* later in 1939, he was completely undermined, said Polanyi, by his own statements at a Planning Conference in 1932. On that occasion, according to Polanyi, Vavilov had mistakenly deprecated the pursuit of science for its own purposes saying, "The divorce of genetics from practical selection, which characterizes the research work of the USA, England and other countries, must be resolutely removed from genetics-selection research in the USSR."[116] Thus, in Polanyi's view, Vavilov had opened himself up to the charges against genetics that it was an idealist bourgeois science, and Vavilov and his colleagues could only grasp at claims for the consistency of their science with the principles of Marx and Engels. No good comes from scientists adapting their science to ideology or popular demands, warned Polanyi. The threat to scientific freedom can be seen even in England, he concluded, citing Hyman Levy's remark to an applauding throng of scientists at a meeting of the Association of Scientific Workers that "science must be marshaled for the people."[117]

Later, Polanyi specifically located his decision to write about the nature of science in the Lysenko affair, although he also attributed the turning point to his meeting with Tansley and Baker and the founding of the SFS in 1941. In the introduction to the second edition of *Science, Faith and Society*, Polanyi wrote that Vavilov's last public defense of Mendelian genetics at the 1939 *Under the Banner of Marxism* conference was marked by an attempt to try to appeal to the authority of Western science in validating his work, but that the defense on these grounds failed. Polanyi said that he came to see the necessity not only to argue against the planning of science for the benefit of society but also the need to define most clearly the very conception of science in the West—the meaning and nature of science—to which Vavilov could not successfully make appeal. What gives science—Western science—its veracity? What distinguishes true science and scientists from impostors and opponents? "How was its general acceptance among us to be accounted for? Was this acceptance justified? On what grounds?"[118]

Polanyi was coming now to a project in the philosophy of science, but the project remained tightly intertwined with his thinking in economics and politics. He remained a member of the SFS until it disbanded in 1962. In 1953, he also became active in the Congress for Cultural Freedom (CCF), which was founded at a meeting in West Berlin in June 1950 by a group that included Melvin Lasky, Arthur Koestler, Sidney Hook, Arthur Schlesinger Jr., and Raymond Aron. A highlight of the meeting was Koestler's speech, in which he denounced totalitarian states in the name of intellectual freedom, to an audience of fifteen thousand in the Funkturm Sporthalle. Hugh Trevor-Roper later said about the occasion, "I felt that we were being invited to summon up Beelzebub in order to defeat Satan."[119]

Alex Weissberg came to Manchester to persuade Polanyi to chair the CCF's Committee on Science and Freedom and to help organize a future meeting in Hamburg.[120] To this end, Polanyi marshaled his resources at the Rockefeller Foundation. In 1945, the SFS had published as its Occasional Pamphlet #3 a letter to the *New York Times* written by Warren Weaver in support of "free science" in postwar America. Weaver's letter was a response to Waldemar Kaempfert's review in the *New York Times* of Vannevar Bush's book *Science—The Endless Frontier*. In the review, Kaempfert attacked what he called Bush's notion of "free enterprise" in science. In rebuttal, Weaver argued that Kaempfert's preference for the Manhattan Project as a model for science was no blueprint for the future of scientific organization in the United States.[121] Polanyi was in touch with Weaver in the spring of 1946 about U.S. Senate hearings on science legislation and thanked Weaver for helping shape policy for the freedom in science movement in the United States. At the time Polanyi was looking forward to his visit to Princeton in September where he would receive an honorary degree and speak on freedom in science at the university's bicentenary celebration.[122]

Polanyi wrote Weaver in February 1953 in support of the CCF's application to the Rockefeller Foundation for funds for the Hamburg conference, which would include sessions such as Ideology and Science, Science and Moral Responsibility, Science and Scientism, the Condition of Science in Totalitarian Countries, and the Condition of Science in Free Countries. Polanyi's letter was important, if not decisive, for Weaver, who worried about the involvement in the CCF of scientists whom he considered pinko, emotional, or naïve, unlike Polanyi whom he trusted and admired. Weaver's worries were magnified by ongoing committee hearings in the U.S. Congress in the early 1950s to investigate whether foundations were promoting Socialism and undermining national security. Before making the final positive decision on CCF, Weaver received assurance by telephone from Caltech's Joseph Koepfli that the State Department was definitely interested in the meeting, considered it constructive and would assure that American delegates were sound defenders of democratic liberties. What Weaver did not know was that the Fairfield Foundation, which paid half of the expenses for Hamburg, was a CIA-funded organization.[123]

In July 1953, 119 scientists convened in Hamburg at the Science and

Freedom conference chaired by Polanyi and Weissberg. Polanyi delivered a paper that distinguished pure from applied science and likened the structure of the scientific world to that of the free market. Edward Shils wrote in the *Bulletin of Atomic Scientists* that there was a division at the conference between "idealists" and "positivists" who failed to recognize that they could have found common ground in the notion of a scientific community with its own powers of self-maintenance and self-regulation. Conflicting claims for "planning" and "laissez-faire" should be reconciled by letting the community of scientists, not individual scientists, do the planning.[124]

After Hamburg, Weaver, in agreement with Dean Rusk, who was president of the Rockefeller Foundation, considered making a three-year grant of $30,000 for Polanyi's Committee on Science and Freedom. The committee had its office in Oxford, where Polanyi's son George served as secretary and his wife, Priscilla, served as staff. Rusk checked with the State Department again to make sure that Rockefeller Foundation support of CCF was in accord with public policy, but the foundation ended up never making the grant. In early 1957, Warren Weaver hosted a lunch for Polanyi and Shils at the Rockefeller Foundation, inviting him to make the case for the CCF committee, but the role of Polanyi's son and daughter-in-law made Weaver and others uneasy, as did an increasing concern that Polanyi's pamphlets simply were not effective in the broad goal of defending academic freedom at universities worldwide. Weaver was also adamant that the Rockefeller Foundation needed to avoid clearly political projects. In June 1957, Polanyi withdrew the application for the $30,000 grant.[125]

As discussed by Frances Stonor Saunders, the CCF, which was based in Paris, included many distinguished intellectuals and artists across all parts of the political spectrum. The American Committee for Cultural Freedom, established in early 1951, was headed by Sidney Hook and Irving Kristol in New York City. Among its members were J. Robert Oppenheimer, W. A. Auden, Thornton Wilder, and Arthur Koestler. Major periodicals such as *Encounter* were linked to the CCF, and it served as a clearing house for the distribution of a wide range of cultural journals such as *Partisan Review, Sewanee Review, Kenyon Review, Journal of the History of Ideas,* and *Daedalus* (the journal of the American Academy of Arts and Sciences). Edward Shils's journal *Minerva* also received funds from CCF. Paul Goodman first made the public charge in 1962 that CCF and *Encounter* were instruments of the CIA, and the *New York Times* published articles to this effect in April of 1966. George Kennan wryly commented in 1967 that "this country has no Ministry of Culture, and CIA was obliged to do what it could to try to fill the gap."[126] Polanyi resigned from the CCF in October 1967 after Mike Josselson, its executive secretary, was expelled from CCF following the admission that he had secretly accepted CIA funding for seventeen years. The American committee had suspended its activities a decade earlier, in 1957, for lack of funds. Despite the scandal, Polanyi had genuine sympathy with Josselson's political aims.[127]

While Polanyi's name was indelibly associated politically with the issues boxed under the rubrics of central planning and freedom of science, British scientists throughout the political spectrum, whether writing in *Nature* or in the circulars of the SFS, all emphasized the necessity for truth and freedom in science. There really was no argument on this issue. Polanyi was asked to address the Manchester section of the left-wing AScW in May of 1945 and his talk was well received. As for the right of scientists to be able to follow research paths of no obvious social utility, Blackett was among those who cautioned that socialist thinking could go too far in reacting against the attitude of science-for-science's-sake. A progressive Socialist society had the obligation, he insisted, to devote an appreciable fraction of its resources to pure science, just as to music and art.[128]

There was some effort at planning for British universities right after the war. The Royal Society was enlisted to furnish a report titled "The Balanced Development of Science in the United Kingdom," and Blackett was one of the committee's members during 1945 to 1946. Polanyi submitted a document arguing that universities should be left to fill professorships with the most eminent candidates available and that the needs of industry, medicine, and defense should be considered only as subsidiary factors in the funding of various branches of pure science. In the end, the committee report, known as the Darwin Report for its chairman Charles G. Darwin, supported the "natural" development of science by its most "distinguished" leaders.[129] Blackett was president of the AScW from 1943 to 1947, and under his presidency the organization recommended against the centralization of governmental science offices into a single Ministry of Science. The 1950s organization of science in the UK, then, turned out to be not much different than in the prewar period, although the situation would change in the next decades.[130]

## J. D. Bernal and *The Social Function of Science*

Let us return now to Bernal and his book *The Social Function of Science* by way of examining the elements in the book that positioned it at a social turn in the study of science. Polanyi had known Bernal and his scientific work, as well as his political views, for many years. John Desmond (J. D.)

Bernal was a pioneer in studies of molecular structure using X-ray crystallography. After taking his degree in 1922 and doing further research on a theory for analyzing the arrangement of atoms in space, Bernal moved from Cambridge to the Royal Institution in London in 1923, where he worked with William Henry Bragg. In London, Bernal developed the method of determining the structure of graphite by X-raying the graphite crystal as it rotated.[131] He drew up a chart to simplify the process of classifying the reflections, publishing a paper in 1927 that Dorothy Hodgkin praised fifty years later as still valuable for beginners in X-ray crystallography.[132]

After returning to Cambridge as a lecturer in 1927, Bernal worked on the structure of vitamin B1, vitamin D2, pepsin, sterols, and the tobacco mosaic virus. Among his students and collaborators were Hodgkin, Rosalind Franklin, Aaron Klug, and Max Perutz. It was well known that Ernest Rutherford, who died in 1937, could not abide Bernal. As Gary Werskey puts it, Bernal was a "boundary rider," "an Irish Catholic [whose Sephardic Jewish ancestors reached Ireland in 1840] in an Anglican society; a middle-class intellectual on the far Left of a working-class movement; a theorist in a decidedly experimentalist community; a man of science who doubled as a man of letters; and a married man who flaunted social conventions."[133] Rutherford objected to Bernal's open commitment to Communism, his ideology of sexual freedom and promiscuity, his speculative theorizing, and his tendency to hand over research projects to students.[134] Collaborators and students had a different view of Bernal, finding him inspirational and appreciating his insistence that they publish papers from his laboratory in their names alone.[135]

While Bernal was at Cambridge during 1927 to 1938, Rutherford assigned him to a sub-department where Bernal did his groundbreaking structural studies of sterols, proteins, and viruses. He worked in what were described as "ill-lit and dirty rooms on the ground floor of a stark, dilapidated grey brick building" with inadequate studentships and technicians.[136] No wonder Bernal addressed himself to the social and economic conditions for scientific work in Britain.

Bernal moved to Birkbeck College in 1938, succeeding Blackett after Blackett came to Manchester. C. P. Snow contributed to Bernal's fame by portraying him as the brilliant X-ray crystallographer Constantine in his 1934 novel *The Search*. After Polanyi moved to England, he and Bernal were in touch in the mid- and late 1930s not only about political questions and the showing of Polanyi's economic film but also about scientific matters, corresponding, for example, in 1935 on data on interatomic distances and bond energies in the water molecule. That letter had a political

Figure 6.2. John Desmond Bernal. (Photograph by Ramsey and Muspratt. © Peter Lofts Photography / National Portrait Gallery, London. Used with permission.)

dimension, too, with Polanyi writing that he had just returned from Moscow and would welcome a discussion with Bernal about his experiences.[137]

At the beginning of *The Social Function of Science*, Bernal introduced the two vying conceptions of science as pure thought and as power. In introducing the notion of pure science, Bernal referred to James Jeans, Arthur Eddington, and Alfred North Whitehead, among others, and he quoted George Sarton, "the great historian of science," for the view that science "is essentially human in its origin and growth" and that it is part of intellectual culture and general humanism. This idealist view, wrote Bernal, dates back to Plato's *Republic* and to the belief that natural knowledge is a category of pure thought concerned with contemplation of truth.[138] A different view than Sarton's, wrote Bernal, characterizes science by its utility and its power. Neither view alone is sufficient, however. "Modern science derives both from the ordered speculation of the magician, priest, or philosopher, and from the practical operation and traditional lore of the craftsman. Until now, far more attention has been given to the first rather than the second, with the result that progress of science seems more miraculous than it was in fact."[139]

The second chapter of Bernal's book is a historical survey, written self-consciously as something different from what Bernal called "histories of science [that] are little more than pious records of great men and their works, suitable perhaps for the inspiration of young workers, but not for understanding the rise and growth of science as an institution."[140] Bernal's history laid out the general lines that Blackett, Crowther, and others used in their lectures and writings on the history of science, including the description of the rise of agriculture and urban centers, the Hellenistic revival, the birth of science in the Italian Renaissance, the new academies, scientists as craftsmen with patrons, and the development in Leeds, Manchester, Birmingham, Glasgow and Philadelphia of a culture of gentlemen who furthered the science of the industrial revolution.[141] Bernal recognized that Marxist ideological jargon alienated most British readers, like Dorothy Singer at the 1931 Congress, but he still adopted it.[142] Bernal conceded that the appeal to Marxism likely immediately influenced many British academics "not to listen to the arguments"—but he persisted.[143]

Only a minor part of Bernal's *Social Function of Science* had to do with the history of science. He divided the book into two large sections, the first titled "What Science Does" and the second "What Science Could Do." His historical chapter appeared in the first section, followed by chapters on the current organization of scientific research in Britain, science in education (including the role of scientists and journalists in popular science), the efficiency of scientific research, the application of science (in which he demonstrated agreement with Polanyi's view that patents stifle research), uses of science in war past and present, and statistics and analysis for science worldwide (including Asia, Fascist Germany, and Socialist Russia). In assembling all his information, he sought help from colleagues, including Polanyi, from whom he got a reprint of Polanyi's 1930 article in *Deutsche Volkswirt*.[144]

Further chapters in Bernal's book focused on measures of change and reform in the training of the scientist; the organization of research; scientific communication, including science and the press and the importance of personal contacts; the finance of science; strategies for scientific advance; ways that science can serve practical needs of man; science and social transformation; and a last chapter called "The Social Function of Science." As already noted, Bernal proposed a major reorganization of research, including the formation of a central endowment based in government and industrial funds. He cited as an example the Centre Nationale de la Recherche Scientifique (CNRS), which had just been established under the Popular Front in France. It was a national endowment for scientific research separate from scientific education.[145] He argued for the social responsibility of scientists and praised the left-wing Association of Scientific Workers.[146]

As for government funding of science, Bernal marshaled figures showing that Great Britain was far behind other nations in funding research in the mid-1930s. Only 0.1 percent of national income supported scientific research, compared with 0.6 percent in the United States and Germany and 0.8 percent in the Soviet Union. In a new British agency for scientific research, he proposed, the distribution of funds would be determined by needs both for the internal development of science, according to scientists' own estimates, and the need for particular developments in sciences on account of urgently required applications in the public interest. This was hardly unmitigated central planning by government bureaucrats, as his detractors claimed. Nor did Bernal advocate adopting organizational structures and methods modeled on business or civil administration. In contrast, he explicitly wrote, regarding the finance of science, that the organization of science must be largely built up by the efforts of scientists themselves: "To submit science to such discipline and routine [as found in business or civil administration] is certainly to kill it."[147]

As for the argument that scientists should not mess in politics, Bernal expressed strong objection to the position enunciated by A. V. Hill that science must adhere to the founding documents of the Royal Society and refuse to meddle with, or be dominated by divinity, morals, politics or rhetoric.[148] This attitude, Bernal warned, was the view that scientists had adopted during the rise of German Fascism and it had led to no good. Adopting Marxist language, Bernal wrote of the failure, as he saw it, of nineteenth-century liberalism in which social relations are expected to operate through market forces. According to the theory of liberalism, every man is free to do as he pleases, but in fact he is tied by the "iron laws of economics," which are socially produced but erroneously stated to be laws of nature. In a new society, this liberal conception of freedom would be replaced by "freedom as the understanding of necessity" and by freedom to take part in a common enterprise for the common good with the means to do so. This kind of freedom, Bernal wrote, "can only be appreciated to the full by living it."[149] In Bernal's view, guarantees of scientific freedom can be achieved only if there is public confidence that public interest lies in scientific advance and if scientists revert to the habits of their seventeenth- and eighteenth-century predecessors of extolling the usefulness of their work, whether that utility lies in the glory of God or the benefit of mankind or personal satisfaction, but not science for science's sake.[150]

## Marxism and Liberalism Meet at the Social Turn

Polanyi, like other members of the Society for Freedom in Science, could not overcome his hatred of Marxist jargon and his anger at Bernal's wide-eyed admiration for Soviet Socialism. The elision of pure and applied science was a sin that Polanyi equally could not forgive, made all the more execrable by a passage in *Social Function of Science* that characterized pure science as potentially pure self-indulgence. In this line of argument, Bernal quoted at length from one of the characters in Aldous Huxley's 1928 novel *Point Counter Point*, who reflects that in his view the real charm of the intellectual life is its easiness. "It's the substitution of simple intellectual schemata for the complexities of reality," the character remarks.[151]

In a statement that Werskey judges "monumentally indiscreet," Bernal went on to say: "This attitude, though rarely admitted, is actually extremely widespread among scientists, particularly those in the safer and more comfortable positions . . . 'Whenever I look out at the world,' a professor once remarked, 'I see such misery and mess that I prefer to bury myself in my own work and to forget about things that in any case I could do nothing about.'"[152] Ironically, Bernal's description is precisely the kind of attitude that Loren Graham finds was a lifesaver for scientists working in the Soviet Union during the Stalinist era.[153]

In his book review of February 1939, Polanyi took special offense at this passage and what he characterized as Bernal's view that the ideal of pure science "was a form of snobbery" and that no economic system would pay a scientist just to search for truth.[154] On the other hand, writing in 1951 in a SFS Occasional Pamphlet, the biochemist and former president of the Royal Society, Henry Dale, conceded that scientists could no longer stick to a rhetoric of pure science and deny public accountability:

> However much we may desire that government advisory councils, charitable foundations or a benevolent industry should be ready to show their faith in individual scientists by simply endowing them to do whatever research they fancy, this cannot be expected as a regular policy.[155]

A final area in which Polanyi took special offense to *Social Function of Science* lay in Bernal's statements about freedom. Bernal argued that freedom must recognize necessity. Toward the end of his essay review, Polanyi wrote:

> Dragooned into the lip service of a preposterous orthodoxy, harried by the crazy suspicions of omnipotent officials, arbitrarily imprisoned or in constant danger of such imprisonment, the scientist in Soviet Russia is told, from England, that the liberty which he enjoys can only be appreciated by living it.[156]

Yet, for all the radicalism of Bernal and the Bernalists, some of Bernal's core assumptions about science did not much differ from most scientists and historians and philosophers of science writing about science in the 1930s and for the next two decades. Bernal and Blackett no less than Sarton and Singer were absolutely confident that science results in the advancement of knowledge and human progress. For all of them, science was a positive good, and it was a good not only in the sense of human advancement but in its dedicated search for the truth. Bernal and Polanyi, just like Blackett and no less than Jeans or Eddington, subscribed to a broadly realist view of what they were doing in their work—what Arthur Fine in his study of Albert Einstein has called "motivational realism." Scientists would hardly subject themselves to the rigors of scientific discipline unless they had the enthusiasm and confidence that they were on the path of discovery.[157]

For both Bernal and Polanyi the important matter, then, in their contrasting views on state support of science, was to teach the public how science works *in order* to demonstrate its dependability and stability. There was a need to demonstrate that science is not simply a procession of discarded wrong ideas, but rather it is valid knowledge acquired through well-established traditions and values. The endless talk about scientific method by historians and philosophers of science was fruitless, in their view, and failed to convince the public that scientific work is well worth public support and even sacrifice. What is required is an understanding of how scientists uniquely go about their work.[158]

For both Bernal and Polanyi, science is a social network and a social activity maintained within a framework of institutions that includes universities, scientific societies, scientific journals, and communities of expert peers. Bernal and Polanyi were men who had learned their trade through the personal dedication and disciplined apprenticeship that they described in their writings. Science, for them, is a matter of personal enthusiasm and discipline, not machine rationality. Things get done because "a handful of the more important scientists in the country know one another," as Bernal put it, and workers in a field "arrange among themselves, when they are on friendly terms, the kind of work each of them intends to pursue and the relation of one man's work to another." "To some it is a game against the unknown where one wins and no one loses, to others, more humanly minded, it is a race between different investigators as to who should wrest the prize from nature."[159] Note the similarity in Bernal to Polanyi's emphasis on the master-apprentice relationship, science as tradition and social practice, the competitiveness and personal devotion in scientific work, and the intimacies of independent actors in the community of science.

As for the moral values and ethical conduct that result in scientific knowledge, Bernal and Polanyi again fundamentally agreed. In what Gary Werskey characterizes as the most widely quoted passage in all of Bernal's writings, Bernal wrote in 1939 of what Robert K. Merton would define in 1942 as the four scientific norms: Merton's universalism, disinterestedness, organized skepticism, and what Merton initially called "communism."[160] Here is Bernal in 1939:

> In its endeavour, science is communism. In science men have learned consciously to subordinate themselves to a common purpose *without losing the individuality* of their achievements. Each one knows that his work depends on that of his predecessors and colleagues, and that it can only reach its fruition through the work of his successors . . . Each man knows that only by advice, honestly and disinterestedly given, can his work succeed, because such advice expresses as near as may be the inexorable logic of the material world, stubborn fact. *Facts cannot be forced to our desires, and freedom comes by admitting this necessity* and not by pretending to ignore it.[161]

Waddington's book *The Scientific Attitude*, so despised by Hayek, was one that praised Bernal for his "very fine statement of the aims and method of science, whether or not one agrees that in acting this way scientists are behaving like communists."[162] Merton changed his own characterization of the scientific community's "communism" to "communalism" in order to make clear the non-Marxist meaning of the common ownership of scientific information, discoveries, and theories. The notion that scientists relinquish intellectual property rights in exchange for recognition and esteem is precisely the argument that both Bernal and Polanyi made against patent rights.

Polanyi identified Bernal's notion of freedom—whether scientific freedom or civil freedom—with the limited concept of freedom in Nazi Germany and the Soviet Union where it was rationalized on grounds of historical necessity. With respect specifically to science, Polanyi wrote, "scientific life" is a system in which independent scientists freely cooperate in a system rooted in common tradition and longstanding institutions. The distribution of authority within the system is not equal, but activities and decisions are naturally coordinated by a mutual adjustment of independent initiatives resulting in what Polanyi called a "General Will" that governs the society of scientists in such a way that "scientists recognize that . . . the resultant decisions of scientific opinion are rightful."[163]

Polanyi had unmitigated confidence in the social processes of scientific opinion. Bernal, too, had a great deal of confidence in scientific expertise, seeing a necessary role for scientists in the planning of science, without reaching Polanyi's conclusion that the state had no right to intervene in the internal working of the republic of science. It is striking how much they agreed in ways that Polanyi could not see at all because of his political differences with Bernal. Philip Mirowski, Charles Thorpe, and David Hollinger are among scholars who have noted the authoritarian tone of Polanyi's insistence on internal and external deference to scientific authority. Mirowski, for example, regards as hypocritical Polanyi's ridicule of Bernal's German-Marxist historicist appeal to "freedom as the understanding of necessity," while Polanyi himself was counseling internal and external acceptance of scientific "authority" and praising the "'naturally' self-optimizing market" of the republic of science.[164]

Polanyi's argument against Bernal that science requires the structure of a liberal republican framework in order to thrive is a clearly political argument and one that was common among natural scientists and social scientists in the West and especially in the United States, as has been well studied by scholars. Vannevar Bush, Edward Shils, and Robert Merton, for example, all insisted on the necessity of a free society for the flourishing of science. A counter-argument appeared toward the end of the twentieth century, as historians of science retold the histories of science under Nazi Germany and the Soviet Union in order to highlight real achievements in many scientific fields despite extreme limitations on scientists' political freedoms and demands of ideological orthodoxy in describing scientific results. In a special issue of *Osiris* titled *Politics and Science in Wartime*, for example, with comparative perspectives on the Kaiser Wilhelm Institutes, Carola Sachse and Mark Walker explicitly reject the belief "that science flourishes best, or even can only really flourish, in a Western-style democracy."[165]

Bernal and British Left scientists of the 1930s had the potential to invigorate and transform the history and philosophy of science for the next generation. In the short run, they failed largely because of Marxist jargon or socialist ideology, pro-Soviet politics, and controversial proposals for central funding of science and technology. The broad social turn in history and philosophy of science was not taken at the time of Bernal and Polanyi's engagement because of its political minefields, even though, paradoxically, these two scientists arrived at the same fork in the intellectual map from the political extremes of the Marxist Left and the liberal Right. Polanyi's decision to write a book on scientific life followed directly from his combat with the Leftist social relations of science movement.

When the broad social turn later occurred around 1970, its advocates rejected the single most fundamental assumption shared by the two scientists Bernal and Polanyi. Neither Bernal nor Polanyi doubted the privileged nature of scientific knowledge and the scientists' notion of truth. Bernal may have been doubly protected against doubt by Marxist realism, as well as by his experience of a scientific life. Polanyi's belief in the transcendence of scientific knowledge was similar to that of his elder Berlin colleagues Max Planck and Albert Einstein, and it required no logical argument. Polanyi believed that the way in which the genuine scientific community functions guarantees its claims to moral virtue and natural truth. In the next chapter, the political underpinnings of the philosophical writings of Karl Popper and Thomas Kuhn are examined in comparison with the political roots of Polanyi's philosophy of science.

# SEVEN

# Political Foundations of the Philosophies of Science of Popper, Kuhn, and Polanyi

If political concerns and social values played a role in the writings on science by Polanyi and Bernal in the 1930s, these influences equally infused the writing of other historians and philosophers of science in the mid-twentieth century, with Karl Popper and Thomas Kuhn foremost among them. By the mid-1950s, Popper, Kuhn, and Polanyi all had parted company with the then-dominant scientific philosophies of empiricism, inductivism, and logical positivism. They rejected the philosophical traditions that used empiricism to differentiate the sciences from the arts, from religion and metaphysics, and from ideology. They each published a magnum opus around 1960: Polanyi's *Personal Knowledge: Towards a Post-Critical Philosophy* (1958); the English translation of Popper's *Logic of Scientific Discovery* (1959); and Kuhn's *Structure of Scientific Revolutions* (1962).

By the twentieth century's end, Kuhn's *Structure of Scientific Revolutions* had sold over a million copies in two dozen languages.[1] This was a resounding success for a scholarly book with footnotes, and it proves the appeal of Kuhn's interpretation of the nature of science to a very broad public audience, as well as to specialists in the natural and social sciences and in historical and philosophical disciplines. Geophysicists were among the first natural scientists to adopt Kuhn's catchy phrase of "paradigm change," applying it around 1968 to their "current Scientific Revolution" and to their emerging consensus about continental drift and plate tectonics.[2] Geophysicist Allan Cox, for example, noted that he preferred to talk about "paradigm change" in geology and geophysics, rather than to engage in "sterile" debate about whether plate tectonics should be described as a hypothesis or a theory.[3]

Like Kuhn, Popper had a broad audience. Many scientists regarded Popper as a giant among philosophers. The philosopher of science Peter

Godfrey-Smith recounts the instance of attending a lecture by a Nobel Prize winner in medicine only to hear mostly about Popper rather than about virology.[4] The virologist in question might have been Sir Peter Medawar, a great admirer of Sir Karl's. Medawar often is quoted as having said, "I think Popper is incomparably the greatest philosopher of science that has ever been."[5] Popper, like Kuhn, had his admirers among geophysicists, including the British-Canadian geophysicist Edward Irving, one of the pioneers in continental drift. Irving participated in a tiny seminar led by Popper in Canberra in 1963. Irving later said that Popper's emphasis on testability and falsifiability fortified his own determination to demonstrate that the opponents of continental drift were wrong. A quotation from Popper appears as the epigraph for Irving's *Paleomagnetism and Its Application to Geological and Geophysical Problems* (1964).[6]

In comparison to Kuhn and Popper, Polanyi was less well known at the end of the twentieth century in the general public. He also was less influential than Kuhn and Popper among professional philosophers of science despite his effort to reach the philosophical community. Godfrey-Smith does not once mention Polanyi in *Theory and Reality* (2003), a popular introductory textbook for philosophy of science. In contrast, Polanyi's name has iconic status among sociologists of science who routinely cite Polanyi's *Personal Knowledge* for its anti-positivism, formulation of the idea of tacit knowledge, and originality in recognizing and describing science as social practice, rather than as formal method.[7]

David Hollinger, George A. Reisch, and Steve Fuller are among scholars who have described aspects of the politics of scientific culture within the intellectual establishment of the United States during the 1940s and 1950s when the philosophies of Polanyi, Popper, and Kuhn were first widely welcomed.[8] A close examination of the political origins of Popper's and Kuhn's philosophical work, with comparisons to Polanyi, demonstrates how their scientific philosophy, like scientific life, was partly embedded in political ideology during this period. Further, the political commitments of the three men were fundamentally the same, although Kuhn was an American intellectual of a younger generation who matured in a society that suffered none of the radical and violent political shifts and anti-Semitic persecutions of Popper's and Polanyi's generation in central Europe from the 1920s to the 1940s. Nor was Kuhn a staunch antitotalitarian warrior as were Popper and Polanyi. In contrast, Kuhn's political purview was couched more narrowly within the American scientific and philosophical community, and his philosophy did not bear the moral fervor of Popper's and Polanyi's writings.

This chapter focuses on the early experiences and writings of Popper and Kuhn before the publication of the *Structure of Scientific Revolutions* and makes comparisons of Popper's and Kuhn's epistemological arguments and political assumptions with Polanyi's published views prior to *Personal Knowledge*. The argument of this chapter is that the initial influence of Popper's and Polanyi's truly original ideas in philosophy of science occurred in their explicitly political writings that appeared well before the publication in the United States and Great Britain of their big philosophical books. Kuhn's publications, too, carry a political agenda that asserts the reciprocal relationship of science and democracy, but the message is implicit rather than explicit in his philosophy of science. Polanyi's early writings laid out an argument for the social nature of science that can be found later in Kuhn, while Popper found Polanyi's and Kuhn's portrayal of the stability of belief and dogma in a scientific community antithetical to his valuation of skepticism in science.[9]

## Karl Popper versus Positivism and Historical Determinism

Polanyi, born in 1891, was almost a decade older than Karl Popper, who was born in Vienna in 1902. Kuhn, more than thirty years younger than Polanyi, was born in Cincinnati in 1922 and spent his early years in New York City and Croton-on-Hudson before going to private schools in Pennsylvania and Connecticut.[10] All three men were from originally Jewish families of some financial means and intellectual ambitions. None of the families was religiously observant. Popper's parents converted to Lutheranism in 1900.[11] Like Polanyi's wife, Magda Kemeny, Popper's wife, Anna Josefine Henniger, was Catholic.[12]

Karl Popper was the youngest of three children and the only boy. His father headed a law firm located adjacent to the large family apartment, which contained a library of some ten thousand books. Simon S. C. Popper was a scholar, social reformer, Freemason, and radical liberal, characterized by his son as a man "of the school of John Stuart Mill" whose philosophy Karl Popper later criticized. Friends of the family were Sigmund Freud's sister Rosa Graf and the psychiatrist Joseph Breuer. Among Karl's friends were the ethologist Konrad Lorenz and the pianist Rudolf Serkin.[13] Young Karl left home at sixteen, toward the end of the Great War, and tried out some classes at the University of Vienna. He ruminated on philosophical problems, concluding, he later said, that words and their meanings need not be taken seriously, but only questions of fact and assertions about facts.[14]

The end of the war brought events that appear pivotal for Popper's mature orientation. Sympathetic with the goals of socialism, Popper briefly turned to the Communists in 1919, until he witnessed the police shooting of unarmed young Socialists during a demonstration orchestrated by the Communists. Popper later said that he was horrified by Marxist demands that this kind of conflict should be intensified in order to bring about the historical destiny of the rule of the proletariat. He abandoned Marxism and the Communists for the Social Democrats.[15]

Postwar inflation wiped out his father's savings, so Popper had to find work while attending university lectures. He was in the audience at a university lecture by Albert Einstein whose explanation of relativity dazed him. Max Elstein, a mathematics student, helped Popper to better understand Einstein's general theory of relativity. "But what impressed me most," Popper later wrote, "was Einstein's clear statement that he would regard his theory as untenable if it should fail certain tests, such as the predicted redshift of spectral lines due to the gravitational potential." This insight, concluded Popper, is the essence of the real scientific attitude.[16] Popper took it upon himself to learn about standards of truth through studying mathematics and physics, attending university lectures during the 1920s given by Felix Ehrenhaft, Ernst Lecher, Hans Thirring, and Arthur Haas.[17] On the practical side of things, Popper apprenticed himself to a master cabinet maker from 1922 to 1924, then took up social work for a year in the psychologist Alfred Adler's child guidance clinics, and passed qualifying exams for teaching in primary school and for teaching mathematics, physics and chemistry in secondary school.[18]

Popper began taking classes in 1925 at Vienna's new Pedagogic Institute, and he defended his thesis, "Method in the Psychology of Thought," in 1928 before a jury that included Moritz Schlick. In 1924, Popper met Karl Polanyi at one of Karl's seminars on Socialism, and they had mutual friends in Vienna's circles of adult education and evening seminars, including a study group on cooperation among the sciences that was run by Rudolf Carnap.[19] Karl Polanyi introduced Popper to Heinrich Gomperz, who directed Popper's studies at the Pedagogic Institute, where Popper's interests were shifting from the psychology of discovery to the logic of discovery.[20]

In discussions with Karl Polanyi and with Gomperz, Popper found himself reconsidering his own views on the roles of nominalism and realism in the sciences—an old debate in which nominalists deny the existence of universal or real entities corresponding to particular named objects. Karl Polanyi told Popper to keep in mind that what Popper called methodological nominalism was characteristic of the natural sciences, but not of the social sciences. Gomperz told Popper that his views were closer to realism than to nominalism.[21] At Gomperz's suggestion, Popper began reading works by Ludwig Wittgenstein, Rudolf Carnap, Hans Reichenbach, and other contemporary philosophers, some of whom he began meeting, at Schlick's invitation, when he occasionally attended Schlick's Thursday evening private seminar, known as the Vienna Circle.

Schlick, whom Popper met at the Pedagogic Institute, had received his PhD in 1904 in Berlin under Max Planck and published a philosophical defense of relativity theory in 1917. He was appointed in 1922 to the chair in philosophy of the inductive sciences, at the University of Vienna, which previously had been held by Ernst Mach and Boltzmann. During the academic term, Schlick hosted the Vienna Circle at the Mathematical Institute on Boltzmann Strasse. The meetings were attended by the economist and philosopher Otto Neurath, the mathematician Hans Hahn, the Prague physicist Philipp Frank, the logician Rudolf Carnap, the philosopher Herbert Feigl, and the historian of science Edgar Zilsel. Karl Popper and Ludwig von Mises's younger brother, the physicist Richard von Mises, were among occasional visitors. Neurath, Frank, and Zilsel were committed Socialist activists who viewed scientific method as a tool for social change, while Schlick and Richard von Mises were politically disengaged liberals. What they all held in common were commitments to an empiricist and positivist definition of knowledge and to an emphasis on logical analysis. In November 1928, the group went public in founding the Verein Ernst Mach and publishing a programmatic pamphlet coauthored by the circle's more leftist members: Carnap, Hahn, and Neurath.[22]

Popper sometimes met, too, with a group that assembled in the apartments of Vienna Circle members Zilsel or Victor Kraft.[23] Kraft taught philosophy at the University of Vienna, while Zilsel, a Marxist, was teaching in the Volkshochschule. Zilsel completed a critical study of the "cult of genius" in 1926, using sociological explanation, but the work was rejected as a *Habilitationsschrift* by Vienna's philosophical faculty, despite support from Gomperz, Schlick, and Ernst Cassirer.[24] Reichenbach, too, made an argument in 1929 that the exaggerated status attributed to men of the past could not but cripple initiative in current science, an argument consistent with emphasis on the importance of properly defining the methodology of scientific induction rather than ascribing scientific achievement to genius.[25]

Popper conceived his ideas to be very different from the legacy of Ernst Mach and from the program of logical positivism that admitted only two

Figure 7.1. Karl Popper in Vienna in 1930. (Courtesy of Karl Popper Papers, envelope B, Hoover Institution Archives, Stanford University.)

kinds of general knowledge: the purely formal (logico-mathematical) and the factual (empirical). In contrast to Reichenbach, Zilsel, Kraft, Carnap, Neurath, Schlick, and others associated with the Vienna Circle and with the Ernst Mach Verein, Popper was skeptical about theories that identified meaning with verifiability, and therefore with induction. He did not like Carnap's *Der logische Aufbau der Welt* (The logical structure of the world; 1928), in which Carnap developed a rigorous formal version of empiricism with phenomenological definitions of scientific terms. Indeed, Carnap's approach left itself open to the charge of solipsism, or idealism, leading Hans Reichenbach to a preference for the term "logical empiricism" rather than "logical positivism" for the Vienna Circle collaboration.[26]

Popper also questioned the validity of the logical positivist argument for demarcation of science from metaphysics on the basis of verifiability and induction. In the tradition of the eighteenth-century skeptic David Hume, Popper regarded objections to inductivism as insuperable. Unlike most members of the Vienna Circle, he doubted that widely accepted scientific theories have forfeited their hypothetical character, and he worried about the efficacy of empiricism for distinguishing true scientific laws, such as the universal law of gravitation, from pretended scientific laws, such as historical determinism.[27] By the spring of 1932, Popper settled on falsifiability as a criterion for a true scientific law.[28]

Feigl, working with Gomperz, had completed his doctoral thesis in 1927 on probability and natural law. He encouraged Popper to map out his arguments about the problems of induction and falsification in chapters that became Popper's long manuscript *Die beiden Grundprobleme der Erkenntnistheorie* (The two fundamental problems of a theory of knowledge; 1932).[29] Karl Polanyi tried to help Popper publish the manuscript, enlisting the aid of his brother Michael in Berlin, but to no avail.[30] It is unlikely that Michael Polanyi actually saw Popper's full manuscript. In 1928, Schlick and Frank, who held Einstein's former chair of physics in Prague, began editing the book series *Schriften zur wissenschaftlichen Weltauffassung* (Papers on the scientific worldview), and they decided to publish a short version of Popper's manuscript under the title *Logik der Forschung*. Popper also published articles in the journal *Erkenntnis*, which Carnap and Reichenbach began editing in 1930.

In *Logik der Forschung*, Popper engaged directly with the arguments and texts of members of the Vienna Circle. He rejected at the outset of his manuscript what he called the "naturalistic" approach of studying the actual behavior of scientists in favor of a more critical and rational approach. This was Popper's "critical rationalism."[31] Popper made his case that falsifiability, not verifiability, is the proper definition of the scientific method, and it is the means of demarcation of science from nonscience. Popper wrote a long chapter on degrees of falsifiability in which he argued that the empirical content of a theory increases with its degree of falsifiability. He rejected pragmatic and aesthetic ideas of "simplicity" in favor of simplicity as the degree of falsifiability of a theory.[32]

Popper outlined a theory of probability as a frequency theory, along lines used by Feigl and some other members of the Vienna Circle, and criticized John Maynard Keynes's 1921 *Treatise on Probability*. Popper then tried to demonstrate that Werner Heisenberg's principle of indeterminism contradicts the "formulas" of quantum theory if the formulas are interpreted statistically. A concluding chapter on "corroboration" of hypotheses and theories shows the way to go about avoiding the concepts of "true" and "false" in philosophy of science, while conceding that the search for truth is the strongest motive for scientific discovery.[33] In response to these arguments, Reichenbach told Neurath that Popper's *Logik* was a naive, senseless book marred by mathematical mistakes. Neurath countered that Popper had his right of speech, and Carnap, despite some basic disagreements, applauded Popper's highly original criticisms, as reported to Feigl. Carnap, Neurath, and Carl Hempel all came to call Popper their "official opposition," although his loose affiliation with the Vienna Circle ironically resulted in the assumption outside the Circle that he was one of the logical positivists.[34]

The arguments and examples used in Popper's *Logik* are rooted in his reading in philosophy of science, the physical sciences, and mathematics. About the time *Logik* appeared in German, Popper completed the main outline of another book, *The Poverty of Historicism*, and summarized its arguments in discussions at a friend's house in Brussels and at Friedrich von Hayek's London School of Economics seminar in the winter of 1936.[35] By this time, Popper was hoping for a job in England. Following the demise of republican governments in both Germany and Austria in the spring of 1933, many Vienna Circle intellectuals were seeking refuge and new positions elsewhere. Feigl already had left for Iowa in 1931; Reichenbach departed Berlin in 1933 for Istanbul and then Los Angeles; Karl Polanyi went to London; Carnap left Prague in 1935 for Chicago; Neurath delayed until 1938 and then fled to Holland, which was the same year that Frank left Prague for Harvard University. Zilsel went to London and then to New York. Schlick, who continued lecturing in Vienna, was murdered in 1936.[36]

Canterbury University College at Christchurch in New Zealand offered Popper a refuge in spring 1936, two years before German troops marched into Austria in March 1938. He completed the *Poverty of Historicism* in his new home. It first appeared in three parts in *Economica* during 1944–1945 and then as a book in 1957. *The Open Society and Its Enemies* (1945), written in New Zealand, was the other arm of what Popper called his "war effort" after he was denied enlistment in the New Zealand armed forces.[37] *Open Society* contained Popper's critiques of Plato, Hegel, and Marx that were earlier adumbrated in the *Poverty* manuscript. It was these political books that made Popper's critique of scientific positivism and inductivism first widely known, supplemented later by the English translation of The *Logic of Scientific Discovery* in 1959 and by Popper's collection of essays and lectures titled *Conjectures and Refutations: The Growth of Scientific Knowledge* published in 1962.[38]

Popper's main target in these political writings was the doctrine of historical determinism, which he characterized as sheer superstition. While he thought it was legitimate to interpret history in metaphors or themes of the struggle of classes, races, or ideas, he rejected the notion that one can arrive at historical laws based in observations of historical facts and the discovery of historical rhythms or patterns.[39] In order to argue that history is not a science, Popper had to describe what science is and what scientists do, and in this he used the same arguments that appear in the German manuscripts of *Die beiden Grundprobleme* and *Logik*.[40]

Science is not a method of beginning with observations, wrote Popper, and so there cannot be historical laws based in induction. Descartes was partly right in his description of scientific knowledge as deductive and systematic in nature, but Descartes was wrong in taking the foundations of scientific knowledge to be secure and self-evident initial premises. Rather, argued Popper, the starting point of scientific thinking is conjecture or hypothesis, and it must be refutable in principle, unlike Descartes's fundamental axioms. Contrary to the arguments of philosophical conventionalists, whom Popper identified as Pierre Duhem and Henri Poincaré, crucial experiments can refute theories, although they cannot verify theories. Progress in science depends on free competition among rival conceptions of nature and thus on political freedom that guarantees such competition. All this was argued in *The Poverty of Historicism*, written around 1935.[41]

Falsifiability as a criterion of scientific method and as a criterion of demarcation of science from pseudoscience appears clearly in the *Poverty of Historicism*:

> We make progress if, and only if, we are prepared to *learn from our mistakes*: to recognize our errors and to utilize them critically instead of persevering in them dogmatically. Though this analysis may sound trivial, it describes, I believe, the method of all empirical sciences. This method assumes a more and more scientific character the more freely and consciously we are prepared to risk a trial, and the more critically we watch for the mistakes we always make . . . All theories are trials; they are tentative hypotheses . . . and all experimental corroboration is simply the result of tests undertaken in a critical spirit, in an attempt to find out where our theories err.[42]

Popper's main political target in the mid-1930s, when he wrote *Poverty of Historicism*, was fascist totalitarianism and the claims of historical destiny in Mussolini's Italy and Hitler's Germany. In *Poverty*, as in *Open Society*, Popper criticized by implication the historical claims of Soviet Communism, but his critique of Marx was considerably more muted than his critique of Plato's politics and epistemology. As noted by both Malachai Hacohen and Michelle-Irène Brudny in their biographies of Popper, he withheld judgment on Soviet Russia, as did most Democrats and Socialists, and he praised Marxist intellectuals for wanting to solve urgent social problems.[43] In his autobiography Popper wrote that he demurred from publishing anything anti-Marxist until 1935 because anti-Marxism was so closely associated in the 1920s and 1930s with Fascism. Popper's anti-Fascism in the 1930s outweighed his skepticism about the Soviet regime.[44]

Popper was less patient in *Poverty of Historicism* with his contemporary Karl Mannheim than with the historical Marx, and Popper especially singled out for criticism Mannheim's enthusiasm for central planning and social engineering in the 1935 book *Mensch und Gesellschaft im Zeitalter der Umbaus* (Man and society in an age of reconstruction). Mannheim, who had taught in Frankfurt until 1933, now was at the London School of Economics. Popper identified Mannheim's utopian vision with a closed society that is fundamentally antithetical to an open society based in the free dialectic of making mistakes and correcting them.[45] Equally erroneous in Popper's view was Mannheim's sociology of knowledge, and, in his criticism, Popper struck a political theme, which is not at all emphasized in *Logik*, on the social aspects of scientific method:

> [Mannheim's] naïve view that scientific objectivity rests on the mental or psychological attitude of the individual scientist, on his training, care, and scientific detachment, generate as a reaction the skeptical view that scientists can never be objective . . . This doctrine, developed in detail by the so-called *sociology of knowledge* . . . entirely overlooks the social or institutional character of scientific knowledge . . . if we had to depend on [the scientist's] detachment, science . . . would be quite impossible. *What the "sociology of knowledge" overlooks is just the sociology of knowledge*—the social or public character of science. It overlooks the fact that it is the public character of science and of its institutions which imposes a mental discipline upon the individual scientist, and which preserves the objectivity of science and its tradition of critically discussing new ideas.[46]

This was a theme repeated in a short chapter on the sociology of knowledge in *Open Society*.[47]

This theme sounds remarkably like Michael Polanyi's writings of the early 1940s, and it has led Ian Jarvie to argue that Popper anticipated the later social turn in philosophy of science.[48] The social aspect of science was not Popper's legacy in philosophy of science, however, because of his stress instead on the critical rationalism of skepticism and falsifiability. Popper refers in *Logik* to the necessary intersubjectivity of scientific testing but clarifies that he is not leaving room for belief or commitment. Rather, for Popper, "subjective experience, or a feeling of conviction, can never justify a scientific statement, and . . . within science it can play no part."[49]

Just after the *Open Society* appeared, Popper was appointed in 1946 to be reader at the London School of Economics, where he became professor of logic and scientific method in 1949. He owed this good luck in part to Friedrich von Hayek, who would leave the London School of Economics in 1950 for a position at the University of Chicago. Hayek's 1944 book *Road to Serfdom* was a topic of heated discussion in the mid-1940s as a result of its controversial argument that the inevitable consequence of any degree of socialist planning is a gradual death spiral into totalitarianism. Explicitly criticizing British Left intellectuals who favored socialist and welfare ideas, Hayek charged that the British Left intelligentsia seemed no longer to recognize "any good in the characteristic English institutions and traditions." "A foreign background," wrote the Viennese émigré, "is sometimes helpful in seeing more clearly to what circumstances the peculiar excellencies of the moral atmosphere of a nation are due."[50]

After reading Popper's *Open Society and Its Enemies*, Carnap wrote Popper from Chicago: "I was somewhat surprised to see your acknowledgment of von Hayek. I have not read his book myself; it is much read and discussed in this country, but praised mostly by the protagonists of free enterprise and unrestricted capitalism while all leftists regard him as reactionary."[51] This was not to be Popper's only acknowledgment of Hayek. They had met for the first time during Popper's visit to England during 1935–1936 when Popper presented an early version of *The Poverty of Historicism* at Hayek's seminar at the London School of Economics. Hayek helped Popper publish *The Open Society and Its Enemies* by talking with the editors at Routledge, who had published *The Road to Serfdom*, after Cambridge University Press turned down Popper's manuscript. In 1962 Popper dedicated *Conjectures and Refutations* to Hayek.[52]

Once settled in London, Popper's seminar at the London School of

Economics became well-known for its court of admirers, and also for Popper's authoritarian and egocentric demeanor, which many people found insufferable. Paul Feyerabend, who came from Vienna in 1952 to spend a year with Popper, recalled that his "lecture started with a line that became widely known: 'I am a Professor of Scientific Method—but I have a problem: there is no scientific method.'" By the time Feyerabend returned to England to teach at the University of Bristol from 1955 to 1957, he found that Popperian orthodoxy was required of seminar participants. In his autobiography, Feyerabend recalled dinners in London in 1957 with Popper's protégé John Watkins. Watkins "would welcome me at the door, take me up to his study, and invite me to take a seat. Walking up and down with a stern face he would chastise me for having been a bad Popperian: too little Popper in the text of my papers, no Popper in the footnotes."[53]

Michael Polanyi presented a paper in London in 1952 at a meeting of the Philosophy of Science Group chaired by Popper. (The group was the predecessor of the British Society for the Philosophy of Science.) In the paper "The Stability of Beliefs," Polanyi drew upon studies by Edward Evans-Pritchard and compared belief-systems of the African Azande people to modern belief systems, such as Marxism and science. As in his 1951 article "Scientific Beliefs," Polanyi emphasized the difficulty of breaking out of any belief system in which one is educated, including what he called the belief system of our "apprenticeship to science." Further, Polanyi dismissed what he called the "positivistic" story that a scientist immediately drops a hypothesis or theory the moment that it conflicts with experience, using the counterexamples of Bohr's theory of spectra and the periodic system of the elements which were not abandoned when confronted with anomalous data.[54] According to John Watkins, Polanyi emerged from the meeting gravely offended at the way that Popper had chaired the session and furious at Popper's treatment of him.[55] Edward Shils later wrote of Popper and Polanyi's relationship, "As far as I can recall, whenever I mentioned the one to the other, I was answered in silence."[56]

In 1950, Popper gave the William James Lectures at Harvard University, where he met George Sarton, James Bryant Conant, I. Bernard Cohen, Gerald Holton, and Thomas Kuhn.[57] Kuhn found Popper's insistence "too positivist" that later scientific theories embrace earlier theories.[58] Popper and Kuhn may have seen each other again when Popper spent the 1956–1957 academic year at the Stanford Center for Advanced Studies in the Behavioral Sciences, just at the time when Kuhn started teaching at the University of California at Berkeley. They met at Kuhn's home in Berkeley in 1962 and at a 1965 conference at Bedford College in London which was organized as a response to Kuhn's publication of *The Structure of Scientific Revolutions*.[59] Popper did not care for Kuhn's views, saying, in particular, of Kuhn's notion of normal science that, if accurate, it was a recent phenomenon and "a danger to science."[60] We turn now to Kuhn and to the political foundations of Kuhn's philosophy of science.

## Thomas Kuhn and the Goals of General Education in a Democracy

Thomas Kuhn's father, a consulting engineer, had studied at Harvard University, and his son entered Harvard in the fall of 1940 with the aim of majoring in physics. Young Kuhn graduated in three years and immediately went to work in radar for the war effort, joining John Van Vleck in the spring of 1943 at the Radio Research Laboratory on the Harvard campus. During 1944 and 1945 Kuhn was assigned to Great Malvern in Worcester and then to the 9th Air Force in Rheims. He was in Paris on August 25, 1944, Liberation Day.[61]

Had the war not ended in August 1945, Kuhn would have been assigned to a bomber base in the Pacific. Instead, in September he entered the physics graduate program at Harvard, working with Van Vleck in solid-state theoretical physics and completing a PhD thesis in 1949 on electronic binding energies of metallic solids.[62] In the meantime, Kuhn took philosophy courses. At a cocktail party, a young woman asked him what he did, and, as John Heilbron relates, the party chanced to fall silent, so that everyone heard Kuhn's answer: "I just want to know what truth is."[63] He was not sure, however, that physics was his route to the truth. He was finding group theory and solid-state physics dull in comparison with the work of contemporaries who had been at Los Alamos. By his own account, he was not convinced that he could shine in his current field.[64]

As an undergraduate writer and editor for the Harvard student newspaper, the *Harvard Crimson*, Kuhn had voiced strong support for the policies of Harvard's president, the organic chemist and wartime science defense administrator James Bryant Conant. In particular, Kuhn applauded Conant's calls for using American higher education as a tool in total war, and he found appealing Conant's argument for the importance in a democracy of public understanding of science.[65] In June 1945, the Harvard Committee on the Objectives of a General Education in a Free Society, established by Conant in 1943, published its report, *General Education in a Free Society*. The goal was to set up a program of undergraduate education that would give students an understanding of the texts and achievements of western civilization. The "red book" promoted not only literature, history, and philosophy, but the role of science in a free and democratic society as the foundation of "spiritual values" for citizens who could develop the "scientific attitude": the ability to form "objective, disinterested judgments based upon exact evidence."[66] When objections surfaced at a Harvard faculty meeting, Conant disarmed critics by announcing that he would teach one of the courses, as did Edwin C. Kemble, the molecular spectroscopist who was chairman of the Physics Department.[67]

In 1947 Conant followed up the red book with his own book, *On Understanding Science: An Historical Approach,* based on materials in his course Natural Sciences 4.[68] In the course and in the widely read book, Conant argued for the importance in a democracy of a lay understanding of science, and he suggested the positive role of historical case studies in showing "in the clearest light the necessary fumblings of . . . intellectual giants," the false starts and blind alleys in scientific research, and the inefficacy of glib formulas for scientific method.[69]

Four different courses were initially set up to incorporate historical elements into teaching the physical sciences to undergraduates. Conant taught Natural Sciences 4 for three years in the late 1940s, collaborating with younger colleagues, all of whom completed their PhDs at Harvard in the 1940s. They included the physical chemist Leonard K. Nash, whose research field was thermodynamics; George Sarton's student I. Bernard Cohen, whose dissertation focused on Benjamin Franklin's experiments; and Kuhn.[70] Conant invited Kuhn to assist in Natural Sciences 4 in the spring of 1947, and Kuhn saw this as an opportunity to do something interesting and important while finishing his physics dissertation.[71]

Nash had acquired a reputation as a gifted teacher in Harvard's Chemistry Department, and he had a strong interest in incorporating the history of chemistry into both his teaching of chemistry and into general education. Nash and Conant, along with another new graduate student in history of science, Duane H. D. Roller, and Roller's father, the physicist Duane Roller, all cooperated in writing the *Harvard Case Histories in Experimental Sciences* that were published as pamphlets in the early 1950s and then as a two-volume set by Harvard University Press in 1957. Nash's book *The Nature of Natural Sciences* appeared in 1963.[72]

Similarly to Nash's historical project for chemistry, Kemble taught a physics and history–based general education course titled Natural Sciences 2, enlisting the assistance of Percy Bridgman's student Gerald Holton, whose research focused on high-pressure physics. Holton incorporated the history of physics into his 1952 textbook *Introduction to Concepts and Theories in Physical Sciences,* which reappeared in new iterations afterward. Cohen developed a mimeographed text, which was published as *The Nature and Growth of the Physical Sciences* (1954) for Natural Sciences 3. In 1960, Cohen's very short and very popular book *Birth of a New Physics* appeared.[73]

Cohen and Nash were among Harvard faculty members who participated and contributed to a workshop on science in general education held in July 1950 at the Harvard Summer School. So was former Vienna Circle member Philipp Frank, who now taught physics and philosophy of physics in Harvard's Physics Department. Frank's biography of Albert Einstein appeared in 1947, followed in 1949 by *Modern Science and Its Philosophy,* based on his philosophy of physics course.[74] Frank became president in 1949 of the newly established Institute for Unity of Science, which carried on the agendas of the original Vienna Circle in association with the American Academy of Arts and Sciences in Boston. He emphasized in these years that historical and sociological studies were badly needed in order to show how political and religious ideologies can misappropriate the meanings of scientific theories to their own ends, as well as in order to investigate how social and cultural values may affect scientists' decisions when evidence underdetermines a logical choice between competing theories.[75]

With Conant's support, Kuhn received a three-year fellowship from 1948 to 1951 in Harvard's Society of Fellows, where he could read and study whatever struck his fancy without embarking on a formal curriculum in history of science.[76] As Kuhn studied texts for a unit in the natural sciences course on the history of mechanics from Aristotle to Galileo, he experienced the joy of discovery. "To my complete surprise, that exposure to out-of-date scientific theory and practice radically undermined some of my basic conceptions about the nature of science and the reasons for its special success."[77] For Kuhn, who also was reading psychology, it was a Gestalt switch in his conceptual framework.[78] Kuhn decided to follow the history of science as a profession.

At Conant's request, Kuhn and Nash agreed to take full responsibility for Conant's course together during the 1950–1951 academic year. They renamed it *Research Patterns in Physical Science,* a title that reflected recent literature on the role in the sciences of recurrent ideals or patterns. Frank, for example, taught that analogies or patterns, such as the notion of a living organism or a simple mechanism, guide scientific thinking for a time until a change or revolution occurs. The British historian Robin G. Collingwood's 1945 book *The Idea of Nature* had used this trope of explanation for distinguishing ancient (organismic) and early modern (mechanistic) ideas of nature from current evolutionary patterns of explanation.[79]

Kuhn missed the July 1950 Harvard workshop on general science education because he was in Europe. Before the fall term began, Kuhn and his wife, Kay, spent four months in France and England, where he visited Alexandre Koyré in Paris, whom he had previously met at Harvard and whose work he had been reading. In London, Kuhn made the acquaintance of historians and philosophers of science who included Alistair Crombie, Mary Hesse, Herbert Butterfield, and Stephen Toulmin.[80] His wide reading as a Junior Fellow prepared him for these meetings, with his range extending from the histories of early science by Hélène Metzger and Anneliese Maier to the psychological theories of Jean Piaget, the language theories of B. L. Whorf, and the philosophical logic of Willard Van Orman Quine, who had been one of Harvard's first Junior Fellows in the mid-1930s and taught in the Philosophy Department.

A reference in Reichenbach's *Experience and Prediction* led Kuhn around 1950 to Ludwik Fleck's then obscure monograph *Entstehung und Entwicklung einer wissenschaftlichen Tatsache* (Genesis and development of a scientific fact; 1935), a work that, as Kuhn said in his preface to the *Structure of Scientific Revolutions*, made him realize that ideas that he had developed before reading Fleck should be set within the sociology of the scientific community. What attracted Kuhn to the book originally was the title's implication that a fact does not simply exist but is created or developed, a view with which Kuhn agreed, although he later said that he did not thoroughly read the book because of the difficulty of Fleck's German and the unfamiliarity of concepts from medicine and biochemistry.[81]

Like Michael Polanyi, Fleck was educated in medicine and, unlike Polanyi, his scientific career centered on medical research, specifically bacteriology, which he pursued in a private laboratory in L'viv until the German occupation of Poland in 1941. He returned to research in Lublin and Warsaw after surviving deportation to Auschwitz and Buchenwald during the war. Fleck emigrated to Israel in 1961.[82] In his 1935 book, Fleck developed the notion of the "thought-collective" and "thought-style" (*Denkkollektiv*, *Denkstil*) while referring to the work of the French sociologist Émile Durkheim and the Viennese philosopher Wilhelm Jerusalem. As Mary Douglas has noted, Fleck's "thought collective" is similar to Durkheim's "social group," and the "thought style" is similar to Durkheim's "collective representations." Durkheim was talking, however, only about primitive cultures and their religious and magical beliefs, whereas Fleck extended the analysis to modern science, using as an example the history of syphilis and the discovery of the Wasserman test for syphilis.[83]

Every field of knowledge, including natural science, Fleck wrote, requires a period of apprenticeship in which the thought-commune is bonded together through statutory and customary arrangements that result in limitations on what problems are deemed significant and what kinds of answers are possible. The thought-style of science rests in a distinct tradition, and all empirical discovery can be construed as a supplement, development, or transformation of this traditional thought style.[84] The thought-collective does not rule out the role of the individual, however. Fleck suggested in a striking analogy: "If the individual may be compared to a soccer player and the thought collective to the soccer team trained for cooperation, then cognition would be the progress of the game."[85]

It should be noted, too, that Fleck laid out some of these ideas around 1930 in articles in the "General Questions and Philosophy" (*Allgemeines und Philosophie*) section of the widely read weekly periodical *Die Naturwissenschaften*. Responding to Kurt Riezler's essay about the crisis of reality ("Die Krise der Wirklichkeit") as a philosophical result of relativity theory and quantum mechanics, Fleck described the social character of intellectual activity and praised the creation in modern physics of a new thought style and a new reality. The mathematical physicist Gustav Mie singled out this article by Fleck for criticism in his book *Naturwissenschaft und Theologie*, and he explicitly rejected Fleck's view that reality is dependent on time, place, culture, and individual personalities. Under the guise of a review of Mie's book in *Die Naturwissenschaften*, Fleck replied that his own views had no implications for theology, but that he sought to reshape "the concepts of the objective and the subjective as well as of perception and being."[86]

Thus, while Fleck's views may have been relatively uninfluential in the 1930s, they were part of an ongoing debate in Austrian and German circles about rapid changes in the physical sciences and a perceived crisis in the security of scientific knowledge. For Kuhn in 1950, as he later reminisced, Fleck reinforced his own thinking about changes in the various Gestalts "in which nature presented itself" and the difficulties of making "fact" independent of "point of view." As for the thought-collective, Kuhn later said that he was troubled in Fleck's book by the thought-collective's seeming to function as an individual mind writ large simply because many people possess it. His reading of Fleck, however, confirmed Kuhn's commitment to his own developing point of view.[87]

In laying the groundwork for history of science in the new general education curriculum at Harvard, Conant had been using some of Michael Polanyi's work. Kuhn later recalled, "I liked it quite a lot—I don't remember just what it was, except that I kept feeling terrible at those points where [Polanyi] sort of spoke as though extrasensory perception was the source of what the scientists did."[88] Michael Polanyi's little book *Science, Faith and Society* had been published in 1946, and it is not surprising that Conant would have noticed it, since Polanyi was a fellow colleague in chemistry. As noted in chapter 6, Polanyi also was writing essays for the Society for Freedom in Science, and his pamphlets "Rights and Duties of Science," "The Planning of Science," and "The Foundations of Academic Freedom" can be found in Harvard's Widener Library. They echo the themes of science and democracy to which Conant was committed. Similarly, too, Conant expressed his concern in his 1951 book, *Science and Common Sense*, that the pendulum was swinging too far in the postwar era toward organized programmatic research directed toward solving practical problems in an expanding field of government laboratories and military research.[89] Kuhn may have read the pamphlets; in any case, he recalls in a letter to William Poteat that he read *Science, Faith and Society* and the collection of essays *The Logic of Liberty*.[90]

In the spring of 1951, while Nash taught their Harvard course, Kuhn gave a series of eight lectures at Boston's Lowell Institute. In his opening lecture, he announced that his lectures would show that the empiricist view of the "ideal scientist [as one who] abandons all prejudice so that he may proceed first to a dispassionate analysis of all the facts and then to the formulation of the immutable laws which govern them . . . is altogether wrong." As outlined by historian Karl Hufbauer, Kuhn argued that firm preconceptions are woven into patterns of scientific research, and that this is a good thing. He denied the possibility of dispassionate neutrality in the working scientist, but not the objective validity of scientific results. By the fifth Lowell lecture, Kuhn tackled the changing roles of what he called "orientations" or "points of view" in the evolution of a scientific field, and in a subsequent Lowell lecture he explored the similarities between science and the perceptual or behavioral worlds studied by psychologists. According to Hufbauer, Kuhn used the word "paradigm" in these lectures. Indeed the word was entering general philosophical debate in Great Britain and the United States as a consequence of Ludwig Wittenstein's work and, in the natural sciences, the book *On Understanding Physics* (1938) by Wittenstein's student William H. Watson.[91]

In spring of 1952, Kuhn accepted a five-year appointment as assistant professor in general education and history of science at Harvard. Planning ahead for a pre-tenure leave from teaching, Kuhn applied in 1953 for a

Guggenheim Fellowship with a proposal for two books, one on the Copernican revolution and another on the structure of scientific revolutions. The Copernican revolution proposal grew out of a segment in the course with Nash, and it expressed Kuhn's growing preoccupation with the problem of scientific revolutions as shifts in intellectual systems. This was the theme of Alexandre Koyré in his *Etudes galiléennes* (1939), which examined Galileo's achievements through the lens of the history of ideas and history of philosophy. Koyré emphasized revolutionary breaks in scientific thought and the impact on Galileo's scientific work of Plato's philosophical ideas, while denying the role in Galileo's mechanics of any active experimentation. In general, Koyré ignored the role of practical or technical dimensions in scientists' work, associating any historical emphasis on technique or applied science with Marxism and the political left. The American historian of chemistry Henry Guerlac recalled that Koyré once objected to Guerlac's writing about Lavoisier's interest in practical matters, such as street lighting, by cautioning Guerlac that this point was interesting, but "un peu Marxiste."[92]

In his Guggenheim proposal, Kuhn outlined his intention to argue in a second book, after he completed *The Copernican Revolution*, that "established scientific theories [function] as *ideologies* which . . . serve the practitioner of sciences both as directives for future work and as restraints upon the creative imagination."[93] He had a contract for this second book with the *International Encyclopedia of Unified Science*, the series envisioned in the 1930s by Otto Neurath as the Vienna Circle's new counterpart to the eighteenth-century *Encyclopédie*. The series was being published by the University of Chicago Press under the active editorship of Charles Morris and Rudolf Carnap at the University of Chicago and Philipp Frank at Harvard. Bernard Cohen had recommended Kuhn to Morris after deciding himself not to write the history of science volume for the series.[94]

While Kuhn worked on *The Copernican Revolution*, which would be published by Harvard University Press in 1957, he was pessimistic about prospects for tenure at Harvard. There was no History of Science Department—only a program—and some colleagues regarded his Copernican revolution book as too popular in its approach and analysis.[95] As a consequence, he accepted a joint position in the departments of philosophy and history at the University of California at Berkeley in the fall of 1956.[96] The young historian of American science A. Hunter Dupree joined the History Department at the same time as Kuhn, in order to help establish a history of science program, and Stanley Cavell, who had been in Harvard's Society of Fellows in the mid-1950s, joined the Philosophy Department. It was Cavell who introduced Kuhn to Wittgenstein's notion of language games.[97] After Paul Feyerabend moved to Berkeley from Bristol in 1958, he, too, became an intellectual sparring mate for Kuhn as they engaged in lively arguments in the Café Old Europe on Telegraph Avenue.[98]

Kuhn spent the 1958–1959 academic year near Berkeley at the Center for the History of the Behavioral Sciences at Stanford, on the recommendations of Robert Merton and Edward Shils. Quine was another of the resident Fellows that year. At Stanford, Kuhn was struck by the inability of the center's resident social scientists to agree on the fundamental problems and practices of their discipline. It was here that Kuhn worked out a fuller concept of the paradigm, a term that also appeared in Norwood Russell Hanson's *Patterns of Discovery: An Inquiry into the Conceptual Foundations of Science* in 1958 and in Stephen Toulmin's work, for example, the 1961 *Foresight and Understanding*, with the meaning of an ideal or pattern of natural order.[99]

Early in Kuhn's academic year at the center, in November 1958, Polanyi visited Palo Alto, and Kuhn heard him give a talk on tacit knowledge. Kuhn later reminisced, "I liked the lecture all right, and it's possible that it helped me to get to the idea of paradigm, although I'm not sure."[100] Polanyi's new book, *Personal Knowledge*, had just appeared in June. Kuhn looked at it, he later said, but decided at the time that "I *must* not read this book now. I would have to go back to first principles and start over again, and I wasn't going to do that." Similarly, he decided not to read Toulmin's *Foresight and Understanding* when it appeared. In this book, Toulmin discussed the role in science of what he called explanatory paradigms and offered a naturalistic analogy from evolutionary biology for scientific development.[101]

Later, sometime before the publication of *The Structure of Scientific Revolutions*, Kuhn did try to read Polanyi's *Personal Knowledge*, but he did not like it, although he mentioned Polanyi in a footnote for "brilliantly" developing a theme congruent with Kuhn's view that much of the scientist's success depends on tacit knowledge that is acquired through practice rather than through articulated rules.[102] More to Kuhn's liking was Hanson's *Patterns of Discovery* with its anti-positivistic argument about the role of thematic interpretations in observations. Hanson had left the University of Cambridge and established the Department of History and Philosophy of Science at Indiana University in 1958. In the late 1950s, Gerald Holton also was working out a notion of thematic interpretations in science and lecturing and writing on the subject by 1962.[103]

Some of Kuhn's writing in Palo Alto included revisions of an essay on the role of measurement that he first had given in 1956 as a talk to the

Social Sciences Colloquium in Berkeley. He also wrote a paper on tradition and innovation for a conference in Salt Lake City.[104] In the 1956 paper Kuhn argued that "the normal practice of science" is a "mopping-up operation" and that "finer and finer investigations of the quantitative match between theory and observation cannot be described as attempts at discovery or at confirmation." Far, from it, "failure to solve one of these puzzles counts only against the scientist" and not against the theory.[105]

In the Salt Lake City paper of 1959, titled "The Essential Tension: Tradition and Innovation in Scientific Research," he argued that scientific education is a "dogmatic initiation in a pre-established tradition" and an "apprenticeship" that inculcates "commitment" to the tradition. Taking up the theme of normal science, Kuhn suggested that almost none of the research by even the greatest scientists is meant to be revolutionary, and thus great discoveries are the product of scientists who "simultaneously display the characteristics of the traditionalist and of the iconoclast."[106] These arguments in the "Essential Tension" paper, as in the *Structure* book, later struck Polanyi as strongly similar to his own views not just in *Personal Knowledge* but in his earlier books *Science, Faith and Society* and *The Logic of Liberty*. Polanyi later complained to some colleagues that he had not received proper recognition from Kuhn.[107]

By the late spring of 1961 Kuhn had finished a full draft of the *Structure* manuscript. He mimeographed copies and gave them to colleagues for comments and criticisms. He sent a copy to Conant with a letter asking for criticism and for an endorsement, and he requested comments from the theoretical physicist H. Pierre Noyes at the Lawrence Radiation Laboratory, Ernest Nagel at Columbia University, Paul Feyerabend at Berkeley, and his student John L. Heilbron.[108] Conant was not pleased with all that he read, commenting in letters on Kuhn's "use (and abuse)" of the term "paradigm" as a "magic verbal word to explain everything!" and "a word you seem to have fallen in love with!" Kuhn's notion of every significant scientific advance as a "new world view" struck Conant as "far too grandiose a characterization of most of the revolutions you cite as examples," and Conant accused Kuhn of "taking refuge in the word 'paradigm'" in order to avoid the dilemmas created by his vague treatment of the scientific community as a "community with a single point of view."[109]

In the meantime, Kuhn attended a symposium on scientific change at the University of Oxford in July of 1961, shortly before *The Structure of Scientific Revolutions* was to appear. This symposium was organized as a major event by the Division of the History of Science in the International Union for the History and Philosophy of Science. Polanyi, who had been a Senior Research Fellow at Merton College in Oxford since 1959, was invited to comment on Kuhn's paper, along with A. Rupert Hall of the University of Cambridge. Kuhn summarized his ideas on paradigms, normal science, and scientific revolutions in his paper "The Function of Dogma in Scientific Research." Polanyi followed up by saying, "The paper by Mr. Thomas Kuhn may arouse opposition from various quarters, but not from me," since, Polanyi said, he had been trying for years to call attention to scientists' deep commitment to established beliefs. A commitment to a paradigm, Polanyi continued, "has a function hardly distinguishable from that which I have ascribed to a heuristic vision, to a scientific belief, or a scientific conviction."[110]

Kuhn explicitly referred in his Oxford paper to Polanyi's *Personal Knowledge* and *Logic of Liberty* as arguments for the "importance of quasi-dogmatic commitments as a requisite for productive scientific research." In participants' discussion following Kuhn's paper, which included comments by Bentley Glass, Stephen Toulmin and E. F. Caldin, Kuhn firmly distinguished his point of view from at least one aspect of Polanyi's philosophy of science. Just as Kuhn later differentiated his notions of paradigm and of normal science from Fleck's "thought-collective," Kuhn cautioned that Polanyi, too, put too much emphasis on the individual scientist and his individual responsibility. In contrast, in Kuhn's view, "It is not, after all, the individual who decides whether his discoveries of theoretical inventions shall become part of the body of established science. Rather it is his professional community . . . And on what ground should we trust *their* judgment?"[111]

## Michael Polanyi and Scientific Tradition

Earlier chapters have focused on Michael Polanyi's migrations among scientific centers, his scientific career, his classically liberal views in economics and politics, and his writings against Socialism, central planning, and the social responsibility of scientists. His invitation to give the Riddell Lectures, which were sponsored to address spiritual and moral issues, followed his lecture at Leeds in January 1941, in which he argued that a better understanding of the nature and limits of scientific method might "re-vindicate the ideal of truth in all aspects, which jointly constitute the heritage of Christian civilization . . . Science forms an integral part of the way of life which the citizens of the West are pledged to defend and carry on."[112]

The Riddell Lectures, delivered in March 1946 and published later that year as *Science, Faith and Society*, began with the question of "what is the nature of science" and then moved to the question that had perplexed members of the Vienna Circle and Karl Popper. What are the legitimate grounds for demarcating magical or metaphysical explanations from naturalistic interpretations of events? Polanyi found empiricism and observation too weak a reed upon which to base the demarcation. The method of skepticism, too, was inadequate as a demarcation, he concluded, and he reiterated the argument of his 1944 article on John Dalton in the *Manchester Guardian* that, while modern science may have taken its origin in a spirit of skepticism, this doubt was of a very modest kind.[113]

Polanyi told his audience that the part played in the process of scientific discovery by new observations and experiments in the process of discovery is usually overestimated. Another popular misconception, he continued, is the notion that a "working hypothesis" is immediately abandoned in the face of conflicting observational evidence. In contrast, he explained, scientists spend most of their time in fruitless efforts to find support for a strongly felt intuition, taking "beating after beating for months on end" followed by fresh outbursts of confidence in a scientific hypothesis deemed too good to relinquish.

Even in the case of well-established theories, he argued, data in conflict with theory usually are ignored:

> In my laboratory I find the laws of nature formally contradicted at every hour, but I explain this away by the assumption of experimental error. I know that this may cause me one day to explain away a fundamentally new phenomenon and to miss a great discovery . . . Yet . . . if every anomaly observed in my laboratory were taken at its face value, research would instantly degenerate into a wild-goose chase after imaginary fundamental novelties.[114]

Science is characterized, he reiterated from earlier essays, by "schools of research" in each of which there is a master-pupil relationship within a wider institutional and material domain, consisting of periodicals and textbooks, research grants and salaries, and facilities for teaching and research. Decisions on research worthy of publication and on employing scientists in their posts are reached through scientists' valuations of each other's work. The system is coherent and self-governing, with no central authority. The pupil learns from the master in the way that a child learns a language or a student learns how to play the piano. The master chooses the problems, selects a technique, reacts to clues and difficulties, and keeps speculating all the time. It is a system of apprenticeship rooted in tacit knowledge that often cannot be articulated and constitutes a tradition passed from mentor to apprentice.[115]

No revolutionary scientist can succeed without appealing to this tradition and authority in which a few fundamental standards are used to judge scientific results. In *Science, Faith and Society* Polanyi called them validity, profundity, and intrinsic interest.[116] By the time he wrote "The Republic of Science" in 1962, the standards were rechristened plausibility, scientific value, and originality. A contribution in science must be plausible within the canons of current theory or else, as he said in 1946, scientific journals would be beset by cranks, frauds, and bunglers.[117] A scientific contribution is judged for its value in terms of accuracy, its systematic importance, and the intrinsic interest of its subject matter. Finally, even with plausibility and value attributed to a reported result, merit also is determined by originality, or the degree of surprise which the result or the theory occasions among scientists.[118] Plausibility and value tend to enforce conformity within the scientific community, while originality encourages dissent. This is the "internal tension" always operating within science.[119] The process of choosing one system over another, where competing claims are based on different premises, is a process of "conversion"—a loaded term most often used in discussing religion—influenced by intuition and "finally conscience." The outcome has the sanction of the scientific community: "Scientists recognize that, inasmuch as each scientist is following the ideals of science according to his own conscience, the resultant decisions of scientific opinion are rightful."[120]

Thus, science has a social nature, in which well-established values and rules are adjudicated. But, what then, Polanyi asks, is the claim of this science on truth? The scientist's search for the solution of problems, he writes in the lectures that became *Science, Faith and Society*, is analogous to the "prayerful search for God . . . the urge to make contact with a reality, which is felt to be there already . . . waiting to be apprehended."[121] The "novice" scientist gives himself to the service of an ideal. To accord validity to science is to express a "faith" that is upheld within the community. For Polanyi, this faith in science is the wellspring of liberal Western society.

There are significant echoes in *Science, Faith and Society* from some of Polanyi's earlier readings from the Berlin period. In particular, in his diary of 1928, Polanyi took notes on Julien Benda's new book *Le Trahison des Clercs* (*The Treason of the Intellectuals*). Benda, like Polanyi an assimilated Jew, condemned the racial, class, and national passions sweeping Europe, including Zionism.[122] Also like Polanyi, Benda lamented the preeminence of what he called the "religion of the material" and the romanticized positivism that makes a cult of facts and claims to have discovered historical laws according to which political movements are merely carrying out the spirit of history.[123] So far, this might well have been written by Karl Popper. But then there is a turn. Benda wrote that there existed until the last half century

> [a] class of men whom I shall designate "the clerks" [*clercs*, in the medieval sense of the word], by which term I mean all those whose activity essentially is *not* the pursuit of practical aims, all those who seek their joy in the practice of an art of a science or metaphysical speculation . . . and hence in a certain manner say: "My kingdom is not of this world."[124]

The clerks had fallen prey to the game of political passions, wrote Benda, and had become members of the bourgeoisie with the rare exceptions of men who had no passion "but the passion for thought"—Einstein an exemplar among them.[125] What is needed, wrote Benda, quoting the nineteenth-century French philosopher Charles Renouvier, is faith: "The world is suffering from lack of faith in a transcendental truth."[126]

Benda's reflections bring to mind Max Weber's well-known Munich lecture of 1918, "Science as a Vocation," in which he decried the waning of asceticism and devotion in German scientific life, saying that self-interested professionalism was coming to characterize the social and natural sciences.[127] Yet Weber had warned that the qualities that make a man an excellent scholar and academic teacher are not the qualities that make him a leader in practical life or, more specifically, in politics. The scientist must be devoted to his calling.[128] For Polanyi, the lesson was clear that the scientist must resist practical demands or political interests in order to remain committed to the goals of transcendent science.

Truth is not demonstrable, Polanyi continued in *Science, Faith and Society*, but it is knowable, "and I have said how."[129] For Kuhn, in his reading of these Riddell essays, Polanyi voiced a kind of mysticism or "extra-sensory perception" that made Kuhn uneasy. Further, in Kuhn's view, Polanyi failed to work out fully the sociological implications of his ideas, whereas Kuhn intended to "take the 'social implication'" of his own historical account of normal science and paradigm change to its logical end of a denial of a single true world of universals. He said this to Polanyi in Oxford in 1961.[130] For Popper, in contrast, as revealed during Polanyi's visit to his London School of Economics seminar in the spring of 1951, Polanyi's emphasis on the "stability of beliefs," like Kuhn's notion of "normal science," was anathema to any plausible defense of the privileged position and credibility of scientific knowledge.[131]

Like *Science, Faith and Society*, Polanyi's *Logic of Liberty* resulted from an invited series of lectures, which he presented at the University of Chicago when he was invited there for a visiting professorship during the spring term of 1950. The lectures reiterated themes now familiar: the role of scientific convictions in scientific practice, the need for self-government of science, the dangers of planning in science, and the working of polycentricity in scientific and economic life.[132] While in Chicago, Polanyi met the philosopher Marjorie Grene, who had studied in Freiberg and Heidelberg, as well as at Harvard, where Alfred North Whitehead had confirmed her anti-Cartesian bias and dissatisfaction with logical positivism. Grene, who took her doctoral degree in 1935 at Radcliffe College, had become Carnap's teaching assistant in his seminar at the University of Chicago and then, briefly, an instructor in the Philosophy Department.

With funding from the Rockefeller Foundation, Grene officially was Polanyi's research assistant in Manchester in the spring of 1952, as she helped him with the writing of the Gifford Lectures, which he gave at Aberdeen during 1951–1952. These lectures went into the final version of the book *Personal Knowledge*. As explicitly noted by William Scott and Martin Moleski, Grene became the dominant figure in the development of Polanyi's philosophical skills.[133] As we will see in the detailed discussion of *Personal Knowledge* in chapter 8, Grene never liked the Christian overtones in the final section of the 1958 book nor did she care for Polanyi's use of the analogy between religious faith and scientific faith.[134]

Like Popper's *The Logic of Scientific Discovery* and Kuhn's *The Structure of Scientific Revolutions*, Polanyi's *Personal Knowledge* uses historical examples, as well as sources from philosophy, psychology, and language theory, to further develop ideas that earlier prefigured his magnum opus. Like them, he rejected inductive empiricism as the modus operandi of science, using some of the same historical examples as they used, but differing elements of interpretation. Negotiating a middle way between objectivity and subjectivity in characterizing scientific epistemology, Polanyi settled upon what he called "personal knowledge" by way of emphasizing the personal commitment, passion, and faith in the experience of the scientific calling. He insisted on what he called the "fiduciary" nature of the scientist's quest for truth. Without faith in a transcendent truth, science cannot be done, in Polanyi's view, a conception, as noted earlier in looking at Bernal's faith in scientific progress, which expresses a kind of motivational realism for an intellectual activity that has both ethical and political dimensions.[135]

## The Political Dimensions of Post-Empiricist History and Philosophy of Science

The normative and political dimensions inherent in scientists' and other intellectuals' accounts of the history and philosophy of science have been discussed by a number of scholars. In a widely discussed lecture titled "Science as a Weapon in *Kulturkämpfe* [Culture Struggles] in the United States during and after World War II," historian David Hollinger examined the anti-Soviet and anti-centralization political views of natural scientists, social scientists, and other commentators as the basis for a shift from emphasis on scientific method to scientific community in descriptions of the nature of science during the 1940s and early cold war period. Hollinger especially noted the roles of social scientists such as Robert K. Merton, natural scientists such as James Bryant Conant, and public journalists such as Walter Lippman in setting up a reciprocal relationship in which not only are freedom and democracy essential for scientific development but the scientific community itself is an exemplary model of organization for a free and democratic society. In this view, the self-governing autonomy of the scientific community is crucial for the advancement of science.[136]

From Hollinger's perspective—a point of view taken up by other scholars such as the historian George Reisch and the sociologist and philosopher Steve Fuller—Conant's introduction of the history of science into the Harvard general education curriculum, Popper's popular writings on the open society and science, and Polanyi's essays in support of freedom in science and the pursuit of pure science all explicitly used the history of science as a an illustration of the history of liberal humanism. It was an Anglo-American point of view that was shared with anti-Marxist and liberal-humanist historians and philosophers of science in Great Britain following the war. At Cambridge, for instance, British historian Herbert Butterfield, theologian Charles Raven, English professor Basil Wiley, and members of the Society for Freedom in Science became leaders of the university's History of Science Committee. They rejected materialist and economic interpretations of history, and they emphasized liberalism as the embodiment of science and vice versa. They taught the history of science as the history of ideas, placing emphasis on civic notions of truth and virtue.[137]

At the time it was easy to see that politically Popper and Polanyi took the side of liberal humanism in their writing on the nature of science. They adopted the role of public intellectuals in explicit campaigns—their "war efforts"—to defend science against philosophies or ideologies that might endanger the autonomy of the scientific community and the trust of the public in the scientific attitude. In contrast, a political view is not explicit in the writings of Kuhn. He did not set out on a political mission to become a public intellectual, and he tried to avoid political readings of his work. In his biographical study of Kuhn, Fuller engaged in a wholesale critique against Kuhn's philosophy of science on these very grounds, condemning Kuhn for an alleged subservience to dominant political values in the United States and for an alleged failure to take overt political stands. Fuller argued, perhaps paradoxically, that, on the one hand, Kuhn's *Structure of Scientific Revolutions* is a document of the cold war and, on the other hand, that Kuhn was entirely silent on contemporary politics.[138]

*Structure of Scientific Revolutions* is consistent with the aims of Conant's general education program and with the liberal humanist politics of elite and mainstream American intellectuals of the postwar period. Kuhn's philosophy provides an implicit argument against central planning of science, since Kuhn characterizes science, as did Polanyi, as a closed, or at least normally closed, community of self-governing peers. Kuhn is mainly silent on the matter of the pursuit of science for practical applications, but he by and large describes only pure or fundamental science and its internal technical problems. Kuhn made a sharp distinction between pure and applied science in his essay "The Essential Tension" in 1959, saying:

> It seems likely, for example, that the applied scientist, to whose problems no scientific paradigm need be fully relevant, may profit by a far broader and less rigid education than that to which the pure scientist has characteristically been exposed . . . [T]he personality requisites of the pure scientist and of the inventor may be quite different, perhaps with those of the applied scientist lying somewhere between.[139]

In a section in *Structure* on science as a "special kind of community" in which only members of the group are entitled to evaluate each others' achievement, Kuhn explicitly justified the insulation of the scientific community from social demands, in order that the scientist may concentrate on problems he believes he can solve. In this respect, Kuhn claimed that there is a contrast between natural scientists on the one hand and social scientists on the other, since social scientists often defend their choice of a research problem by its social importance.[140]

Like Polanyi, Kuhn recognized the inconsistency of Popper's emphasis on skepticism and falsification with the behavior of scientists whom he knew. Nor did Kuhn accept Popper's demand that scientists abandon their theories when faced with falsifying evidence or Popper's skepticism about scientific procedures for verification. Sounding much like Polanyi, Kuhn wrote in 1959 that "anomalies . . . could conceivably be the clue to a fundamental innovation in scientific theory or technique, but the man who pauses to examine them one by one never completes his first project." "If any and every failure to fit were ground for theory rejection . . . all theories ought to be rejected at all times."[141]

Explicitly criticizing Popper, Kuhn firmly distinguished the role of anomalies in normal science from the role attributed by Popper to falsification in the scientific process. If verificationists had found it necessary to devise a criterion of probability for a statement in light of available evidence, so falsification needs a criterion of improbability, which Kuhn did not think Popper had satisfactorily supplied. History shows us, argued Kuhn, that there is a joint verification-falsification process going on in theory choice.[142] Sounding a bit like Toulmin, but citing Ernest Nagel, Kuhn suggested that "verification is like natural selection: it picks out the most viable among the actual alternatives in a particular historical situation."[143]

Kuhn's view on the truth of scientific theories lay neither with Popper nor with Polanyi. For Kuhn, the unit of scientific achievement is the solved problem rather than a theory or paradigm that is closer to the truth.[144] Science is progressive, because new paradigms increase the numbers of problems solved regardless of the premises of the new paradigm. For Popper, confidence in truth of a theory can be increased, but never achieved, and judgments of "true" should be avoided unless carefully qualified. For Polanyi, no scientist would continue working without faith that there is a reality and truth to be known, and the scientist's calling is the search for that reality.

Some critics have noted an element of theomachy in Kuhn's philosophy of science. His history is told as the conflict of the gods, of dogma and heresy, of paradigm and revolution, resulting in the triumph of a new order that becomes a stable regime for a while but yet will be overthrown.[145] Polanyi's history of science is less dramatic, but it rests in tension between tradition and innovation which is resolved by individual conversions and community authority. Popper offers a scheme of conflict, except that the crisis is an everyday matter of laboratory life that Kuhn and Polanyi found absurd. Popper denied that there are patterns or laws for historical change. Polanyi talked more about his experiences of practice than about detailed historical cases.

In contrast to Polanyi and Popper, does Kuhn's framework for under-standing scientific change constitute a historical theory? Kuhn explicitly denied that he offered a *theory* of scientific change. He did not ask students to test his interpretation, he said, nor did he intend to write history himself as a verification of *The Structure of Scientific Revolutions*.[146] Nor did Kuhn intend for his history and philosophy of science to alter scientific practice. Like Polanyi, Kuhn thought that he had described science as it actually had been done in the past and as it works today, and that the scientific com-munity is a triumph of Western civilization that should be left alone to run its own affairs.[147]

Before it appeared in print, Kuhn's point of view was severely criticized on political grounds by Paul Feyerabend when he read the draft of *Structure of Scientific Revolutions* that Kuhn circulated in the spring of 1961. After the Second World War, Feyerabend had been part of the Viennese philosophy club organized by his teacher Viktor Kraft, who had been a member of the Vienna Circle in the 1920s and 1930s and was a proponent of logical em-piricism as scientific philosophy. Feyerabend's studies in London with Pop-per in 1952 and his translation into German of Popper's *Open Society and Its Enemies* thoroughly acquainted him with Popper's views, which Feyera-bend found still persuasive in the early 1960s. In coffeehouse discussions with Kuhn on Telegraph Avenue and in detailed written critiques, Feyera-bend disputed some key parts of Kuhn's interpretation in the manuscript of *Structure*.

Most centrally, Feyerabend accused Kuhn of writing "ideology covered up as history" like Hegel and all historical determinists. Feyerabend agreed with Popper that there are no historical laws, a point of view disputed at the time among philosophers. Carl Hempel was among those who de-fended the existence of historical laws fully analogous to laws in the natu-ral sciences. Such laws, Hempel argued, are indispensable instruments of research. Hempel had studied with members of the Vienna Circle in the early 1930s and taken a doctorate in philosophy in Berlin in 1934 before emigrating to the United States from Belgium in 1937. He briefly was Car-nap's assistant at Chicago and then taught in New York and at Yale Uni-versity before moving to Princeton in 1955, where Kuhn later became his colleague.[148]

Like all advocates of historical laws, Feyerabend said, Kuhn was badly mistaken in describing what has happened in the past as if these events were following a course of inherent reason rather than consisting merely of a "series of accidents combined with struggle for power."[149] Feyerabend strongly objected to what he saw as the political implications in Kuhn's work, made explicit in one brief section of the manuscript in Kuhn's specific analogy between political revolution and scientific revolution.

In a chapter on the nature and necessity of scientific revolutions, Kuhn wrote that political revolutions aim to change political institutions in ways that those institutions themselves prohibit. Society divides into polarized camps for which political recourse fails. Only techniques of "mass persuasion, often including force," can resolve the conflict. "Like the choice between competing political institutions," wrote Kuhn, "that between competing paradigms proves to be a choice between incompatible modes of community life." Thus the choice cannot be determined by the usual evaluative procedures of normal science such as logic and experiment alone.[150]

For Feyerabend, the domination of science by a single dogma or paradigm was an unacceptable conclusion that Kuhn was reading into his allegedly historical accounts. There were profound political consequences. Thus, "in the domain of politics your rule not to consider alternatives unless the existent paradigm has got into trouble would mean that nobody should consider alternatives to a tyranny *until it has become very obvious that it cannot work*; this means revisions, or democratic ideas should be considered only after quite a lot of people have been killed and only after it has become apparent that the tyrant cannot keep his power a moment longer."[151]

Feyerabend objected vehemently to Kuhn's view that it is "psychologically impossible for a scientist and, indeed, for any human being to be able to entertain various alternative hypotheses and to discuss them impassionately."[152] Anticipating Popper's later objections, Feyerabend warned Kuhn that "normalcy" is undesirable: . . . better live in permanent revolution than in the state of normalcy. *Revolution in Permanence* should be the battle cry of every empiricist."[153] What is unique in science, Feyerabend insisted to Kuhn, is that scientists regard as valuable what is testable and refutable. Kuhn's argument that falsifiability is a naïve and a detrimental view of scientific procedure was itself a naïve interpretation of Popper's methodology that ignored the compatibility of the idea of falsifiability with Kuhn's historical accounts.[154]

By the 1970s, Feyerabend parted company, too, with Popper and with Popper's protégé Imre Lakatos. In *Against Method* and in *Science in a Free Society*, Feyerabend argued that Kuhn and Polanyi had been correct in combating the idea of a rationally definable scientific method, but that they were wrong in their insistence that the scientific process is delicate machinery that requires special treatment. Scientific rationalism, argued Feyerabend, is only one tradition among many traditions and not one to which all others must conform: "A free society is a society in which all traditions

Figure 7.2. Thomas S. Kuhn at MIT in 1980. (Photograph by MIT
News Office staff. Courtesy of the MIT Museum.)

have equal rights and equal access to the centers of power . . . because
[a tradition] gives meaning to the lives of those who participate in it."[155]
Feyerabend's anarchism caused great unease for his colleagues. Kuhn later
conceded in an interview that *Structure of Scientific Revolutions* was a profoundly conservative book, in the sense "that I was trying to explain how it
could be that the most rigid of all disciplines, and in certain circumstances
the most authoritarian, could also be the most creative of novelty."[156]

Like Kuhn, Polanyi proposed a view of the nature of science and the scientific community that was fundamentally politically conservative and sociologically elite. Feyerabend saw this clearly, and it was a criticism made as
well against Polanyi by Imre Lakatos and others in Popper's inner circle.[157]
In lectures at the London School of Economics in 1973, Lakatos characterized Polanyi's and Kuhn's views on science, along with Merton's, under the

rubric "elitest authoritarianism," in which only wise scientists can interpret the difference between good science and bad science or pseudoscience.[158] Polanyi's freedom in science was not Popper's "open society," as Polanyi explicitly wrote in *The Logic of Liberty*, because the community of science is "one fully dedicated to a distinctive set of beliefs" in which the scientist "risks defeat but never *seeks* it" in a quest for discovery that is personal and passionate.[159]

Where Polanyi profoundly disagreed with Kuhn was in the matter of a theory of scientific knowledge, rather than in analysis of the scientific community and scientific practice. As discovered by Martin Moleski, Polanyi's annotations of photocopies of the last three pages of *Structure* include the comment "Truth.!!!!!!!! This *really* needs analysis" next to Kuhn's sentence that "we may . . . have to relinquish the notion, explicit or implicit, that changes of paradigm carry scientists and those who learn from them closer and closer to the truth." In a notebook, Polanyi wrote that changes in dogma in science seemed to suggest to Kuhn "that science may not be moving toward greater truth; rather 'one damned thing after another' as Darwin has taught us to understand organic evolution."[160] Indeed, among Kuhn's final remarks in *Structure* was the statement that the entire process of scientific development "may have occurred, as we now suppose biological evolution did, without benefit of a set goal, a permanent fixed truth."[161]

Kuhn reiterated this interpretation in 1991 in his Rothschild Lecture at Harvard and surprised some members of the audience by the expanded scope of his analogy between natural selection and the scientific process. "Scientific development is like Darwinian evolution," he argued, and the "episodes that I once described as scientific revolutions are intimately associated with . . . speciation." There are more specialties after a revolutionary change than there were beforehand. There is no one big mind-independent world to be discovered as the truth, but a "variety of niches within which the practitioners of these various specialties practice their trade." The niches are not independent of mind and they are not independent of culture, but they are as "solid, real, resistant to arbitrary change as the external world was once said to be." As in Darwinian evolution, scientific development is driven from behind rather than pulled toward a fixed goal.[162]

Polanyi's *Personal Knowledge* and Popper's *Logic of Scientific Discovery* are dense, difficult books whose basic arguments were first made and clearly stated in their earlier political writings. It was Polanyi's early writings that had an impact on Kuhn and many readers rather than *Personal Knowledge*. When Toulmin reviewed Popper's English-language *Logic of Scientific Discovery* for the widely read periodical *Scientific American*, he mentioned the public's familiarity with Popper's work on political theory and the phi-losophy of history. This is what they knew about Popper.[163] Popper's and Polanyi's politically inspired work was widely disseminated through news-paper articles, intellectual periodicals, and the BBC, and it became iden-tified with the ideology of a free and democratic society and the role of freedom and autonomy in scientific progress. Karl Popper was knighted in 1965 for his defense of freedom and rationalism. Michael Polanyi received an honorary degree at Princeton University in 1946 with a citation that de-scribed him as "a physical chemist who has devised new tools to determine how fast atoms react; a veteran campaigner against those who would take from science the freedom she requires for the pursuit of the truth."[164]

In contrast, Kuhn's philosophy of science received no honors tied to political values, although his philosophy was consistent with liberal hu-manist political values and with the argument that the scientific commu-nity flourishes best by ruling its own affairs, even in times of intellectual anomaly and crisis. Among the three books that appeared around 1960, it is *Structure* that is the most readable and the most read. Indeed, it is tempt-ing to argue that it is precisely the absence of an impassioned and explicit political agenda that is the key to the power, adaptability, and longevity of Kuhn's short, synthetic interpretation of the nature and practice of science. It is paradoxical, too, that, among the three philosophers, Kuhn perhaps best embodies, in personal circumstance, temperament, and intellect, the *clerc* who does not involve himself in daily political matters and whose passing Max Weber and Julien Benda and Michael Polanyi had lamented.

Of course what might be politically neutral in one milieu could be revolutionary in another, as Alan Richardson has noted in a discussion of philosophy of science and the cold war. The Columbia University philoso-pher of science Ernest Nagel was in Vienna in 1935, where he observed Vienna Circle leader Moritz Schlick lecture, one year before Schlick's assas-sination. Schlick spoke in a huge auditorium to an audience packed with students. Schlick's analytic philosophy, wrote Nagel, was politically radical in mid-1930s Vienna because it was formally ethically neutral when what was expected was a philosophy of indoctrination into dogmas of religion, metaphysics, and patriotism.[165] After the migration of Vienna Circle refu-gees to the United States in the 1930s, some people thought that the po-litical views of the leftist members of the group became considerably more muted, especially in the early 1950s when anti-Soviet politicians punished leftists with public hearings and dismissals from their jobs. It was partly the Vienna Circle's politics of scientific rationalism and social planning that Popper and Polanyi had found objectionable in logical positivism. In its new milieu, logical positivism gave way to an analytical philosophy that had less ambitious aims.[166]

Something important happened in the philosophy of science in the 1960s. Of this, Kuhn said in his 1990 presidential address to the Philosophy of Science Association, "I was, if you will, present at the creation, and it wasn't very crowded. But others were present too: Paul Feyerabend and Russ Hanson, in particular, as well as Mary Hesse, Michael Polanyi, Stephen Toulmin, and a few more besides. Whatever a Zeitgeist is, we provided a striking illustration of its role in intellectual affairs."[167] As we will see in the next chapter, Polanyi's philosophy of science, with its origins in his experiences in the European milieu of the 1920s and 1930s and its first public presentations in the 1940s and 1950s, was adopted and adapted in the late 1960s and the 1970s to very different meanings from his original intentions. Like most authors, Polanyi lost control of his texts. Chapter 8 examines in detail the philosophy of *Personal Knowledge*, its limited success with philosophers of science, and its meaning for two very different camps in modern American and British culture with which Polanyi engaged: religious thinkers on the one hand and sociologists on the other hand.

# EIGHT

# Personal Knowledge

## Argument, Audiences, and Sociological Engagement

The immediate incentive for publication of Polanyi's *Personal Knowledge: Towards a Post-Critical Philosophy of Science* was his invitation to give a set of Gifford Lectures at the University of Aberdeen. The mandate of the lectures is one of discussing natural religion or natural theology as knowledge "without reference to or reliance upon any supposed special exceptional or so-called miraculous revelation."[1] Polanyi gave his first round of Gifford Lectures later than originally planned, in late spring of 1951, and the second set in late autumn of 1952.[2]

The lectures and their revision in *Personal Knowledge* owed a considerable debt to the philosopher Marjorie Grene, who worked with Polanyi as his assistant and critic throughout the 1950s. The multifaceted origins and audiences for the book in its religious, political, humanistic, and sociological dimensions are demonstrated in the names of those whom Polanyi asked to read the manuscript: the Christian ecumenist J. H. Oldham, the neoconservative journalist Irving Kristol, the poet and novelist Elizabeth Sewell, and the sociologist Edward Shils. Funding for Grene's salary and other expenses came from the Rockefeller Foundation, the Congress for Cultural Freedom, and the Volker Fund for promotion of free-market economics.[3] Many people were interested in seeing this book completed. Polanyi dedicated the volume to his wife, Magda, who was completing her own book on textile chemistry at the time.[4]

Certainly the need to meet the general mandate of the Gifford Lectures for a demonstration of the existence of God within the tradition of natural theology influenced Polanyi's greater emphasis on God in *Personal Knowledge* than is found in his briefer lectures of the early 1960s given at Yale University and elsewhere and published as *The Tacit Dimension*.[5] This little book likely has had a far greater readership than *Personal Knowledge*, particularly among philosophers, sociologists, and historians, because of its clarity, brevity, and less explicit religious message. The immediate reaction to his Gifford Lectures and *Personal Knowledge* did not meet Polanyi's hopes, although reactions were very different among distinct communities of readers (who are not always so easy to distinguish by discipline).[6]

On the whole, as we will see, philosophers were not friendly and found Polanyi's new category of nonsubjective and nonobjective personal knowledge unpersuasive. The reaction of Karl Popper was contemptuous, but Popper's once-faithful acolyte and Hungarian disciple Imre Lakatos responded in a considerably more constructive fashion. Polanyi's realism appealed to many scientists who found his accounts of scientific life and scientists' behavior more recognizable than most philosophers' or historians' analyses. The religious tone of the realism was also congenial to many scientists. The spiritual dimensions of *Personal Knowledge* found favor among many Christian believers, and his discussion of cosmic evolution proved useful to proponents of teleology and intelligent design in arguments against mainstream evolutionary biology.

Paradoxically, although Polanyi's intellectual targets included the social-relations-of-science position defended since the 1930s by J. D. Bernal and the sociology of knowledge first developed by Karl Mannheim in the 1920s, sociologists were among Polanyi's most fervent allies and adepts. Like Polanyi, American sociologists largely rejected Mannheim's approach, usually identifying it with Marxism, and by the 1940s, Robert K. Merton's sociology intersected in important ways with Polanyi's writings on scientific life. Merton, like Thomas Kuhn among philosophers, Edward Shils among sociologists, and James Bryant Conant and Harvey Brooks among policy-oriented scientists, helped create an intellectual climate in the United States in which Polanyi's notions of scientific tradition, apprenticeship, tacit knowledge, and creativity found a congenial home, along with Polanyi's insistence on the necessary relationship between science and democracy in a free-market culture. The view that scientists should govern their own affairs independently of government direction came to dominate American science policy in the 1960s.

By the late 1960s, however, the relationship that had developed between big science and the military-industrial complex during the first decades of the cold war and the early years of the Vietnam War became a matter of concern not only among political leaders and science critics but also among some influential scientists. The unprecedented expansion of science and technology also triggered discussions about the funding of science, science education, and the relationship between the sciences and the humanities. These concerns played a role in the creation of academic programs for study of the relationship between science and society. A result was the emergence of a new academic movement in the third generation of the social construction of science which both incorporated and rejected basic assumptions and arguments of its predecessors.

## The Argument of *Personal Knowledge*

Polanyi's *Personal Knowledge* is a dense and difficult book. It is a great oversimplification to say, as Polanyi did, that it is based on his Gifford Lectures of 1951–1952. The book had been in the making since the late 1930s. Polanyi had hired Olive Davies as his part-time personal secretary in November 1941 to aid him in his writing projects. At the time, he began gathering a large collection of books from which he made notes on Aristotle, Locke, Machiavelli, Edmund Burke, George Trevelyan, Alfred North Whitehead, and many other authors. His notebooks of the early 1940s describe what he later called "tacit knowledge," using examples from grammar, scientific method, and the arts and crafts. He began thinking of knowledge in terms of Gestalt psychology, and he drafted essays and potential book chapters on the characteristics of scientific truth, the autonomy of science, the distribution of functions in scientific development, and the meaning of the western conception of science. He told friends that he was going to write a book on scientific life, and in the spring of 1943 he wrote his sister Mausi that he wanted to bring all his ideas together into a "magnum opus to deal with science and everything else in the world." This was the formidable—and ultimately unachievable—aim for what became *Personal Knowledge*.[7]

An opening statement in *Personal Knowledge* has a familiar ring: "This is primarily an enquiry into the nature and justification of scientific knowledge" with the aim "to show that complete objectivity as usually attributed to the exact sciences is a delusion and a false ideal."[8] Yet he also intended something bigger, namely, to treat science as a subset of systems of knowledge that prevail through personal persuasion, conversion, and commitment. Following a statement toward the end of the book that the broad scientific community is one of "mutual trust and confidence between thousands of scientists of different specialties," Polanyi writes, "Though each may dissent . . . from some of the accepted standards of science, such heterodoxies must remain fragmentary if science is to survive as a *coherent system of superior knowledge, upheld by people mutually recognizing each other as scientists, and acknowledged by modern society as its guide.*" "This superior knowledge will be taken to include . . . beside the systems of science and other factual truth, all that is coherently believed to be right and excellent by men within their culture."[9]

*Personal Knowledge* is divided into four parts, which treat in turn the "art" of knowing, the "tacit component" in knowing, "justification" of "personal knowledge," and "knowing and being." As Marjorie Grene reflected, the first two parts are primarily descriptive of skillful doing and of informal or tacit arts of understanding, while the third part self-reflexively asserts the element of personal commitment in a philosophy of science. The fourth part concludes by placing the search for knowledge within the context of natural history and evolutionary biology.[10]

In the book, Polanyi moves back and forth between talk of scientific norms, organization, apprenticeship, skills, and interpretive frameworks on the one hand, and Stalinist science, Freudian psychoanalysis, Rhine's parapsychology, Lysenko's anti-Mendelian biology, Azande belief systems, and the Roman Catholic Church on the other hand. He identifies the empiricist and positivist Enlightenment and what he calls the "Laplacian goals" of reductionism and materialism with a post-Kantian "critical" period in history, which had ended in the last decades in the horrendous "moral inversion" that took place when science was harnessed for supposedly righteous social ends by the totalitarian and inhuman regimes of Stalin and Hitler.[11]

Polanyi analyzes scientific knowledge as a stable system of belief in which an interpretive "framework" or "paradigm" or "tradition" plays a crucial role. He identifies as one of these paradigms what he calls the "Laplacian Delusion," which led the early-nineteenth-century mathematical physicist and astronomer Pierre-Simon Laplace to state that all the phenomena and events of the world could be known once sufficient data was acquired about the positions of bodies and forces in the universe. Taking the Laplacian reductive program as a guide to human affairs resulted in political systems, argues Polanyi, that were totalitarian and left "no justification to public liberties" while demanding that natural and social scientists serve the power of the state. In Polanyi's view, these developments discredited science as an end in itself and as a token of universal truth.[12]

Polanyi's difficulty was that he was convinced that science—as an autonomous system independent of externally imposed political ends—was a system of knowledge that results in true knowledge of the natural world even though it builds its interpretive framework through social mechanisms that overlap with other ways of life and with ideological systems that may be wrong or evil. His description of the operation of conservative social arrangements in favor of stability is balanced against his discussion of the role of individual freedom and initiative in effecting change.

Polanyi portrays scientific discovery as a phenomenon that puzzles or worries the scientist or relieves the community of scientists from the burden of a problem. Discovery is not a strictly logical operation, and he describes the obstacle to be overcome as the "logical gap" between the dominant interpretive framework and a new way of seeing things, since in conflicting systems of belief, members of different systems live in different worlds. Their divisions can only be resolved by "a heuristic process, a self-modifying act, and to this extent a conversion."[13]

The elation at the moment of a discovery is not a simple psychological byproduct but registers the passion that distinguishes between demonstrable facts which are of scientific interest and those which are not. The value of the discovery lies in its certainty or accuracy, its relevance or profundity, and its intrinsic interest, as Polanyi had argued elsewhere.[14] "Having made a discovery, I shall never see the world again as before." "I have crossed the gap," he writes. "Major discoveries change our interpretive framework . . . We have to cross the logical gap between a problem and its solution by relying on the unspecifiable impulse of our heuristic passion."[15] The framework at which we arrive, like the framework which we have abandoned "is relatively stable, for it can account for most of the evidence which it accepts as well established, and it is sufficiently coherent in itself to justify . . . the neglect for the time being of facts, or alleged fact, which it cannot interpret." Proponents of any new system must in turn work through a process of winning their opponents' intellectual sympathy and conversion.[16]

The skills of scientists, which are used in discovery and in their everyday judgments of scientific life, are achieved by following a "set of rules not known as such to the person following them" just as a cyclist or pianist learns his craft. The art is passed from master to apprentice in the finer points of connoisseurship that are communicated by example, not by precept, and in specific locales rather than by learning universal abstract rules. As demonstrated in Gestalt psychology, in Polanyi's view, the parts and the whole of what is learned are in tension with one another, since subsidiary awareness and focal awareness are mutually exclusive, as demonstrated by the pianist's confusion when he begins paying attention to what his fingers are doing.[17] "The large amount of time spent by students of chemistry, biology and medicine in their practical courses shows how greatly these sciences rely on the transmission of skills and connoisseurship from master to apprentice."[18]

This kind of knowledge is the "tacit knowledge" at the heart of scientific and all forms of knowledge. In contrast to the views of the early Ludwig Wittgenstein and logical positivists, Polanyi writes that there is a meaningful domain of knowledge that is ineffable rather than empirical. He uses the term "indwelling" to denote the incorporation of knowledge that incorporates both formal (objective) and nonformal (subjective) understanding into "personal knowledge."[19]

Polanyi uses historical examples, but none in detail and none explicitly from his own personal chemical researches. There are references to Copernicus and Ptolemy, to Johannes Kepler, Tycho Brahe, Isaac Newton, and, more contemporaneously, to Friedrich Kohlrausch, Max von Laue, Jean Perrin, Ernst Mach, and Albert Einstein. Among chemists and controversies associated with their names, he mentions John Dalton, Stanislao Cannizzaro, Friedrich Wöhler, Justus von Liebig, J. H. Van't Hoff, Hermann Kolbe, Louis Pasteur, and Harold Urey. He reiterates the basic theme of his 1936 article on the inexact in chemistry in a passage in which he argues the existence of emergent properties in biology and chemistry that cannot be explained merely in terms of constituent parts. In this he implicitly draws upon his past work on the semi-empirical method in chemical kinetics.

In this context, Polanyi writes that there is a logical gap between any "topography" or field of vision and some pattern that is derived from it. "No Laplacean [sic] mind schooled in quantum mechanics could replace the science of chemistry. For chemistry answers questions regarding the interaction of more or less stable chemical substances, and these questions cannot be raised without experience of these substances and of the practical conditions in which they are to be handled . . . [W]hile quantum mechanics can explain in principle all chemical reactions, it cannot replace, even in principle, our knowledge of chemistry."[20]

Polanyi's most detailed historical example in *Personal Knowledge* was intended to demonstrate the absurdity of strictly empirical explanations of scientific discovery through refutation of textbook clichés and historical accounts that routinely claimed that Einstein conceived the special theory of relativity in order to explain the anomalous negative result of the Michelson-Morley experiment, which had aimed to detect a physical effect from the presence of a universal ether. Polanyi asked their mutual friend Nándor Balázs to make an inquiry of Einstein in 1953 about the accuracy of the interpretation and reported Einstein's answer that the Michelson-Morley experiment had a negligible effect on development of his relativity theory. In addition, as he did rarely in *Personal Knowledge*, Polanyi invoked his own personal memories of reactions in the 1920s to D. C. Miller's claim at that time that he had achieved a positive result for the Michelson-Morley experiment. According to Polanyi:

[By 1920, scientists] had so well closed their minds to any suggestion which threatened the new rationality achieved by Einstein's world-picture, that it was almost impossible for them to think again in different terms. Little attention was paid to the experiments, the evidence being set aside in the hope that it would one day turn out to be wrong.[21]

Examples of dogmas, conflicts, and breakthroughs in the history of science were important in Polanyi's *Personal Knowledge* but they were not worked out as detailed historical case studies, because his aim was not a historically based philosophy of science but a sociologically and psychologically sound philosophy of science that preserved the validity and dependability of the scientific process as long as it operated with freedom and autonomy. How could he be so sure, however, of the soundness of science so operating? Midway through *Personal Knowledge*, he explains, "The principal purpose of this book is to achieve a frame of mind in which I may hold firmly to what I believe to be true, even though I know that it might conceivably be false."[22] To the question "Who convinces whom here?" Polanyi answers simply, "I am trying to convince myself."[23]

Turning to the teachings of Saint Augustine, Polanyi recommends the reader to Augustine's adage that all knowledge is a gift of grace received through the guidance of belief. For seekers of knowledge in a new postcritical age, Polanyi writes, "We must now recognize belief once more as the source of all knowledge. Tacit assent and intellectual passions, the sharing of an idiom and of a cultural heritage, affiliation to a like-minded community . . . shape our vision of the nature of things on which we rely for our mastery of things."[24] We enter into this community "in the hope of achieving thereby closer contact with reality" and "self-satisfaction . . . as a token of what should be universally satisfying."[25]

Polanyi's method of achieving this self-satisfaction takes the form in the book's last chapters of a critique of neo-Darwinian evolutionary biology. In this analysis, he reasserts the irreducibility of life and mind to physics and chemistry, and he marshals arguments for ascending, rather than random, evolutionary processes in a biological evolution that Polanyi calls a history of achievement. After an attack on scientific and philosophical accounts of a "neurological model" of mind as a computing machine, he returns to the problem of knowledge and asserts that "biology can be extended by continuous stages into epistemology and more generally into the justification of my own fundamental commitments."[26] An understanding of the relation between biology and knowledge can "re-equip men with the faculties which centuries of critical thought have taught them to distrust" so that men come to realize "the crippling mutilations imposed by an objectivist framework."[27] Thus his epistemology is meant to be a step in a process of conversion from the critical to the postcritical understanding of science.

Among the few philosophers whose views Polanyi mentions in *Personal Knowledge* are Henri Bergson and Teilhard de Chardin, whose writings he sees as support of his notion of personal knowledge in a process in which "we hope to be visited by powers for which we cannot account in terms of our specifiable capabilities. This hope is a clue to God."[28] In Polanyi's view, human biology can be extended into an epistemology in which "the human mind has been so far the ultimate stage in the awakening of the world." He ends the book with two sentences that perhaps perfectly fit the occasion of his Gifford Lectures at Aberdeen but baffled many later readers of *Personal Knowledge*:

> We may envisage then a cosmic field which called forth all these centres by offering them a short-lived, limited, hazardous opportunity for making some progress of their own towards an unthinkable consummation. And that is also, I believe, how a Christian is placed when worshipping God.[29]

## Response of the History and Philosophy of Science Community

Polanyi's book was received by a philosophical community that was pre-occupied with verification and justification in the philosophy of science. The central core of philosophy of science in 1958 lay in the logical empiricist camp of Rudolf Carnap, R. B. Braithwaite, and Ernest Nagel, which was rooted in turn in prohibitions by the late-nineteenth-century German philosopher Gottlob Frege against the introduction of historical or psychological explanations into analysis of the foundations of scientific laws.[30] Among those sympathetic with Polanyi's rejection of the kind of positivism that "denies the scientist any title to being an imaginative thinker" was Stephen Toulmin, whose book *The Uses of Argument* appeared in 1958.

Toulmin, born in 1922, the same year as Thomas Kuhn, had taken a degree in mathematics and physics at King's College in London before serving in the Ministry of Aircraft Production during the Second World War. He earned a doctorate in philosophy in 1948 from Cambridge, where he studied with Ludwig Wittgenstein and John Wisdom. Polanyi, Marjorie Grene, and Elizabeth Sewell all took a train to Leeds during the 1955–1956 academic year to hear Toulmin lecture on the topics that became part of *The Uses of Argument*. Toulmin criticized formal logic as a misleading and inadequate representation of how humans really argue, and he questioned the claims of formal logic for universality and certain truth. By 1972, his book *Human Understanding: The Collective Use and Evolution of Concepts* laid out a fully articulated Darwinian model for conceptual change in the sciences.[31]

Toulmin reviewed *Personal Knowledge* in 1959 and was dismissive of the book as a "post-critical" work that could take "the next major step on from Kant" or contribute to the mainstream of philosophy.[32] Like most philosophers in the late 1960s and 1970s, Toulmin was considerably less interested in Polanyi's approach than in Kuhn's. The explanation for this reaction is not hard to find. Gerd Buchdahl, the first lecturer in the history and philosophy of science at Cambridge, wrote an influential review of Kuhn's *Structure of Scientific Revolutions*, in which he praised the implications of Kuhn's approach for the notion of scientific change "against logical positivists, old-fashioned empiricists, verificationists (probabilistic or otherwise) and falsificationists." In Buchdahl's view Kuhn's work was novel and important, because it showed that science is never a tidy structure but "ramshackle," and it argued that paradigms speak different languages and are "incommensurable."[33]

All this could be found, too, in Polanyi. What worried Buchdahl in Kuhn, however, was the notion of conversion, an idea that was central to Polanyi's *Personal Knowledge*: "surely Kuhn does not want to say that Pascal's espousal of the conceptual scheme of the 'sea of air' (against the 'horror vacui') is of the same kind as those religious experiences which led to his famous conversion."[34]

The central aim of Polanyi's philosophy of science was the destruction of logical empiricist characterizations of science as a system of detached objectivity, but he also wished to demonstrate the roles of ineffable, passionate, and personal factors in an everyday scientific life to which he attributed moral virtue. Most reviewers recognized that aim clearly but were not at all satisfied with Polanyi's densely argued attempts to define and illustrate "personal" knowledge. A reviewer in *Scientific American* complained, "It is less of a philosophical inquiry than a terribly long and passionate affirmation of a creed, with supporting evidence drawn from every corner of human activity, from quantum theory to Marxism. The main target of Polanyi's attack is the notion that science is objective and impersonal . . . To know is to make an intellectual commitment . . . Knowing is an art . . . a passionate contribution . . . He is, however, quite unable to pin this element down."[35]

The Northwestern University philosopher William Earle thought that Polanyi's arguments against scientific objectivity were effective with respect to his demonstration of personal elements in acts of scientific appraisal, choice, and accreditation and in personal choices of frameworks for the operation of science. In Earle's view, however, Polanyi failed to persuade the reader that personal commitments are not subjective or "whimsical" or to distinguish between science and superstition: "I do not see that Polanyi has provided us in the end with any means whatsoever for distinguishing truth from error, the personal from the subjective, science from superstition." "If, as Polanyi argues, we must always dwell *within* a framework of belief within which there are such things as 'facts' and 'truths' but outside of which there are none, then indeed we have no right to adjudge any other belief whatsoever 'subjective,' except insofar as we simply do not share that belief." For Earle, who wanted a more rigorous logical argument, Polanyi ends up with "radical subjectivism" and only gets out of this conundrum by concluding the book with God.[36]

The idealist and liberal philosopher Michael Oakeshott, who taught at the London School of Economics from 1950 to 1969, wrote a review of *Personal Knowledge* for *Encounter*. He commended Polanyi's criticism of empiricism, his denial of moral neutrality in scientific investigation, and his demonstration of personal elements in scientific knowledge. Oakeshott found the book, however, "a jungle through which the reader must hack his way" that failed to avoid the pitfall of subjectivity, which could have been achieved by a bit of skepticism. At the edge of Polanyi's argument, wrote Oakeshott, "there is a suspicion of philosophical innocence."[37]

The University of Minnesota philosopher May Brodbeck was considerably less kind than Oakeshott, similarly seeing Polanyi's comparison of the scientist's personal belief to religious faith as a failed attempt to avoid solipsism:

In place of the "objectivist" prejudice . . . Polanyi's "post-critical" philosophy explicitly invokes a return to St. Augustine's doctrine that belief precedes understanding. Doubt must give way to a critical acceptance of belief . . . Polanyi's doctrine is basically subjective; in biology it is vitalist and teleological (for "clues to God"); in religion, it is the existentialist mumbo-jumbo by which one can consistently deny that God "exists" and yet "affirm" and find it sensible to worship him. This is indeed the epiphany of the Higher obscurantism. I have not been gentle with Polanyi who is, I am sure, a fine and noble gentleman as well as a distinguished chemist . . . It is late in the day for these tender-minded assaults on reason. They have had and still have far too much company . . . It is not necessary to be a scientific materialist, as I am not, to recognize that it is time once again to stand up and be counted against the forces of irrationalism wherever they appear, in no matter how benign a guise.[38]

Perhaps Polanyi inevitably had to fail in any attempt to convince philosophers who were looking for formal argument to surrender to his belief in personal knowledge. Most philosophers were not persuaded, Karl Popper foremost among them. Popper included a strong rejection of Polanyi's view at the end of his preface to the 1959 translated edition of *The Logic of Scientific Discovery*. Popper wrote that it was his own aim in *Logic* to "save the sciences and philosophy from narrow specialization and from an obscurantist faith in the expert's special skill and in his personal knowledge and authority; a faith that so well fits our 'post-rationalist' and 'post-critical' age, proudly dedicated to the destruction of the tradition of rational philosophy, and of rational thought itself."[39] For Popper, Polanyi's philosophy was not only antirational; it upheld a dangerous ideal of a closed scientific community that underwent only infrequent and transitory dissension in a system controlled by a small expert elite.[40] Popper's view was replicated in the work of his student Alan Musgrave who titled his 1969 doctoral dissertation "Impersonal Knowledge: A Criticism of Subjectivism in Epistemology," and Popper's objection would be repeated later by Popper's admirer Steve Fuller.[41]

Among Popper's acolytes in the early 1960s was the Hungarian-born philosopher of science Imre Lakatos, who fled to Vienna following the failed 1956 Hungarian revolution and then studied at Cambridge with George Polya in mathematics and Braithwaite in philosophy of science. Lakatos's dissertation focused on the logic of mathematical discovery. He joined Popper at the London School of Economics in 1960, but Popper fell out with him after Lakatos began questioning Popper's emphasis on falsification, demarcation, and justification in the philosophy of science. It was partly his reading and correspondence with Polanyi, as well as with Kuhn, that realigned Lakatos's attention on the process of scientific discovery.[42]

In 1965, Lakatos organized a symposium for Popper around the "Popper-Kuhn debate," in which Lakatos was to summarize Popper's position and make a reply to Kuhn. Lakatos informally invited Polanyi to participate, but Popper insisted that Lakatos rescind the invitation.[43] The symposium, on the subject criticism and the growth of knowledge, took place on July 13, 1965, as part of the International Colloquium in the Philosophy of Science held at Bedford College from July 11 to 17. As it turned out, Lakatos was unable to attend and did not deliver a paper on that day, but he completed the essay in 1969 that became his best-known publication, "Falsification and the Methodology of Scientific Research Programs." This long essay appeared in the volume *Criticism and the Growth of Knowledge* (1970), which also included papers by Toulmin, Popper, Kuhn, John Watkins, L. Pearce Williams, Margaret Masterman, and Paul Feyerabend, although Feyerabend, too, had been absent from the symposium.[44] In late fall of 1969, Lakatos sent some of his own essays to Polanyi with the note, "I guess, I just came to understand you, really. This new recognition will appear in my future publications."[45]

In his symposium essay, Lakatos defined scientific progress in terms of scientific research programs, each of which includes a series of theories, to which adjustments are made by a process considerably more complex than naïve falsification. A negative heuristic forbids removing the hard core of a research program if a prediction is not confirmed, while a positive heuristic uses auxiliary hypotheses as a protective belt that can be adjusted and readjusted. A successful research program guides scientists in their choice of problems. As long as a research program leads to new knowledge, it is a progressive enterprise. Conceptual frameworks are not fixed, as Kant argued, but they can be developed and replaced by new and better ones: "it is *we* who create our 'prisons' and we can also, critically, demolish them."[46] Since the history of science does not bear out a theory of scientific rationality, it must be relinquished. Polanyi's and Kuhn's prescription is to explain changes in 'paradigms' in terms of social psychology, but a preferred alternative is to develop a more sophisticated version of falsificationism in which we say that a theory is acceptable and scientific if it "has corroborated excess empirical content over its predecessor (or rival), that is, only if it leads to the discovery of novel facts." Nor is there a possibility of falsification before the emergence of a better theory with the consequence that scientific progress advances through a proliferation of theories or competing research programs.[47]

Lakatos, who died in 1974, succeeded Popper in 1970 at the London School of Economics, and he wrote Polanyi that "our students now read your books with great profit and interest." In December 1969, he met Polanyi for dinner and organized a presentation at the school entitled "What Is A Painting?" Later, he invited Polanyi to give a lecture, which Polanyi titled "Genius in Science."[48] Lakatos objected, however, to what he saw as Polanyi's elitist notion of validation in science, and he thought that Polanyi, like Kuhn, failed to explain the common origin of rival scientific schools or how scientific revolutions come about.[49]

Scientists, like philosophers, often found Polanyi's *Personal Knowledge* tedious going, but they mostly responded positively to Polanyi's depiction of the scientist's personal passion for discovery and to Polanyi's discussions of skill, connoisseurship, and tacit knowledge. Hugh Taylor, Polanyi's longtime friend at Princeton, expressed enthusiasm for the book but encouraged colleagues to read instead some of Polanyi's shorter essays of the 1960s, especially his March 1957 article in *Science*.[50]

Among scientists who themselves turned to the history and philosophy of science, the physicist Gerald Holton, like Thomas Kuhn, knew Polanyi personally, and he shared Polanyi's distaste for extreme forms of positivism. By the 1960s, Holton, like Mary Hesse in England and Russell Hanson in the United States, was arguing that theory comes first, not observation, and he was proposing the addition of a third dimension—what he called "themata"—to the logical-analytic and phenomenomic-empirical dimensions of the growth of science. Holton's first historical paper, published in 1956 on Johannes Kepler's physics, stressed the importance of Kepler's fiercely held presuppositions about the universe as simultaneously a physical machine, a mathematical harmony, and a theological order (Holton's "themata"). Holton, however, stood firm on empiricism, logic, and themata as the defining public characteristics of science while also following up in his later historical work with discussions of creative play, passionate invention, and emotional intensity in the life of science.[51]

Holton found *Personal Knowledge* "maddening in spots," almost as if its genre were the confession literature of Augustine or Rousseau. Given his own historical and philosophical interests in Einstein, Holton noted Polanyi's attempt to use a report that the Michelson-Morley experiment had played no role in the foundation of relativity theory as a refutation of logical empiricism. Polanyi's hearsay argument, writes Holton, "was not found convincing either by philosophers of science or by historians of science" although Holton's own later historical work unearthed evidence for the validity of Polanyi's hunch. Polanyi overreached when he declared that Einstein's theory was framed "on the basis of pure speculation, rationally intuited" but it was borne out on the whole because, Holton suggests, Polanyi had "*internalized* how scientists think, and had observed how others do their work, in finished publications as well as in conversations, and in debates, for example during his time in Berlin, when Einstein was also there." In short, concludes Holton, if there is such a thing as apperception, personal or tacit knowledge, and in-dwelling, Polanyi's intuition about Einstein itself exemplifies his concept of personal and tacit knowledge.[52]

And what of Marjorie Grene? Like May Brodbeck, Grene did not mince words on most occasions. She had moved from Illinois to Ireland in 1952, where she and her children put together the index for *Personal Knowledge*.

Beginning in 1960, she accepted an appointment in philosophy in Belfast that became one of several short-term appointments until she returned to the University of California at Davis, where she taught for thirteen years, eventually moving to the philosophy department at Virginia Tech. Grene died in Blacksburg, Virginia, in March 2009 at the age of ninety-eight as one of the leading philosophers of biology in the United States. She was the first female philosopher to have a volume in the Library of Living Philosophers dedicated to her work.[53]

Following the publication of *Personal Knowledge*, Grene clarified or defended many of the views in the book. An early intervention was her response to Michael Oakeshott's review with her own essay in *Encounter* laying out briefly the argument of Polanyi's book.[54] She continued to read and advise Polanyi in his later work, writing him in 1963 of her worries about his original versions of some of the Terry lectures: "Michael, you must not publish the . . . [first Terry lecture] without very considerable revision. The brief version was brilliant; the present text is so bogged down and cluttered up with stuff that it obscures what you had before made clear."[55]

Speaking in 1977 at a conference after Polanyi's death, Grene wrote that his primary problem had been how to analyze the life of science in order to display the social and political conditions necessary to sustain it, and, secondly, to resolve the question of how we can justify dubitable beliefs, for example, the belief that the Soviet government was wrong in 1935. Since, as Hume had shown, all scientific allegations outrun their evidence, how do we achieve cognitive certainty and the ability to distinguish good science from bad? Polanyi was putting on record "what he felt himself as an educated European scientist in the mid-twentieth century, called upon to believe."[56] She regarded his most significant philosophical contribution as the development of the theory of tacit knowing in order to show the relationship between tacit and explicit knowledge.[57] The emphasis on ineffable and tacit knowing came from the problem of the relation of freedom to science. "That is: when you have a central organization, you tell people what to do and it's supposed to be all explicit. On the other hand, when people are actually working in a good lab or a theoretical situation, there's a lot that goes on that isn't explicit."[58]

She also supported Polanyi's realism, writing later that she was an "unrepentant" realist for whom realism was an "epistemological consequence of an evolutionary metaphysic." In Grene's view, positivists and logical empiricists had created a pseudoproblem by their insistence in rooting scientific knowledge in observables and arguing about the reality of unobservables. It is "glaringly false," she wrote, "that scientists are always chasing observations only . . . What all these busy people are doing is trying, through experiment, to find out how something in the real world really works."[59]

In Grene's view, however, Polanyi did not have serious historical studies and examples in mind in his philosophical work. She assisted him with historical information when it was needed, "but he thought of history from a scientist's point of view—as a source from which to cull tidbits, but no more."[60] She had severe misgivings about the fourth section and especially the last chapter of *Personal Knowledge*. She did not at all like the theistic hints and Christian overtones, including the use of the Fall and redemption scheme and the analogy between Augustine's faith and the scientist's faith. For Grene, a sense of historical contingency and fallibility was lost at the end of *Personal Knowledge* when Polanyi slipped into "ontological dogmatism" and a "hopelessly anthropocentric evolutionism" followed by his closing Christian apologetic.[61] The notion of a "stratified universe" was unconvincing, she thought, as was his cosmology of emergence.[62] In 2005, she told an interviewer, "He hadn't a clue about evolutionary theory. He didn't think that neo-Darwinism could be right at all . . . He just had *no* understanding of evolutionary theory . . . I started looking through the literature to help him out."[63]

## Realism, Religion, and Design

It has been said that theologians more than scholars in other academic disciplines were attracted immediately to Polanyi's thought and to his volume *Personal Knowledge* because of its religious dimension. Polanyi remarked in a conversation in 1970 with Walter B. Mead, one of the presidents of the Polanyi Society, that he was disappointed at his scant recognition from British academic philosophers but gratified at his reception on American campuses—at least outside philosophy departments.[64] At a conference focused on Polanyi in 1992, Holton noted how Polanyi's views, including his commitment to a kind of vitalism in the origin of living things, found a more interested audience outside the laboratory than in it and "may account in part for the fact that we are holding this meeting in recognition of Michael Polanyi not in our Mallinckrodt Laboratory of Chemistry, but in the Sperry Room of the Harvard Divinity School."[65] Polanyi's close friend Elizabeth Sewell, herself a practicing Catholic, said that she was puzzled at his co-optation after his death by Christian thinkers, remembering that Polanyi once remarked on how scientists were liable to write long books and then put God into the last sentence, fully aware that she knew that God is the last word in *Personal Knowledge*.[66]

The religiously oriented Polanyi Society, founded in 1972, began publishing the journal *Tradition and Discovery* in 1973 with the aim of encouraging scholarship in Polanyian thought and to "open, broaden, and integrate our understanding and thereby, to bring us into closer appreciation
of, and oneness with, that which is transcendent and, not inappropriately,
perceived as 'holy.'" The Polanyi Society meets annually in conjunction with
the American Academy of Religion and the Society for Biblical Literature.[67]

While there is much in *Personal Knowledge* and in Polanyi's earlier and
later writings that bears upon analogies between religious belief or faith
and scientific belief and faith, two sections in the book are noted especially
by religious scholars: "Dwelling In and Breaking Out" at the conclusion of
the chapter "Intellectual Passions," and "Religious Doubt" in the chapter
"The Critique of Doubt."[68] Polanyi distinguishes between religious worship and religious belief, defining worship as "an indwelling rather than
an affirmation" and as a "heuristic vision" that aligns religion with other
great intellectual systems "which are validated by becoming happy dwelling places of the human mind."[69] The indwelling of the Christian worship-
per is a continued attempt at breaking out and casting off the condition of
man; it "sustains, as it were, an eternal, never to be consummated hunch: a
heuristic vision . . . an obsession with a problem known to be insoluble."[70]
Religious and natural "findings," he writes, "by-pass each other" and it is
illogical to attempt to prove the supernatural by natural tests. Nonetheless,
he writes by way of pointing to the book's final chapter, "The Rise of Man,"
he would "show how we can arrive by continuous stages from the scientific
study of evolution to its interpretation as a clue to God."[71]

These and other passages led some religiously inclined scholars to debate the status of Polanyi's own belief and religious commitment, some
of them asserting that they had personal experiences of him as a Christian
believer, others questioning whether his theological views edged toward
Catholicism or Protestantism, and still others concluding that Polanyi
yearned to acquire religious belief but that the great sadness of his life was
failure to achieve religious conviction.[72] He attended church from time to
time, including services at Saint Aiden's Presbyterian Church in Didsbury
during the 1940s.[73]

From about 1944 to 1960, Polanyi participated in many of the meetings of the Moot, a discussion group organized by J. H. Oldham that met
in St. Julian's in a rural setting near Horsham, south of London. Oldham, a
former missionary and a leader of the ecumenical movement, asked members of the Moot to offer opening and concluding prayers at the meeting.
The group included Manchester's vice-chancellor Walter Moberly, the poet

T. S. Eliot, the writer John Middleton Murry, the philosopher H. A. Hodges, and other guests. It was around the time he joined the Moot that Polanyi began carrying the Anglican Book of Common Prayer.[74]

Some of the papers Polanyi presented at Moot meetings articulated themes that appeared in chapters of *Science, Faith and Society* and *Personal Knowledge* (the latter of which Oldham read while still in manuscript form). A letter Polanyi wrote Oldham in 1948, however, appeared to express Polanyi's religious doubts. While planning a meeting to discuss modern atheism and Christian belief in God, Oldham wrote Polanyi, "We should, if possible, have one or two non-Christians in the groups, provided we can find the right people." Polanyi's response may have surprised Oldham: "I also feel a little at a loss as to how I could contribute to the subject which you discuss. Our meetings leave me increasingly with the feeling that I have no right to describe myself as a Christian. So perhaps I may feel the part of the outsider in the discussion."[75]

It is not the case that scientists were routinely put off by a fellow scientist's making links between the aims of scientific investigation and the aims of religious worship. In the modern Western tradition, the links between science and religion are old and deep, extending from the Roman Catholic Church's patronage of scientific research through the tradition of eighteenth-century natural theology and into religious beliefs in the twentieth century.[76] A survey published in *Nature* in 1997 found that 40 percent of American scientists professed belief in an immortal soul and in a God who answers prayer, a percentage more or less unchanged since 1916. Another survey in 2008 discovered that only 55 percent of American scientists (presumably in many different disciplines) agreed to the statement that man has developed over millions of years from less advanced forms of life with no God participating in this process.[77] In Great Britain, one of Polanyi's colleagues in chemistry was the Oxford theoretical quantum chemist Charles A. Coulson, whose books of the 1950s included *Valence* (1952), *Science, Technology and the Christian* (1953), and *Science and Christian Belief* (1955). In a review of *Personal Knowledge*, Coulson wrote that Polanyi's book was the "germ of a new Christian apologetic, relevant to the twentieth century."[78]

Among the great scientists whom Polanyi knew in Berlin, the religious or spiritual dimension in Einstein's philosophical thought is well known. Einstein rejected the idea of personal immortality but wrote, "To sense that behind anything that can be experienced there is something that our minds cannot grasp, whose beauty and sublimity reaches us only indirectly: this is religiousness. In this sense, and in this sense only, I am a devoutly religious man." And, "That deeply emotional conviction of the presence of a superior reasoning power, which is revealed in the incomprehensible universe, forms my idea of God."[79] Polanyi's Berlin colleague Max Planck was a profoundly religious man and, like Coulson, he gave talks that were sermons. Planck's lecture of May 1937, "Religion und Wissenschaft," spoke explicitly of a drive toward unity that obliges us to identify "the world order of science with the God of religion."[80]

Einstein, Planck, and Polanyi shared confidence in an epistemological outlook that the scientist's knowledge establishes what Polanyi called contact with hidden reality. For Einstein and Polanyi, the existence of this reality was a matter of emotional and passionate belief or commitment on the scientist's part, while for Planck it was a result of expressed confidence in the existence of God. Among scientist-philosophers of the twentieth century, one who came very close to some of Polanyi's views was the University of Bordeaux scientist Pierre Duhem, whose philosophical work was much discussed in the 1960s following publication of Harvard philosopher Willard Quine's 1951 paper "Two Dogmas of Empiricism." Quine made use of Duhem's argument against the possibility of crucial experiments in what became known as the Duhem-Quine Underdetermination Thesis. Within three years, a much-discussed English translation appeared of Duhem's major work in philosophy of science, *The Aim and Structure of Physical Theory*.[81]

Polanyi only once mentioned Duhem in *Personal Knowledge*, using a reference from the historian of science Alistair Crombie and (incorrectly) identifying Duhem, like Henri Poincaré, as a conventionalist.[82] Had Polanyi read Duhem's *The Aim and Structure of Physical Theory*, he would have found insights from an expert practitioner in theoretical thermodynamics and physical chemistry who, like himself, tried to explain scientists' convictions in the validity of their work. As a practicing Roman Catholic who felt besieged by anticlericalism and scientism and who detested the Darwinian "monkey theory," Duhem's philosophical work was affected by his determination to give theology higher ground over science in the matter of transcendental truth, but some of his remarks demonstrate a similarity to Polanyi's way of thinking about the lessons of scientific practice.[83]

Pertinently, as Karen Merikangas Darling has argued, Duhem was convinced that the physical laws he investigated had value beyond mere usefulness and that physical theory is a reflection of a real and logically unified ontological order. "Without claiming to explain the reality hiding under the phenomena whose laws we group, we feel (*nous sentons*)," writes Duhem, "that the groupings established by our theory correspond to real affinities among the things themselves"; "while the physicist is powerless to justify this conviction, he is nonetheless powerless to rid his reason of it." The physicist's conviction in his work is an "intuition" (*une intuition*) and an "act of faith" (*un acte de foi*) which reminds us of Pascal's "reasons of the heart 'that reason does not know.'"[84] Compare this with Polanyi's *Personal Knowledge*: "He is guided by his intimations of a hidden knowledge. He senses the proximity of something unknown and strives passionately towards it."[85] "The physicist is compelled to recognize that *it would be unreasonable to work for the progress of physical theory*," writes Duhem, "*if this theory were not the increasingly better defined and more precise reflection of a metaphysics: the belief in an order transcending physics is the sole justification* (raison d'être) *of physical theory*."[86]

If this kind of realism has a name, it may be akin to what Arthur Fine calls "motivational realism." Einstein, like Polanyi or Duhem, trusted "in the rational character of reality and in its being accessible, at least to some extent, to human reason," Fine argues. This truth "drives the scientific effort with no deliberate intention or program, but *straight from the heart*."[87] Without a doubt, this kind of realism—whether in Duhem, whom Polanyi did not know, or Einstein, whom he knew and revered—is the scientific practitioner's kind of realism that Polanyi announced as his credo in *Personal Knowledge*.

As for the religious reaction to Polanyi's reflections on evolution in the last part of *Personal Knowledge*, its best known manifestation became the co-optation of Polanyi's name by the intelligent design community in the short-lived establishment in 1999 of a Michael Polanyi Center for Complexity, Information, and Design at Baylor University. After an outcry from many members of the Baylor faculty and dissension within the university, an external review committee recommended termination of the Center partly on the grounds that Polanyi's interest in science and religion and his criticisms in *Personal Knowledge* of Darwinian evolutionary theory did not include a claim that an agency such as required by intelligent design must be invoked in order to explain the "growth in complexity of the living world over aeons past."[88] Leading scholars in the Polanyi Society argued that Polanyi's teleological views in evolutionary theory do not support the intelligent design thesis.[89] What may have seemed to intelligent design advocates as crucial to their interests was Polanyi's claim that scientists can distinguish randomness in nature from order in nature.[90]

At the time of publication of *Personal Knowledge*, the book was not identified by reviewers as an example of natural theology in its treatment of evolution, but neither did Polanyi's views on evolution excite enthusiasm. One reviewer who might have been expected to be sympathetic was Joseph Henry Woodger, an experimental biologist at Middlesex Hospital Medical School who was the author of several biological textbooks and a philosopher of biology. His *The Axiomatic Method in Biology* appeared in 1937. Woodger was a member of the Theoretical Biology Club, which was established at Cambridge in the 1930s with an antimechanistic view in biology that emphasized developmental biology and the use of philosophy, mathematics, and the physical sciences as well as biology in understanding the complexity of living organisms. Other members of the Club included Conrad Hal Waddington, Dorothy Wrinch, and Dorothy and Joseph Needham.[91]

The group's members used as a starting point of their biological discussions the concept that living things are organized in hierarchical levels, an argument set out in Needham's *Order and Life* in 1936. The basic concept was that the whole organism can not be fully grasped at any one of the lower levels—the molecular, macromolecular, cellular, tissue, and so forth—and that new modes of behavior emerge at higher levels that can not be interpreted merely in terms of levels below. Needham wrote, "The hierarchy of relations from the molecular structure of carbon to the equilibrium of the species and the ecological whole will perhaps be the leading idea of the future."[92]

On the face of it, much of Polanyi's biological reading and thinking is fundamentally consistent with the approach of this group, which Needham called "organic." In *Personal Knowledge*, Polanyi cites Waddington but not Needham or Woodger in his discussion of the inadequacy of mechanical or reductive biology to account for developmental processes such as regeneration and embryonic growth or to account, for that matter, for evolutionary development in general. Polanyi mentions regulative principles such as that of localized embryonic "organizers" and Waddington's notion of the "competence" of tissues.[93]

Evolution, in Polanyi's view, is a process of fundamental innovations tending to produce ever higher biotic achievements. Evolution makes use of accidental hereditary changes that offer reproductive advantages, but it also is a process that can only be understood as one of upward complexity and achievement. Evolution works with statistical populations, but it also confers individual attributes and identities. Living beings are instances of morphological types and of operational principles subordinated to a center of individuality; no types, no operational principles, no individualities, and no new forms of life are definable in terms of physics and chemistry alone. "No new creative agent, therefore, need be said to enter an emergent system at consecutive new stages of being. Novel forms of existence take control of the system by a process of *maturation.*" The process is neither "predetermined from the start" nor the result of "continuous intensification of an external creative agency."[94] This is nothing like intelligent design. However, it is also nothing like the neo-Darwinian evolution of the Modern Synthesis engineered by R. A. Fisher, Julian Huxley, George Gaylord Simpson, and other pioneering evolutionary theorists of the mid-twentieth century. Polanyi sounds almost like Jean-Baptiste Lamarck but without the mechanism of the inheritance of acquired characteristics.[95]

In Woodger's view, Polanyi's insistence that biology must take into account "the notion of the person" is a perennial problem in the experimental biological sciences that follow closely the example of physics in method and explanatory hypotheses. Woodger offers no detailed description or evaluation of Polanyi's biological chapters but instead writes, with no commitment to Polanyi's views, "On all these topics Polanyi makes interesting suggestions. His book is a challenge to orthodoxy in many fields."[96] As for Polanyi's main theme of personal knowledge, commitment, and the fiduciary principle, Woodger concludes that Polanyi's book is haunted by the rival claims of belief and doubt "and by the inescapability of both."[97]

## Mannheim, Merton, and the Sociology of Science

There is no mention in *Personal Knowledge* or in its briefer summary in *The Tacit Dimension* of any major sociologist of the nineteenth or twentieth centuries. Nothing at all is said of Max Weber or Émile Durkheim or Talcott Parsons or even Karl Mannheim, whom Polanyi had long known and whom he saw from time to time in London before Mannheim's death in 1947. Nor is there any indication in *Personal Knowledge* that Polanyi was familiar at the time with the writings of Robert K. Merton, although Polanyi was a visitor in Cambridge, Massachusetts, and New York City and familiar with Harvard sociology.

In *The Tacit Dimension*, Polanyi refers three brief times to sociology, each time in a negative connotation, saying that his own writings show that he dissents from large tracts of scientific views, "particularly in psychology and sociology" and that he worries that "there is not one higher principle of our minds that is not in danger of being falsely explained away by psychological or sociological analysis, by historical determinism, by mechanical models or computers." All these false sirens are objects of detailed attack in *Personal Knowledge.*[98]

Polanyi's main sociological target in *Personal Knowledge* is what he describes as a sociology that teaches the public to distrust traditional morality and leads to moral inversion. These include sociologies of "aggressiveness" or "competitiveness" or "social stability," as well as sociological acceptance of the inefficacy of the individual in history as argued, he particularly mentions, by the Soviet historian M. N. Pokrovsky.[99] Primarily with Marxist sociology in mind, Polanyi argues that sociology bases its principles in a set of beliefs and then uses those principles in turn to justify the very society it studies: "it is precisely this consistency which renders the universal intent of such declarations suspect, since it shows that they lend support to established powers, after having been instilled in us by the very society which they vindicate."[100] In Polanyi's view, in his final explicit reference to sociology in *Personal Knowledge*, it is possible to build a sociology—and indeed a sociology of knowledge—that escapes determinism and relativism by abandoning objectivism and by acknowledging the role in life of spontaneous and individually unique action. In the spirit of liberal humanism, he writes:

> Objectivism requires a specifiably functioning mindless knower. To accept the indeterminacy of knowledge requires, on the contrary, that we accredit a person entitled to shape his knowing according to his own judgment, unspecifiably. This notion—applied to man—implies in its turn a sociology in which the growth of thought is acknowledged as an independent force. And such a sociology is a declaration of loyalty to a society in which truth is respected and human thought is cultivated for its own sake.[101]

Although Polanyi does not mention Mannheim, unlike Popper in *The Poverty of Historicism* and *The Open Society and Its Enemies*, Mannheim's sociology of knowledge is the shark cruising beneath the surface waters of Polanyi's argument. Polanyi and Mannheim had been discussing intellectual and political issues since their Sunday afternoon meetings at Béla Balázs's home in Budapest during the First World War. Polanyi had kept up with Mannheim's career in Heidelberg and Frankfurt in the 1920s and at the London School of Economics after Mannheim left Frankfurt for London in 1933. Both corresponded regularly with their old friend from Budapest Oscar Jászi, who despised what he called Mannheim's reduction of "individuality to zero, making it strictly determined, thus into a mere effect of social development."[102] To Jászi's criticism, Mannheim responded, "I am a sociologist who wants to discern the secret of this modern world (even if it is a ghastly one), because I believe that to be the only way for us to gain control of the new social structure instead of the new social structure gaining control of us."[103] Mannheim said that he aimed to "overcome the vague, ill-considered, and sterile form of relativism with regard to scientific knowledge which is increasingly prevalent today" in order to reformulate the foundation of the social sciences and to instill optimism in the possibility of rational scientific guidance for political life.[104]

Edward Shils and Louis Wirth translated Mannheim's 1929 book *Ideologie und Utopie* into English, along with essays of 1931 and 1935, for publication as the 1936 English-language book *Ideology and Utopia: An Introduction to the Sociology of Knowledge*. In 1940, Shils translated Mannheim's 1935 *Mensch und Gesellschaft im Zeitalter des Umbaus*, which was published as *Man and Society in an Age of Reconstruction*.[105] Shils and Mannheim corresponded regularly and discussed their disagreements, which were substantial. Shils, like Polanyi, identified Mannheim with a Socialist or Marxist point of view, although Shils conceded in an essay on *Ideology and Utopia* that Mannheim likely was not a Marxist but simply succumbed to the power of Marxism even while thinking that he had transcended it.[106] Indeed, Mannheim subjected Marxism to substantial critique in *Ideology and Utopia* alongside what he called the five most representative ideal-types of political movements: bureaucratic conservatism, conservative historicism, liberal-democratic bourgeois thought, the Socialist-Communist conception, and Fascism.[107]

Mannheim saw his sociological epistemology as one of relationism, not relativism or relativization. Just as art may be definitely dated according to its style, so each form of knowledge, or what Mannheim called a "thought-model," is possible only under certain historical conditions. We need to learn how each historical perspective may be recognized in a thought-model in order to obtain a new level of objectivity in what we know. In addition (and in this, Mannheim in 1929 sounds much like the later Polanyi or Thomas Kuhn), insofar as observers are immersed in the same system with common conceptual and categorical apparatus and a common universe of discourse, they eradicate as error whatever deviates from their unanimity. Their thought-model binds them together. When observers have different perspectives, however, what each has perceived must be understood in the light of the differences in the structure of their varied modes of perception. Thus, a goal for sociology of knowledge in reformulating the foundations of sociology is to establish the means to translate the results of one thought-model into the other and to discover a common denominator for the different perspectives. But "it is natural that here we must ask which of the various points of view is the best. And for this too there is a criterion. As in the case of visual perspective, where certain positions have the advantage of revealing the decisive features of the object, so here pre-eminence is given to that perspective which gives evidence of the greatest comprehensiveness and the greatest fruitfulness."[108] In other words, for Mannheim, a better theory was one that has superior explanatory power and a positive heuristic.

Unlike Marxists, Mannheim did not limit the determinants of thought-models to social class but extended categories of social situation to "generations, status groups, sects, occupational groups, schools, etc."[109] In fact, like Polanyi, Mannheim argued against objectivism and the notion of scientific detachment while attempting to save the possibility of valid knowledge that is not subjective: "it should be regarded as right and inevitable that a given finding should contain the traces of the position of the knower. The problem lies not in trying to hide these perspectives or in apologizing for them, but in inquiring into the question of how, granted these perspectives, knowledge and objectivity are still possible."[110] In Mannheim's opinion, the false ideal of a detached, impersonal point of view must be replaced by the ideal of an essentially human point of view which is within the limits of a human perspective, constantly striving to enlarge itself.[111]

As for right and wrong or true and false, Mannheim drew examples from the recent history of the physical sciences and discussions in relativity theory and quantum mechanics in order to try to establish the meaning of relationism. The stability of what he called the "sphere of truth" in the exact sciences had been disrupted in recent years, he wrote.

> Just as the fact that every measurement in space hinges upon the nature of light does not mean that our measurements are arbitrary, but merely that they are only valid in relation to the nature of light, so in the same way not relativism in the sense of arbitrariness but *relationism* applies to our discussions. *Relationism does not signify that there are no criteria of rightness and wrongness in a discussion.* It does insist, however, that it lies in the nature of certain assertions that they cannot be formulated absolutely, but only in terms of the perspective of a given situation.[112]

Thus, though the differences between Polanyi and Mannheim were profound, some shared thought patterns are striking. Once settled in England, Mannheim had a wide circle of friends, including Oldham. It was at Mannheim's suggestion that Polanyi was first invited to be a guest at the Moot. In the spring of 1944, Mannheim met with Polanyi in London to suggest publication of some of Polanyi's recent essays on the freedom and autonomy of science as a volume in Mannheim's series *International Library of*

Figure 8.1. Karl Mannheim. (Photograph by Elliott and Fry. ©
National Portrait Gallery, London. Used with permission.)

*Sociology and Social Reconstruction.* The book remained unfinished because
of Mannheim's death in 1947, but these essays made up the volume that
Polanyi later published as *The Logic of Liberty.*[113]

In a letter following up on that conversation in London, Polanyi reminisced to Mannheim on his youthful enthusiasm for H. G. Wells, the first
awakening of his religious interests when he was in his early twenties, and
his return to religious devotion only in the past five years. In the letter Polanyi brought up his objections to Mannheim's use of the phrase "planning
for freedom," criticized Mannheim's notion of a scientific politics, and took
issue with Mannheim's views on the limitations of historical knowledge. "As
regards the social analysis of the development of ideas, suffice to say that I

reject all social analysis of history which makes social conditions anything more than *opportunities* for a development of thought. You seem inclined to consider moral judgements on history as ludicrous, believing apparently that thought is not merely conditioned, but determined by a social or technical situation. I cannot tell you how strongly I reject such a view."[114]

Mannheim had argued, however, that the history of ideas is inadequate as history of society or knowledge, and he explicitly rejected the notion that intellectual history follows historical laws, writing that the process of knowing does not develop historically in accordance with immanent laws or from the "nature of things" or from "pure logical possibilities," or from an "inner dialectic." "The older method of intellectual history, which was oriented towards the *a priori* conception that changes in ideas were to be understood on the level of ideas (immanent intellectual history), blocked recognition of the penetration of the social process into the intellectual sphere."[115]

Mannheim responded affectionately to Polanyi's criticisms and expressed his gratitude for the intensity of their intellectual exchange. He demurred from agreeing with what he called Polanyi's too great dependence in accounting for the development of scientific knowledge on "emotional decisions, where still a further confrontation of evidence and argument is feasible."[116] Polanyi's defense of his injection of personal intuition and passion into scientific decision-making employed descriptions of the realities of the scientist's work day similar to what he had written elsewhere, for example, in his telling Mannheim:

> Believe me, the most frequently observed course of events in our laboratories gives little encouragement to such an assumption [of simple relations between evidence and theory]. Failures prevail overwhelmingly over successes, and the lack of reproducibility of phenomena is our daily experience. When you walk through such a place in the morning, you find everyone in a mess and every piece of research has just gone wrong altogether. The most important function of a director of research is to keep up his collaborator's morale in the face of these incessant disappointments . . . Evidence . . . can neither kill nor create fundamental beliefs . . . [but when a new solution] opens up before our eyes, we undergo a conversion. Henceforth we do not doubt the faith to which we have been converted, but rather reject such evidence as may seem to contradict it.[117]

Polanyi published reviews in the *Manchester Guardian* of Mannheim's posthumous 1951 collection of essays, *Freedom, Power and Democratic Planning,* and the 1952 volume *Essays on the Sociology of Knowledge.* The essays in the latter book were written during 1923–1929, a period when Mannheim "had great hopes for the future in Central Europe," according to Polanyi. Mannheim believed, Polanyi claimed, that social forces mold us into what we shall be, they determine what we shall believe, and since they manifest the true meaning of history the outcome will be right. "History has not justified Mannheim's confidence in it," but the present "bears out his analysis of the modern mind which, having consented to regard its own mental processes as determined by the existing social structure, has renounced any standing from which it might pass judgment on an act of violence which transforms the social structure." Thus Polanyi condemned what he elsewhere called moral inversion and Mannheim's self-relativization.[118]

A different reading of Mannheim's sociology of knowledge came from the young Robert Merton in the 1930s and 1940s. Merton would become one of the major figures in American sociology and the founding figure in the sociology of science. As mentioned in chapter 6, Merton studied in the 1930s at Harvard University where he came under the influence of the economic historian Edwin F. Gay, the historian of science George Sarton, and the sociologists Talcott Parsons and Pitirim Sorokin. Merton completed his doctoral dissertation, "Science, Technology and Society in Seventeenth-Century England," between 1933 and 1935. The dissertation, published in *Osiris* in 1938, had an enormous influence in studies of science and seventeenth-century English history through Merton's qualitative and quantitative demonstration of influences from religion, commerce, and industry on English science and its institutions in the age of the early scientific revolution.

Following the publication of *Ideology and Utopia*, Merton was asked to review it in George Sarton's journal *Isis*. Mannheim's book got considerably more attention after Parsons organized a session at the annual meeting of the American Sociological Society focused on discussion of Mannheim's approach.[119] Influenced in part by the writings of Weber and Durkheim, Parsons's own work in the 1930s focused on the notion of the professional "calling" and on the guild-like standards of occupational groups, and he explained medical practice as a paradigmatic case of institutional regulation and social solidarity. Similar to Mannheim, Parsons believed that a future economic order based on planning rather than on market incentives would be possible if personal behavior in the economic sphere could be geared to a commonly accorded definition of the public good and public trust. By the 1960s, Parsons's use of structural functionalism was a leading methodology in the sociology of Neil Smelser, Robert F. Bales, Edward Shils, and other American sociologists who studied small groups as the sites of social, political, and cultural activity. They also developed sociological studies of the ways in which the structure of the academy could serve as a model for the organization of society, an idea that Polanyi developed in his 1962 essay "The Republic of Science."[120]

Following his 1937 review article on Mannheim, Merton wrote essays on the sociology of knowledge in 1941 and 1945.[121] Shils sent Mannheim a copy of Merton's 1941 essay, along with Shils's own comments that had appeared in the same issue of the *Journal of Liberal Religion*. Mannheim thanked Shils in a letter of April 1941, writing of Merton, "I am most grateful for his very conscientious and sound criticism. It has the merit of being really honest . . . [but] he is still too much under the spiritual influence of the Idealistic approach."[122] This "idealistic" undercurrent perhaps refers to the influence of Sarton, but also of Sorokin whose sociology, as Merton himself described it, sought to derive aspects of knowledge not from an existential social basis, but from differing "cultural" or "ideational" mentalities. Indeed, that part of Merton's doctoral dissertation which came to be called "Merton's thesis" on the influence of Puritan values in early English science falls within the methodology of the history of ideas.[123]

As noted in an analysis by David Kaiser, Merton's complaint against the sociology of knowledge in general, whether in Mannheim or in the earlier work of Émile Durkheim on the social origin of categories of knowledge, was that it did not distinguish between different types of knowledge. The sociology of knowledge as so far developed pertains less to "positive knowledge [i.e., the natural sciences and mathematics] than to political convictions, philosophies of history, ideologies and social beliefs," Merton wrote in his 1941 essay. "Had Mannheim systematically and explicitly clarified his positions in this respect, he would have been less disposed to *assume* that the physical sciences are wholly immune from extra-theoretical influences and, correlatively, less inclined to urge that the social sciences are peculiarly subject to such influences."[124] Merton also suggested that Mannheim gave special status to the content of the natural sciences because of his immersion "in the Marxist tradition," a criticism taken up by later pioneers in the sociology of scientific knowledge.[125] More fully, Merton noted "an incipient tendency in Marxism, then, to consider natural science as standing in a relation to the economic base different from that of other spheres of knowledge and belief. In science, the focus of attention may be socially determined but not, presumably, its conceptual apparatus."[126]

For Merton, the perceptive and speculative insights of the early sociology of knowledge now required increasingly rigorous empirical studies and tests. Rather than focusing exclusively on the orientation of "men of knowledge" with respect to their data and to the larger society in which they lived, Merton decided that his units of study would be unique segments of society and their "special demands, criteria of validity, of significant knowledge, or pertinent problems, and so on," as in his earlier analysis of seventeenth-century scientists who organized themselves into scientific societies with audiences very different from the traditional universities. "Searching out such variations in effective audiences, exploring their distinctive criteria of significant and valid knowledge, relating these to their position within the society, and examining the sociopsychological processes through which these operate to constrain certain modes of thought constitutes a procedure which promises to take research in the sociology of knowledge from the plane of general imputation to that of testable empirical inquiry." Merton extended his studies of the 1940s on the internal "norms" and "values" within the scientific community to other social processes internal to the scientific community, such as scientists' emphasis on originality, concern with priority and reward systems, multiple discoveries in science, and changes of foci in interesting problems, all operating within a subset of a larger scientific culture.[127]

In 1952, Bernard Barber's book, *Science and the Social Order*, brought attention to the sociology of science as a field. Polanyi was not much impressed, writing that "its substance is slight." Still, Polanyi continued, the book raises serious issues as a product of the newer sociology, "centering particularly on Harvard," regarding evaluations of the scientist by the public-at-large and in self-evaluations and the social rewards of the scientist" by means of public prestige, financial income, and other honorific symbols.[128] Within twenty years, the sociology of science was well established, as registered in Joseph Ben-David's much-read book, *The Scientist's Role in Society* (1971), in which he praised not only Merton as a founding father in the field, but also Polanyi and Shils for bringing to academic and public attention the concept of the "scientific community" as a collectivity that evolves its own norms and policies.[129]

Merton, too, noted the significance of Polanyi's writings on the "authority structure in science and the social structural basis of scientific objectivity."[130] In a retrospective talk called "Insiders and Outsiders: A Chapter in the Sociology of Knowledge," delivered at the annual meeting of the American Sociological Association in September 1971, Merton praised Polanyi for noting "more perceptively than anyone else I know, how the growth of knowledge depends upon complex sets of social relations based on a largely institutionalized reciprocity of trust among scholars and scientists."[131] At this time, Polanyi was working on the book that

Figure 8.2. Robert K. Merton, ca. 1962. (Courtesy of Robert K. Merton Papers,
Rare Book and Manuscript Library, Columbia University, New York.)

became *Meaning*, but he was beginning to feel that his intellectual powers
were fading. He received an offprint of Merton's article with a handwritten
note from Merton: "Michael—from start (p. 10) to close (p. 44), my debt
is evident and great, affectionately, Bob." Polanyi wrote in his own hand
on the offprint cover: "A message from my chief rival, Robert Merton. New
York."[132]

## The Science and Society Movement of the 1960s

The field of the sociology of science was emerging with its own distinct
identity by the 1960s.[133] In 1963, Yale sociologist Derek de Solla Price's
volume *Little Science, Big Science* stunned readers with its demonstration
of the exponential pace of the growth of science and implications for gov-

ernment spending, social programs, and humanistic culture. Price dramatically warned that the current rate of scientific growth would result in "two scientists for every man, woman, child and dog in the population, and we should spend on them twice as much money as we had. Scientific doomsday is therefore less than a century distant."[134]

Daniel Greenberg's *The Politics of Pure Science* appeared in 1967. The youthful Greenberg had been hired in 1961 at *Science* with the assignment to write about the politics of science rather than the substance of science at a time when the weekly magazine's editors were attempting to attract a broader range of readers and educate them in matters of importance for scientific advance in the United States.[135] At that moment in 1961, President Dwight D. Eisenhower's farewell speech of January 17 was fresh in many minds. Not only had Eisenhower warned against the danger to the American social fabric of the unprecedented conjunction of an immense military establishment and a huge arms industry, but he cautioned that while "holding scientific research and discovery in respect, as we should, we must also be alert to the equal and opposite danger that public policy could itself become the captive of a scientific technological elite."[136]

Some scientists were among those who worried about the postwar expansion of science, including members of the scientific elite such as the Manhattan Project–era nuclear physicist Alvin Weinberg, who was director of the Oak Ridge National Laboratory. He wrote a commentary in *Science* in July 1961 in which he coined the phrase "Big Science" and warned his peers against diversion from scientists' real purpose, "which is the enriching and broadening of human life."[137] He chided fellow scientists for a haughty attitude toward applied research. Weinberg's message was echoed by another prominent Manhattan Project–era physicist, Philip H. Abelson, who was director of the Carnegie Institution's Geophysical Laboratory and editor of *Science*. Abelson cautioned his peers that basic scientists' chauvinism about the inviolability of their work was neither realistic nor beneficial to society.[138]

In Greenberg's *Politics of Pure Science*, he distinguished two schools of thought on the subjects of basic and applied science—what he called the "immaculate conceptionist" view and the "corruptionist" view. The first, he suggested, was exemplified by Harvey Brooks, dean of engineering and applied physics at Harvard, and the second by Robert M. Hutchins, the neo-Thomist humanist who had been president of the University of Chicago from 1929 to 1945. Hutchins viewed scientists with distrust, regarding them not as objective but as self-interested. In contrast, Brooks maintained that basic scientists adhere to a "stern ethical code associated with the question of scientific credit and priority . . . sanctioned within the scientific community" and that basic scientific research "is recognized as one of the characteristic expressions of the highest aspirations of modern man . . . it not only serves the purposes of our society but *is* one of the purposes of our society."[139]

Greenberg quoted from Michael Polanyi's *Logic of Liberty* in order to demonstrate the "immaculate conception" attitude, using Polanyi's statement that the function of public authorities is to provide opportunities for research and to allow scientists to follow their own interests.[140] Reflecting a view, which was to be found in Polanyi and Bernal, on the social nature of science, Greenberg described science as a "community," a "way of life," and a "style of behavior distinct unto itself" in which the common denominator is a "dedication to the understanding of the universe through systematic investigation and measurement, through the harnessing of curiosity, training, discipline, and instruments. (Some would add intuition and good luck.)"[141] In explaining how scientific institutions function to "advance the truth, weed out error, honor the worthy, and reject the crackpots"—a turn of phrase calling to mind Polanyi and Kuhn among others—Greenberg drew upon Bernard Barber's 1961 *Science* article on scientists' resistance to scientific discovery and on Polanyi's 1963 *Science* article on the scientific community's rejection of his early potential theory of adsorption.[142] Greenberg struck a very different note from Polanyi, however, sounding more like Bernal, in writing that "science was once a calling; today it is still a calling for many, but for many others it is simply a living, and an especially comfortable one." In Greenberg's opinion, the financial burden of the Vietnam War and President Lyndon B. Johnson's political preference for "utility in research in the interests of rapid social engineering" was going to provoke a major change in the politics of pure science. As President John F. Kennedy had said shortly before his death in November 1963, "Scientists alone can establish the objectives of their research, but society, in extending support to science, must take account of its own needs."[143]

Reactions from scientists to Greenberg's book often echoed reactions to Bernal's *Social Function of Science* almost thirty years earlier when many scientists, including Polanyi, had taken offense at Bernal's charge that too many scientists were self-indulgent in losing themselves in their own work rather than caring about the social and economic problems of their time. After reading the page proofs of Greenberg's book, the Vienna-born MIT theoretical physicist Victor Weisskopf asked Greenberg to meet him in Cambridge. According to Greenberg, Weisskopf "wanted the text 'toned down' because 'it makes us look bad.' Science is a politically delicate enterprise, Weisskopf explained, and it might be harmed by the book's description of its inner workings."[144] In contrast to Weisskopf, although the physicist Joel A. Snow noted that Greenberg's attack on "well established folkways" would offend many scientists, Snow himself thought Greenberg's analysis was timely, exciting, and generally quite perceptive."[145] This was the favorable opinion, too, of many young mavericks in graduate programs around the country, Steven Shapin among them. Shapin was then a graduate student in genetics at the University of Wisconsin, and he later wrote the foreword for a reissue of Greenberg's book.[146]

On the other side of the Atlantic, another Snow was stirring up discussion about the social nature of science and the social responsibility of science in ways that echoed the Bernal debates of the 1930s and presaged the dichotomy that Greenberg drew in 1967 between the immaculate conceptionist view of the scientist Harvey Brooks and the corruptionist notion of the humanist Robert Hutchins. In the 1950s and 1960s, the physicist C. P. Snow was among a group of scientists, including Bernal, who advised on matters of science and technology for Labour Party leaders Hugh Gaitskell and Harold Wilson. The group argued for greater government support of science and for the establishment of a new Ministry of Technology in what became Harold Wilson's Labour government in 1964. In May 1959, Snow delivered the later much-debated lecture "The Two Cultures and the Scientific Revolution," in the Senate House at Cambridge. One of his targets was literary modernism, personified by his Cambridge colleague F. R. Leavis, who joined T. S. Eliot and other modernists in deploring scientific materialism and literary realism. In his lecture Snow attacked the self-indulgence of literary intellectuals who pined for a premodern world, and he extolled the superior moral values of forward-looking scientists. He also argued for the expansion of scientific education and for greater government investment in science and technology in the interests of social welfare at home and abroad.[147]

Although Polanyi did not like Snow's politics, he admired Snow's narrative novels about scientific life, having written in an essay of 1949 that "the most penetrating and also most moving representation of the young scientist's struggles has come from novelists, like Sinclair Lewis, C. P. Snow and Neville Shute."[148] Polanyi was not keen on Snow's "Two Cultures" lecture, however. He warned in *Encounter* of the dangers of the scientific rationalism extolled by Snow and returned to the theme of the moral inversion of science in the 1930s and 1940s. Polanyi seemed to side with the literary modernists, applauding artistic work for "pulling itself up out of a universal descent into realism and sentimentalism" and repeating his argument in *Personal Knowledge* of the need to revise the claims of science away from the ideal of impersonal and detached objectivity.[149]

Figure 8.3. Michael Polanyi in his study at home on Upland Park Road, Oxford, in the late 1950s. (Courtesy of the Special Collections Research Center, University of Chicago Library, and John C. Polanyi.)

Some years later, while browsing in a Cambridge bookstore, Leavis came across the collection of essays *Knowing and Being*, edited by Marjorie Grene, which included Polanyi's essay on the "Two Cultures." Leavis began referring students and friends to Polanyi and Grene, and he started paying attention to them in his own writings. He found Polanyi's emphasis on the inseparability between knowing and being a support for his own conception that language is not merely descriptive but creative.[150] David Hollinger has argued that Snow hit a nerve in the literary community in the early 1960s which became a factor in its regrouping, "under the cover of Michel Foucault and postmodernism," in order to wrench from scientists the mantle of democratic values and human decency. Guy Ortolano further suggested in *The Two Cultures Controversy* (2009) that assumptions and approaches in the British social construction of scientific knowledge movement in the 1970s echo Leavis's critiques of Snow, scientific rationalism, and technocratic elites.[151]

Among those in Great Britain who paid attention to these debates and began contributing to the burgeoning literature in the sociology of science in the 1960s were John M. Ziman and Jerome Ravetz, both members of Kuhn's transitional generation of intellectuals born in the 1920s. Ziman was a solid-state physicist at Bristol University, and he decided, like Polanyi, that he wanted to explain to a general audience how science works from the standpoint of a "personal witness."[152] He argued that there is no logical method for doing science but that science must be understood in terms of its internal social relations. After reading Hanson, Kuhn, Polanyi, Shils, and the psychologist Patrick Meredith, among others, Ziman settled upon a striking formula, which, he said, meant no antagonism to Polanyi's *Personal Knowledge*, who "goes a long way along the path I follow, and is one of the few writers on Science who have seen the social relations between scientists as a key factor in its nature."[153]

Science, Ziman astutely argued, is *Public Knowledge* within which the principle of freely accepted consensus is fundamental. Science is a community that functions under common social norms—one of Polanyi's insights. Science requires a society in which there is general freedom of speech and comment. Science is not an individual vocation, but a system of deliberate apprenticeship, which was first fully systematized, Ziman argues, in the German university system, where the scientist (such as Polanyi in his days in Berlin) worked not only for the advancement of learning or "his own amusement" but also for personal promotion and reward.[154] Science is a community of coordinated actions where the validation of authority within any one field is mediated by the opinions of authorities in neighboring fields who are competent to make judgments in an overlapping manner.[155] Whereas Polanyi had likened this coordination by mutual adjustments within the scientific community to a free market guided by a "hidden hand," Ziman portrayed interlocking "markets" in which intangible commodities such as "repute" are traded for more tangible currencies such as academic employment. Ziman did not share Polanyi's confidence that the market mechanisms converge on scientific truth, and he preferred the claim that science arrives at reliable knowledge.[156] Ziman opened his book, the first of many that he wrote on the social nature of science, with a Polanyi-like or Pascal-like quotation from Richard Feynman: "A very great deal more truth can become known than can be proved."[157]

Jerome Ravetz's *Scientific Knowledge and Its Social Problems* appeared in 1971. Ravetz had studied with Toulmin at Leeds, and he began giving a short course during 1962–1963 on science and society in the Workers' Educational Association, the organization where Karl Polanyi had taught in the 1930s and 1940s. Following discussion with the historian of science and technology Donald S. L. Cardwell, Ravetz incorporated ideas from Polanyi's *Personal Knowledge* into his WEA course and into his 1971 book. "I am indebted to Michael Polanyi," wrote Ravetz, "for his systematic development of the insight that science is craft work . . . much of this present work derives from an attempt to solve the problems which were raised by his analysis; in particular, how objective scientific knowledge can result from the intensely personal and fallible endeavor of creative scientific inquiry."[158]

In contrast to Polanyi, however, Ziman and Ravetz were concerned not only with the nature of science as a social community but also with the social responsibilities of science and with ongoing interactions between science and technology. In Ravetz's view, "The process of industrialization is irreversible; and the innocence of academic science cannot be regained." Not only is science responsible to society at large for the solution of technical problems that are being set by industry and the State, there also is need for a "critical science" in which scientists address themselves to current and potential threats to human health and welfare that are the unintended results of science and technology.[159] In Ziman's view, as with Ravetz, scientific research should be undertaken for its foreseeable social benefits, although investigators must retain a great deal of autonomy in the performance of their research.[160]

As members of a transitional generation in social epistemology of science, Ravetz and Ziman adopted their predecessors' views on the social nature of science but parted company with those like Polanyi who insisted on the insulation of scientists from the demands of state or industry or on the firm distinction between basic and applied science. Bernal's ideas were newly appealing as it became easier politically to ignore Bernal's Marxist and Stalinist rhetoric in a post-Stalinist era. Similarly, many elements of Polanyi's social epistemology of science could be adopted without using Polanyi's ideology of anti-Stalinism, anti-planning, the free market, and the distinction between pure and applied science. Kuhn's work, however implicitly it carried a political ideology of the status quo, could serve as a bridge, along with the sociologically informed work of his generational peers, into the philosophically radical work of the next generation in the social construction of science.

# SSK, Constructivism, and the Paradoxical Legacy Of Polanyi and the 1930s Generation

Among the three founding generations in the social epistemology of science, David Edge was a member of the transitional second generation in Great Britain. Born in 1932, Edge helped establish the Edinburgh program in sociology of scientific knowledge, which became known as SSK. In discussing the origins of the program, Edge later noted the impact in Great Britain of the kind of argument that Derek de Solla Price was making about big science, as well as the view of Manchester's Lord Bowden (B. V. Bowden), who was briefly minister for education and science in the Labour government, that limitations had to be made on scientific growth. That argument was countered immediately by J. D. Bernal, then in his sixties, who suggested in the *New Scientist* that scientific specialties with the fastest rising curves of growth should be funded as generously as possible.[1]

There was general political agreement at this time in Great Britain, as in the United States and elsewhere, on the need for empirically based science policy and for reform in science education. Snow had spoken to this point in his "Two Cultures" speech. Administrators and academics, including Snow, argued for a more "liberal" and "human" introduction to science in courses that would include study of the social nature of scientific knowledge and of scientists' social responsibility. In Great Britain and America, the arguments for educational reforms were influenced by leaders of environmental, nuclear disarmament, feminist, and anti–Vietnam War movements, often allied in the United States with the civil rights movement, who called into question the influence of scientific elites and the misuse of science and technology for political ends.

Edge's Science Studies Unit at the University of Edinburgh in 1966 was one of the institutional innovations that resulted from these kinds of discussions. The biologist Conrad Hal Waddington, whose notions on the social responsibility of science enraged Polanyi in the 1940s, made the initial proposal in 1964 for a new unit at Edinburgh that would develop an interdisciplinary program of teaching and research to broaden undergraduate education in the science faculty.[2] The unit's first members all were men with scientific training. Edge was a physicist who had worked for the BBC, and he was in charge of setting up the teaching faculty. Barry Barnes had a natural sciences education with postgraduate certification in social theory, and David Bloor had a PhD in psychology and philosophical interests. Gary Werskey, who completed a PhD in history from Harvard University, became another member of the group in the early 1970s. So did Steven Shapin, who completed some graduate work in genetics before finishing a PhD in the Department of History and Sociology of Science at the University of Pennsylvania. The aim of the Edinburgh group's program, as described by Edge, was "to develop an empirically informed view of the social nature of scientific knowledge," drawing inspiration from the work of Thomas Kuhn and, to a lesser extent, J. D. Bernal and Michael Polanyi. In 1970, together with Roy MacLeod at Sussex University, Edge established the journal *Science Studies*, which soon was renamed *Social Studies of Science*. Edge was a founding member of the Society for the Social Studies of Science in 1975, and Robert K. Merton served as the organization's first president.[3]

As emphasized by Michael Friedman in an analysis of the philosophical agenda of SSK, its main thrust aimed for what Harry Collins at the University of Bath called an "empirical program of relativism" in which, as Barnes and Bloor explained, "there is no sense attached to the idea that some standards or beliefs are really rational as distinct from merely locally accepted as such." An equivalence postulate requires that "all beliefs are on a par with one another with respect to the causes of their credibility . . . regardless of whether the sociologist evaluates a belief as true or rational, or as false and irrational."[4] Bloor coined the term "strong program in the sociology of science" in 1976 for an approach that aimed to be causal, impartial, symmetrical, and reflexive in its analysis of science.[5]

On the matter of the correspondence of scientific theories with reality, Shapin explained the issue succinctly in 1982:

> If scientific representations were simply determined by the nature of reality, then no sociological accounts of the production and evaluation of scientific knowledge would be offered . . . It would be pointless to argue against the kind of naïve realism and positivism which has few, if any, philosophical proponents at present. The underdetermination of scientific accounts of real-

ity and the "theory-laden" nature of fact-statements are both quite widely accepted.... the sociology of knowledge is built upon an appreciation of the contingent circumstances affecting the production and evaluation of scientific accounts.[6]

Collins became associated with a more extreme expression than Shapin's on the correspondence of scientific knowledge with reality by writing in 1981 of a new relativism in which "the natural world has a small or nonexistent role in the construction of scientific knowledge."[7]

The strong program in SSK thus brought back into the sociology of science the methodologies of the sociology of knowledge, but SSK ventured an explicit and full analysis of the mathematical and natural sciences that Karl Mannheim had not attempted. Mannheim had adopted instead the assumption that mathematics and natural science are "determined to a large extent by immanent factors."[8] In contrast, SSK adopted a relativistic epistemology in preference to the perspectivist epistemology preferred by Mannheim. SSK also firmly differentiated its methods and aims from functionalist sociology of science that studied scientific norms and institutions in the style of the American school of Robert K. Merton, and SSK denied a clear distinction between science and technology, allying science studies with technology studies. SSK little concerned itself with Marxist or materialist analysis, instead adopting the critique that Marxism perpetuated a naïve and utopian view of the power and promise of scientific method and of scientific progress. At the same time, the SSK impartiality principle undercut the liberal argument that there is a necessary connection between democracy and science because the impartiality principle deprivileged any ideological system over another. Impartiality left open the possibility for the demonstration in science studies of scientific achievements in nondemocratic regimes.[9] In this respect, SSK was directly at odds with the main agenda of Polanyi's life work while drawing upon aspects of that work.

The sociological approach in France of Bruno Latour was initially allied with SSK but it had more radical aims and it had different roots within French anthropology, sociology, philosophy, and politics, notwithstanding Anglo-American collaborations and familiarity with the English-language literature.[10] While SSK concerned itself with "society in science," Latour and his colleagues at the École des Mines in Paris focused on "science in society" in the manner of the philosopher Michel Foucault, who studied how scientific knowledge and technical expertise reproduce and enforce dominant social interests and the existing political order.[11]

Turning John Stuart Mill's nineteenth-century critique against Auguste

Comte's positivist program into a twentieth-century brief against French technoscientific culture, Latour's agenda became one of destabilizing the centralized, elitist, and, in his view, antidemocratic structure of French science and French society. In other words, Latour's agenda did not simply call into question the alliance of science with the state, but the legitimacy of the state itself. From the French point of view, explained Latour and his colleague Geof Bowker in 1987, the work of Bloor, Shapin, and Collins stated the obvious. What was needed was to go further and recognize that science is at the service of reciprocal networks of power relationships in which the state depends upon scientific and technical expertise for its governing discourse and for purposes of coercion. It was necessary to "sociologically reconstruct society" through a critique of the universality of science. An attack on universal and objective science, Latour explained, is an attack on the state, which claims in France to represent rationality and historical transcendence. Latour's views clearly conflicted with Polanyi's commitment to the necessary authority of traditional elites and the role of scientific experts in parliamentary and republican democracy.[12]

Latour adopted a structuralist and constructivist approach, including what he called "actor-network theory" in which there are both human and nonhuman natural actors, and he argued that scientists constitute or construct within this actor network the experimental entities that scientists study in the laboratory rather than discovering or representing real entities.[13] He faulted Foucault for failing to apply his own repertoire of *mentalités*, *épistèmes*, codes, and discourse to the hard sciences. Latour also criticized the senior French sociologist Pierre Bourdieu for backing away from the application to the hard sciences of Bourdieu's concept of the intellectual "field" as a network of agents and distribution of power. Bourdieu had failed, Latour charged, to get full measure from his insight that knowledge is "symbolic capital" which can be "wielded as a weapon, hoarded like a treasure, or re-invested like capital." In Latour's view, Bourdieu made the mistake that Latour also attributed to his "Anglo-Saxon" colleagues in SSK—the error of regarding his own work as empirical and scientific in character and capable of arriving at a neutral or privileged view despite the rhetoric of reflexivity.[14]

Within a few years, Bourdieu replied to Latour's critique in his last lecture course at the Collège de France in the spring of 2001. In these lectures, Bourdieu focused on what he perceived as the problem of the sociology of scientific knowledge and its dangers to modern science. Indeed, he laid out a case that echoes some of Polanyi's fears from half a century earlier. The autonomy that science gradually had won against religious, political, and economic powers and against state bureaucracies was now greatly weakened, Bourdieu warned, and this autonomy is weakened precisely at the time that "external critiques and internal denigration, most recently presented in 'postmodern' rantings" undermine confidence at large in science and especially in the social sciences. Particularly in areas where research results are highly profitable, such as in medicine, genetics, and military research, Bourdieu argued, research scientists are falling under external economic control, and the boundary is blurred increasingly between fundamental and applied research. "Disinterested scientists, who have no program other than the one that springs from the logic of their research and who know how to make the strict minimum of concessions to 'commercial' demands to secure the funding they need for their work risk marginalization in some research areas because of the inadequacy of public support despite the internal recognition that they receive." These conditions, Bourdieu concluded, have made it all the more necessary to analyze science historically and sociologically without relativizing scientific knowledge in order to better understand the social mechanisms that orient scientific practice.[15]

Bourdieu made specific use of what he called Polanyi's "classic and much quoted text" *Logic of Liberty* for statements on the tacit dimension of science, the mastery associated with connoisseurship, the protocols and rules of science, and the force of conviction for a "true idea" that prevails over rivals.[16] Bourdieu explicitly took to task the SSK program and the constructivist program, and he reiterated the decades-old question in the sociology and history of science: "How is it possible for a historical activity, such as scientific activity, to produce trans-historical truths, independent of history, detached from all bonds with both place and time and therefore eternally and universally valid?"[17] Though mindful of the need for reflexivity in his own field, Bourdieu nonetheless concluded that there is truth value or objectivity in scientific knowledge based "not on the subjective self-evidence of an isolated individual but on collective experience, regulated by norms of communication and argumentation" in the "collective validation performed in the quite singular conditions of the scientific field."[18] For Bourdieu, science was social, but, in contrast to Latour, Bourdieu retained belief in the truth value of scientific knowledge.

By the time of Bourdieu's 2001 lectures, SSK and scientific constructivism had become identified with what Bourdieu termed the "postmodern rantings" of anti-science. If C. P. Snow had lamented a gap of misunderstanding between the "two cultures" of the humanities and the sciences in 1959, the gap seemed to have collapsed into a chasm in the so-called science wars or culture wars of the 1990s, when scientists and science-policy advocates squared off against science studies scholars and literary postmodernists.[19] Anti-science was by no means a fair accusation against all science studies scholars, since many of them respected science following their own scientific training or their familiarity with contemporary science and its history. Shapin, for example, explicitly denied the accuracy of the anti-science accusation. His claim was supported by his own historical work demonstrating the virtues of scientists, such as their mutual trust and dedication, and by his argument that many twentieth-century technoscientists regard what they do not as a job, but as a calling, in the spirit of their seventeenth-century predecessors.[20] Echoing Polanyi and Max Weber, Shapin argued the persistent character of science as a form of intensely personal knowledge that depends on people and their personal interactions. Science is a matter of "personal virtue, familiarity, and charisma" and "passion, commitment, and vision."[21]

Shapin had worried as early as 1992 that SSK and scientific constructivism were going off track. He recalled how political culture had undermined trustful acceptance of the so-called value neutrality and impartial objectivity of science for his generation in the 1960s. It had no longer been easy to believe with Merton that science embodies a liberal society and an autonomous marketplace, or with idealist historians that scientific knowledge is the achievement of philosopher-scientists, or with socialists that scientific agendas should meet the needs of society. By way of contrast, Shapin said of his generation, "We are different. The generation which began coming into the history and sociology of science from the mid-1970s has been relatively free of any such commitments." There had been disturbing consequences over the course of the last two decades, however: "The sons and daughters of luxury have tended to be more interested in performing rites of disciplinary purification than in changing the world, while the most luxurious delusion of all has been that disciplinary purification is itself an effective means of changing the world . . . The price of purity is privacy."[22] As Gary Werskey noted, Shapin's remarks, which were written just before the explosion of the science wars, were "positively prophetic."[23]

To be sure, there were those in studies of science in the 1970s and 1980s who adopted a social epistemology of science but who also had political agendas—feminists prominent among them. They saw in the description of science by Merton and Polanyi and Bernal (despite Polanyi's and Bernal's collaborations with major women scientists) a defense of a scientific elite in which women, along with dark-skinned people of the Western and postcolonial worlds, were excluded. Some feminists not only protested the gender and race exclusivity of the scientific community but wondered whether the content of scientific knowledge might be different if women or non-Westerners had been in positions as gatekeepers. They noted, too, that SSK seemed little interested in the idea that gender might be a significant factor in the social production of knowledge. Rethinking the social construction of knowledge need not lead to relativism, however, some argued. Donna Haraway, for example, wrote of "partial, locatable, critical knowledge" within science, and Helen Longino sought to "reconcile the claim that scientific inquiry is value-or-ideology-laden *and* that it is productive of knowledge."[24]

In an article for the first *Handbook of Science and Technology Studies*, published in 1995, Edge focused on the tension in the "Broad Church" of science studies between a "critical" or academic strand, exemplified in SSK, and a "technocratic" or policy-oriented strand. Edge mentioned Steve Fuller's formulation of the high church/low church distinction to describe these two trends, a distinction to which Fuller returned in his book *Thomas Kuhn: A Philosophical History for Our Times* in which the author alleged that SSK had succumbed to what he saw as Kuhn's political neutrality and acceptance of the status quo.[25]

By the early 2000s, Latour was expressing alarm at the ways in which the political deniers of global warming were using his arguments on the uncertainty of science against oceanic, atmospheric, and environmental scientists. To his dismay, Latour found that his critique of science was being used against his own political interests. He had spent years demonstrating *"'the lack of scientific certainty'* inherent in the construction of facts. I intended to emancipate the public from prematurely naturalized objectified facts. Now dangerous extremists are using the very same argument of social construction to destroy hard-won evidence that could save our lives." These are "weapons smuggled through a fuzzy border to the wrong party, [but] these are our weapons nonetheless."[26] In the same vein, Harry Collins and Richard Evans wrote in 2002 that, after teaching people that scientists do not have special access to the truth, science studies scholars now have an obligation to study how decisions are to be made when science and technology intersect with public interests."[27]

The historical wheel had turned. Latour referred in an article in 2007 to the "concerned scientists" of Bernal's generation, writing that scholars in social studies of science must pay attention not just to matters of fact, but to matters of "concern."[28] Just as Polanyi and his generation in the 1930s had developed specific sets of ideas about science in order to fight the ideological battles of their time and had lived to see those weapons used for very different purposes by a later generation in a new era, so it was happening again. The origins of the social construction of science are found in what I have called the 1930s generation. That first generation prominently included natural and social scientists such as Polanyi, Mannheim, Fleck, Bernal, and Merton. They were scientists, and they were "pro-science" in their writings. Their aim was to reinforce confidence in scientists, in the scientific community, and in the validity of scientific knowledge. The writings of this first generation on the social nature of science, the sociology of science, and the social epistemology of science were meant to strengthen public trust in science by demonstrating the stable foundations of science as a consequence of its institutionalized norms, values, and interpretive frameworks. Polanyi and Fleck showed how everyday scientific practice and its traditions or thought styles could be expected to be conservative in dealing with new facts that might break with the dominant interpretative framework. Polanyi aimed to show, too, that science was grossly misrepresented when it was described by philosophical empirical positivists as merely a cold machinery of fact-fed logic. In contrast, he argued that science thrives only because it is a succession of ongoing, everyday acts of personal creativity, passion, and belief.

It is the argument of this book that Polanyi's concern with a new epistemology of science evolved out of the experiences of his changing scientific career in Austro-Hungary, Germany, and Great Britain during the revolutionary and catastrophic decades of the early twentieth century. Polanyi was a member of the twice-exiled Hungarian refugee generation that moved from the intellectual periphery in Budapest to Berlin and Austria and then on to other world capital cities. These central Europeans were of all political persuasions in an era when war, revolution, economic collapse, and everything else was possible—monarchy, democracy, Socialism, liberalism, Communism, Fascism, Nazism. For the physical scientists among them, it was a period of startling intellectual change in the development of electron theory, relativity theory, quantum mechanics, and new tools to study the atom and the molecule. Having grown up in a secular and intellectual family of Jewish origins in the dual monarchy and having been educated in a culture of liberalism, *Bildung*, and science, Polanyi believed in the moral and spiritual value of learning and education and—above all—in science. Late in life, Polanyi wrote of science as the "polestar" that had inspired him since his childhood.[29] The values of his youth were ones that he did not relinquish.

Perhaps it was the career itinerary from Budapest to Karlsruhe to Berlin and then to Manchester that gradually demonstrated to Polanyi the importance of social organization and social networks in science, the significance of disciplinary apprenticeship, situated knowledge and hierarchies of expertise, and the tension between the vying impulses toward innovation and stability in scientific life. His early attempts to build his reputation in physical chemistry convinced him of the necessity to fit scientific results within current popular interpretive frameworks or paradigms. His efforts at building up laboratories and training experimental collaborators reinforced his recognition of the essential part in science of tacit and unarticulated skills of doing and knowing. The sometimes precarious funding of his scientific work and his dependence on private, state, and industrial funds in Berlin and Manchester inculcated in him a fierce loyalty to fundamental, or "pure," science, whether or not it ended up in practical applications that met social needs.

As we have seen, his experiences of the economic collapses of the 1920s and the 1930s, along with his horror at conditions of life in the Soviet Union, the rise of technocratic despotisms, and the particular circumstances of the Lysenko-Vavilov affair led Polanyi into economic and political writings. These writings became the basis of a crusade for a new epistemology of science that would protect science from false pretenders to scientific knowledge and from the estrangement of the sciences and humanities that could result in the enrollment of science for inhuman ends. Polanyi decided that exclusive emphasis on empiricism and logic as the heart of science was not only erroneous, but politically and ethically dangerous.

Polanyi's philosophical writings carried a moral imperative that can be found among his opponents and allies alike. Among the opponents was Popper. Among the allies was Arthur Koestler, whom Polanyi saw regularly from the 1940s to 1960s in London, Wales, and Alpbach, Austria. Particularly important to his philosophical project were conversations with Koestler while on long holidays in Wales during 1946 and 1947, including a two-week stay with their wives at Koestler's rented cottage at Bwlch Ocyn in September 1947. Some of Polanyi's main concerns also inspired Koestler's writings in this period, prominently among them *The Sleepwalkers*. This loosely historical account of Copernicus, Kepler, Galileo, and Newton appeared in 1959 with a preface by historian of science Herbert Butterfield. In describing the heroes of early-modern science as sleepwalkers, Koestler tried to show that scientific progress defies logical explanation and that scientists often cannot explain where they were going or how they got there. Like Polanyi, Koestler believed that reason and faith had been fruitfully linked before the hegemony of objectivity in the post-Enlightenment period and that confidence in meaning and truth needed to be restored.[30]

Neal Ascherson has described the generational obsessions that Koestler and Polanyi shared and how, in Ascherson's view, their era differed from the present.

The whole context in which Koestler fought, survived, preached and rampaged—the epoch of totalitarian dictatorships, millennial mobilisations and total wars—has vanished. And with it have gone (or at least temporarily subsided) the classic moral choices which overshadowed the consciences of so many 20th-century men and women: whether to sacrifice a society's today for a "brighter tomorrow," whether to ally with the lesser evil to overcome the greater, whether to shed innocent blood as the price of breaking humanity's chains.[31]

When Polanyi abandoned his work in physical chemistry in order to launch a crusade on behalf of a liberal and humane science in a free society, he did so without having experienced the apprenticeship and mentorship in social science and philosophy that he knew were necessary for membership in a discipline. As a consequence he was often regarded as an outsider or an amateur except when he could bring to bear his own scientific experience to a question. Raymond Aron reported that the Russian-born Oxford philosopher and historian Isaiah Berlin, who left the Russian Empire in 1921 for Great Britain, offered the opinion on Polanyi that "these Hungarians are strange . . . here is a great scientist giving up the Nobel to write mediocre works of philosophy."[32] Whether or not Polanyi's work outside physical chemistry was amateurish or mediocre, as judged by specialists, there is no question that it was influential, and in no field was it more influential than in the social construction of science.

There is paradox and irony in this outcome. Much of what Polanyi wrote about science, if taken out of context, could be used for ends antithetical to his own aims and values. Polanyi had objected to his brother Karl's economic history, as well as to Mannheim's sociology of knowledge, on the grounds of its historicism and determinism. This objection, as we have seen, was leveled by Paul Feyerabend against Kuhn. SSK, like Kuhn, transformed Polanyi's arguments against objective scientific knowledge into a denial of realism, or Truth, in scientific explanation, a position that was anathema to Polanyi. Polanyi's emphasis on faith and commitment among scientists, along with descriptions by Polanyi, Merton, and Bernal of the authority structure and behavioral norms in science, gave support to the claims of scientific constructivism that science is all about struggles for power in which the winning side convinces others only of its own beliefs.

Scientific theories were merely beliefs. Polanyi's argument against Popper that scientists do not function by acting from doubt and skepticism became practical evidence in SSK of the underdetermination of science by evidence and of the impossibility of arriving at scientific certainty. Despite lip service to Polanyi by SSK, it was Popper's emphasis on the virtues of pluralism that became an antidote to what the third generation perceived as the elitism and dogmatism in science that had been illuminated by Polanyi and Kuhn. Confidence in science, scientific credibility, and the universality of scientific laws was absent in most of science studies in the 1970s and 1980s.

A final irony in the legacy of Polanyi and his generation lies in the renewed focus in SSK and scientific constructivism, as in science studies more broadly, on the social responsibility of science and of social studies of science. Polanyi zealously opposed these kinds of arguments, but the concerns of Bernal, freed in the late twentieth century from the Stalinist Marxism of his own life, became a new preoccupation in what Collins called the "third wave" of the social construction of science, which, from Collins's perspective, began with Kuhn in the 1950s and 1960s.

Each generation of readers can select what it likes from the past. Postmodernism tells us that authors cannot control the uses of their texts in different times and places, but we need only history to teach us this lesson. The ideas and arguments that Michael Polanyi developed in order to fight the battles of the 1930s and 1940s changed their meaning for a generation with very different battles to fight in the 1960s and 1970s, and they were being transformed again at the turn of the twenty-first century. No doubt Polanyi would have been disturbed as much as Latour has been dismayed at later uses of his novel ideas by readers and thinkers whom he could only see as enemies of his beliefs. It is ironic, but not surprising, that Polanyi and other members of his generation have been enlisted for ends with which they might not entirely sympathize, however great the achievement and influence of their ideas and lives.

**BJHS:** *British Journal for the History of Science*

**BP:** Papers of Patrick Maynard Stuart Blackett, OM FRS, Baron Blackett of Chelsea, Royal Society Library, London

**HSPS:** *Historical Studies in the Physical Sciences* (from 1986, *Historical Studies in the Physical and Biological Sciences*)

**JACS:** *Journal of the American Chemical Society*

**MPG:** Max Planck Gesellschaft Archives, Berlin-Dahlem

**MPP:** Michael Polanyi Papers, Special Collections Research Center, University of Chicago Library, Chicago, Illinois

**PKVA:** Protokoll vid Kungl, Vetenskapsakademiens Sammankomster för Behandling af Ärenden Rörande Nobelstiftelsen (Minutes of Meetings of the Royal Swedish Academy of Sciences, for Discussion of Matters concerning the Nobel Foundation), The Nobel Archive of the Royal Swedish Academy of Sciences, Stockholm

**PP:** Linus and Ava Helen Pauling Papers, Special Collections, Valley Library, Oregon State University, Corvallis

**RFA:** Rockefeller Foundation Archives, Rockefeller Archive Center, Pocantico Hills, Sleepy Hollow, New York

**SFS:** Society for Freedom in Science

**SHQP:** Sources for the History of Quantum Physics, Niels Bohr Library, American Institute of Physics, College Park, Maryland

**TFS:** *Transactions of the Faraday Society*

**UMA:** John Rylands University Archives, John Rylands University Library, University of Manchester

**VdpG:** *Verhandlungen der deutschen physikalischen Gesellschaft* (Proceedings of the German Physical Society)

**ZE**:  *Zeitschrift für Electrochemie und angewandte physikalische Chemie* (Journal for electro-chemistry and applied physical chemistry)

**ZP**:  *Zeitschrift für Physik* (Journal for physics)

**Z*p*C**:  *Zeitschrift für physikalische Chemie* (Journal for physical chemistry)

NOTES

INTRODUCTION

1. Forman 1967, 1971. See also Forman 2010.
2. Unguru 1975.
3. Barnes and Shapin 1977, 61–62. See also Zammito 2004, 141–42.
4. Goldman 2006; Fuller 2002, 2009; Bloor 1991, 7.
5. Shapin 2010b.
6. Kuhn 1962, 44n1.
7. For histories of social studies of science and its various subspecialties, including SSK and constructivism, see Bucchi 2004; Sismondo 2004; and Yearley 2005. These books have little to say about events or ideas before Kuhn. Stephen Turner (2008) writes about earlier discussions of the social character of science in historical periods from Bacon to Kuhn.
8. Wigner and Hodgkin 1977; Scott and Moleski 2005.
9. Useful guides to *Personal Knowledge* are Polanyi 1959b; Grene 1958; and Jha 2002.
10. On the post-Kantian and early-nineteenth-century origins of modern notions of objectivity and subjectivity, see Galison and Daston 2007, 30–31.
11. On the prayerful search for God, Polanyi 1964, 34–35.
12. In a notice of the 2009 reissue by the University of Chicago Press of Polanyi's 1966 collection of essays *The Tacit Dimension*, Steven French voices mystification at Polanyi's "cosmic panorama" and references to a fateful conflict between the "moral skepticism of science and the moral demands of modern man." French notes how far distant Polanyi's work lies from mainstream philosophy (French 2010, 157–58, quoting from Polanyi 2009, 57).
13. Golinski 1998, 7.

CHAPTER ONE

1. On the "Hungarian phenomenon," see Palló 1990, 319, 320; 1991, 85. Further, see Frank 2009; Congdon 1991; Somlyody and Somlyody 2003. Quoted material is from Mannheim 1936, 154–55.
2. Hacohen 1999, 2000. Knepper (2005, 285–86) suggests that Polanyi, unlike Popper, drew upon religious spirituality and Protestant social forms of organization in the development of philosophy of science.

3. Quoted in Marton 2006, 11. On Hungarian scientists and patterns of creativity, see Palló 2005. See also Fermi 1968, 53–59.
4. See Scott 1998–1999, 11; Lukacs 1993, 151.
5. Scott 1998–1999, 11; Szapor, 1997, 1. On Paul Polanyi, see Scott and Moleski 2005, 12.
6. Scott and Moleski 2005, 6–10.
7. Knepper 2005, 265; and Kovács 2003, 311.
8. Knepper 2005, 266.
9. On Germanization of names, see Lukacs 1993, 96. On the great-grandfather, see Scott and Moleski 2005, 3.
10. Scott and Moleski 2005, 3–6.
11. Knepper 2005, 265.
12. Ibid.; Lukacs, 1993, 96.
13. Gábor Palló, e-mail correspondence with the author, December 30, 2009.
14. On the successes of Jews in Budapest, see Marton 2006, 17.
15. Knepper 2005, 271.
16. Lukacs 1993, 57–58.
17. Wigner 1992, 77.
18. Scott and Moleski 2005, 15; Szapor 1997, 2. On the schools in Budapest, see Palló 2005, 223; Frank 2009, 55–78.
19. Scott and Moleski 2005, 16; Kovács 2003, 334–35.
20. Litván 2006, 25–26. See also Lukacs 1993, 196–97.
21. Quoted in Lukacs 1993, 199.
22. Mucsi 1990, 27.
23. Quoted in Lukacs 1993, 199.
24. Litván 2006, 25–26.
25. Scott and Moleski 2005, 21–22.
26. Mucsi, 1990, 28; Scott and Moleski 2005, 22.
27. Duczynska, n.d.
28. On Pólya's memories, see Scott 1998–1999, 13; and on Einstein's remark, see Kovács 2003, 312.
29. Scott and Moleski 2005, 24–25. For the hydrocephalic liquid paper, see Polanyi 1911a; for the blood serum paper, see Polanyi 1911b.
30. Scott and Moleski 2005, 33–37; Scott 1998–1999, 15. On Polanyi's early studies, see Palló 1998, 40–41.
31. In the first month of the war, a hundred thousand Austro-Hungarian soldiers were killed in Galicia and a similar number taken prisoner. See Singer 2003, 115, 124–26 for an account of his grandfather's service in Galicia and his physical and mental conditions following that service.
32. Quoted in Scott and Moleski 2005, 46, in letter from Karl Polanyi to Irma Pollacsek, undated.
33. Ibid., 41.
34. Quoted in Nagy 1993, 87, from a letter dated April 14, 1944, from Michael Polanyi to Karl Mannheim.
35. Litván 1990, 31; 2006, 93–96; Scott and Moleski 2005, 47. Also see Gulick 2008, 13.
36. Lendvai 2003, 359 (see p. 358 for casualty figures).
37. Ibid., 361–64.
38. Ibid., 357, 364–65; Scott and Moleski 2005, 48. On Tisza's family, see Lukacs 1993, 211.

39. Polanyi 1917a.
40. Scott and Moleski 2005, 49; Szapor 1997, 3.
41. Hargittai 2006, 26. Hargittai focuses on Kármán, Szilárd, Wigner, Neumann, and Teller, using as his motif a joke that was said to have originated with Fritz Houtermans that his Hungarian friends only pretended to be from Hungary but really were Martians. See Rhodes 1986, 106.
42. Litván 2006, 164.
43. Litván 1990, 33–34.
44. Hargittai 2006, 26.
45. Duczynska n.d., 11–12.
46. Lendvai 2003, 380.
47. Quoted in Duczynska 1977, xiv.
48. Lendvai 2003, 421–26.
49. Egressy 2001, 449.
50. Lendvai 2003, 377–79; Lukacs 1993, 212. See also Ignotus 1961, 4–5.
51. Quoted in Marton 2006, 51.
52. Ibid.
53. Wigner 1992, 42.
54. Scott and Moleski 2005, 51.
55. Lendvai 2003, 378.
56. Scott and Moleski 2005, 50–51; Scott 1998–1999, 17.
57. Egressy 2001, 447n, 447–51.
58. Ibid., 451, 461–62.
59. Ibid., 455.
60. Wigner 1992, 38.
61. Scott and Moleski 2005, 55. On Szilárd, see Hargittai 2006, 28; on Jewish assimilation and religious conversion, pp. 43–50.
62. Quoted in Marton 2006, 52. On Teller's experiences at the Minta, see Teller 2001, 31–41.
63. Frank 2005, 208–13. On the collaboration with Einstein, see Dannen 1997.
64. Eugene P. Wigner, interview by Thomas S. Kuhn, November 21, 1963, at Rockefeller Institute, and December 3–4, 1963, at Princeton University, SHQP.
65. Hargittai 2006, 51.
66. Seitz, Vogt, and Weinberg 1995.
67. Chayut 2001, 59; Marton 2006, 40–41.
68. Wigner1992, 38, 57–59,127–131.
69. On Bródy and Polanyi, see Palló 1996a. Bródy's research laboratory at Tungsram now is called the Imre Bródy Institute for Research. On July 3, 1944, Bródy was dragged from that laboratory and imprisoned at Mühldorf, where he died in December 1944. See Kovács 2003, 341.
70. Kovács 2003, 342.
71. Hargittai 2006, 44.
72. Scott and Moleski 2005, 67, 74, 78, 91, 103; Marton 2006, 65, 72.
73. Hargittai 2006, 73.
74. Ash and Söllner 1996a, 6–7; and Krohn, 1993, 11–12.
75. Krohn 1993, 14.
76. Medawar and Pyke 2001, 58–60. Schrödinger returned to Austria in October 1936 and was dismissed on grounds of political unreliability from his positions at the University of Vienna and the University of Graz in April 1938, despite his having

written a fawning letter to Hitler that was published on March 30, 1938, in German and Austrian newspapers. The *Graz Tagespost* gave the letter the prominent headline "Confession to the Führer." Lindemann and British colleagues were outraged, but Schrödinger received a position in Dublin, where he stayed for eighteen years (Medawar and Pyke 2001, 75–79; Moore 1989, 337).

77. Aaserud 1990; Ash and Söllner 1996a, 11. See also Fleming and Bailyn 1969; Coser 1984; Rider 1984; Siegmund-Schultze 2009.

78. See Krohn 1993; 1996, 191.

79. Medawar and Pyke 2001, 55–57; Szilárd 1969, 97–98.

80. Frank 2001–2002, 8–9. Medawar and Pyke (2001, xii–xiii) write that the AAC assisted about half of 1,700 academic refugees in the United States. Also see Gormley 2006, 123–94.

81. On Wigner and Ladenburg's letter, see Weiner 1969, 215–16; and for the letter, in German and in English, see pp. 229–33. For a list of many notable Hungarian-American émigrés during 1919–1945 in diverse fields, see Frank 2009, 439–52.

82. Szapor 1997, 4–6; Striker 1999; Scott and Moleski 2005, 171–72, 181–82, 188–89. Polanyi's papers at the University of Chicago Library contain letters and affidavits regarding his efforts to hold money in British banks for his relatives and to guarantee their solvency upon emigration to Great Britain or the United States. Examples include an affidavit of December 22, 1937, on behalf of Laura Striker's son Michael (Misi) in his emigration to the United States (MPP 3:10); a letter from Michael Polanyi to Michael Striker in Vienna dated December 28, 1937 (MPP 3:10); and a letter from Michael Polanyi to the German Jewish Aid Committee in London dated July 5, 1938, on behalf of Egon Szecsi (MPP 3:12).

83. Scott and Moleski 2005, 180–81, 190, 193, 195, 198.

84. Letter from P. M. S. Blackett to Michael Polanyi, August 26, 1939 (MPP 4:1). Scott and Moleski (2005, 175) describe the Polanyi family's arrival in Cork on August 15, 1939, and Polanyi's departure alone for Manchester on September 1 after hearing news on the radio of the German invasion of Poland. His family later returned to Manchester.

85. Ash and Söllner 1996b, 12.

86. See Medawar and Pyke 2001, 62. On Millikan, see Kevles 1987, 211–12. That anti-Semitism existed is evident in statistics that Jewish enrollment at Columbia University's medical school declined from about 50 percent in 1920 to less than 7 percent in 1940. At Yale, no Jew held the rank of professor until 1946. Anti-Semitism declined sharply after the Second World War as a result of campaigns against discriminatory practices in revulsion against the Nazi policy of the Final Solution (Hollinger 1996c, 8–9).

87. Weindling 1996, 103; Medawar and Pyke 2001, xvi.

88. Excerpt from telephone diary of Warren Weaver with Hugh S. Taylor, May 24, 1940; letter from Hugh S. Taylor to Warren Weaver, May 28, 1940 (RFA RG 1.1, Series 401D, Box 47, F. 616). (Reproduced courtesy of the Rockefeller Archive Center.)

89. Scott and Moleski 2005, 202. Among those also receiving honorary degrees on this occasion were Niels Bohr, Linus Pauling, and Reinhold Niebuhr.

90. On the Princeton colloquium, see Wheeler 1998, 11–17; Szilárd 1969, 111–14; and more generally, Rhodes 1986, 247–71, 303–14. On Hungarians in the Manhattan Project, see Palló 2005, 228.

91. Medawar and Pyke 2001, 86–87; Badash 1995, 105–6.

92. Quotation from Wang 1999, 85. See also Schrecker 1998.

93. Quotation from Krige 2006, 140; and in Wang 1999, 274. On California, see Kaiser, 2005, 326–27.

94. On Shipley's actions, see Hager 1995, 400–1; Wang 1999, 276–77.

95. Scott and Moleski 2005, 216–17.

96. Letter from Edward Shils to [Dean] Robert Strozier, August 21, 1951; and letter from Edward Shils to Joseph Willitts, September 14, 1951 (RFA, RG 1.1, Series 200S, Box 412, Folder 4880).

97. Krige 2006, 140–41. The three individuals were Frédéric Joliot-Curie, J. B. S. Haldane, and Linus Pauling.

98. Samuel K. Allison was professor of physics and director of the Institute for Nuclear Studies, John U. Nef was professor of economic thought and chairman of the Committee on Social Thought, and Cyril Stanley Smith was professor of metallurgy and director of the Institute for the Study of Metals. Letter to L. A. Kimpton from Allison, Nef, and Smith, December 21, 1951; and Letter from L. A. Kimpton to Mrs. Ruth B. Shipley, January 7, 1952 (RFA RG 1.1, Series 200S, Box 412, Folder 4880).

99. Letter from Michael Polanyi to Joseph Willits, February 11, 1952, on a question about a German refugee society in London (RFA RG 1.2, Series 401S, Box 71, Folder 629); and newspaper clippings from the *Manchester Guardian*, March 3, 1952, and March 8, 1952 (UMA Polanyi 1951–1955). Shils 1952; Polanyi 1952a.

100. Polanyi 1952a, 226.

101. Letter from Michael Polanyi to Joseph Willits, July 22, 1953 (RFA RG 1.2, Series 401S, Box 71, Folder 629).

102. Nye 2004, 93.

103. Letter from Henry Clay to Joseph Willits, January 15, 1952 (RFA RG 1.2, Series 401S, Box 71, Folder 629). (Quotation reproduced by courtesy of the Rockefeller Archive Center.)

104. Letter from Michael Polanyi to Karl Mannheim, May 2, 1944 (MPP 4:11).

105. Knepper 2005, 273, citing Scott 1982, 86–87.

106. After his arrival in Manchester, Polanyi attended Presbyterian services with T. W. Manson, vice-chancellor of the university and a New Testament scholar (Knepper 2005, 284). Charles McCoy spoke of Polanyi's church attendance to William T. Scott; see Scott and Moleski 2005, 289; Moleski 2007–2008, 43. A great deal has been written on Polanyi's religious belief or unbelief. Two short views are Congdon 2005–2006 and Moleski 2005–2006.

107. Knepper 2005, 267–68.

108. On Jellineck, see Knepper 2005, 274. On Polanyi's use of "tribe," see Polanyi 1997b.

109. On Weizmann, see Laqueur 1993; Knepper 2005–2006.

110. Medawar and Pyke 2001, 55.

111. Handwritten draft of a letter from Michael Polanyi to Lewis Namier, May 27, 1934 (MPP 2:16).

112. Knepper 2005–2006, citing Laqueur 1991.

113. See Knepper 2005, 279–80; Knepper 2005–2006, drawing upon Polanyi 1936b (MPP 25:8). On the Quakers, see Scott and Moleski 2005, 162, which also draws from Polanyi 1936b.

114. Knepper 2005–2006, 12.

115. Polanyi 1942 (clipping in MPP 47:2).

116. Polanyi 1943a, 38.

117. This is a running summary of Polanyi's argument in Polanyi 1943a.

118. Quoted in Knepper 2005-2006, 10.
119. Quoted in Marton 2006, 39.
120. Hargittai 2006, 84.
121. Quoted from interview of James Franck by Thomas Kuhn, third session, July 11, 1962, p. 17 in Beyerchen 1996, 72.
122. Willstätter 1965, 418-20.
123. Quoted in Beyerchen 1996, 73.
124. Quoted in Hargittai 2006, 22.
125. Quoted in Hargittai 2006, 22.
126. On Wells, see Rose and Rose 1970, 52-53. Polanyi wrote Kasimir Fajans about Wells's *World Set Free* and confided in Karl Mannheim his early love of Wells. See Scott and Moleski 2005, 41; and letter from Michael Polanyi to Karl Mannheim, May 2, 1944 (MPP 4:11).
127. Quoted in Scott and Moleski 2005, 153.

CHAPTER TWO

1. On the 1860 Karlsruhe Congress, see Nye 1999, 52-53; 1984. On the Karlsruhe TH, see http://www.chem-bio.uni-karlsruhe.de/english/19.php (accessed September 15, 2009).
2. Scott and Moleski 2005, 25-26, 31; Stoltzenberg 2004, 113. On Haber, see Szöllösi-Janze 1998.
3. Anonymous 1966, 402-3.
4. Scott and Moleski 2005, 28, 30.
5. Letter from Georg Bredig to Michael Polanyi, February 12, 1917 (MPP 1:5). Also see the translation in Frank 2006, 10; 2009, 132.
6. Scott and Moleski 2005, 46-47, paraphrasing from a letter from Michael Polanyi to Kasimir Fajans, sometime between March 25 and June 26, 1918 (Kasmir Fajans Papers, 1912-1987, Bentley Historical Library, University of Michigan).
7. Palló 1998, 42. See also Palló 1996a; Scott and Moleski 2005, 55.
8. Scott and Moleski 2005, 55-57, 67.
9. Emeléus 1960, 227, 231.
10. Paneth and Hevesy 1914; Fajans 1914. See also Scerri 2000, 65; Van der Vet 1977, 288-98.
11. See discussion in Scott and Moleski 2005, 40. For bibliography of relevant articles, see Emeléus 1960, 244-45. On the decision in 1923 by the International Committee on Chemical Elements, see Aston et al. 1932.
12. A later version of the resume is Michael Polanyi, "curriculum vitae," composed June 1933 (MPP 2:12). See also Cash 1977; Wigner and Hodgkin 1977, 413-15. On his earliest employment at the KWG, see a letter from Michael Polanyi to Ernst Telschow at the Max Planck Society (Göttingen), June 27, 1952 (MPG Abt. II, Rep. 1A, PA Polanyi 4 L:1952 B:07).
13. Polanyi 1913a, esp. 157; Polanyi 1913b, discussed in Scott 1983, 282-83. See also Wigner and Hodgkin 1977, 416.
14. On scientific life in Berlin, and Polanyi's experiences of it, see, more briefly, Nye 2007b.
15. See Turner 1982; see also Meinel 1983.
16. Merz 1965. See also, Nye 1999, 8-11; Coen 2007, esp. 23, 69; Carson 2010, esp. 33-35. On Prussian reforms and their impact, Vom Bruch 2001; Schwinges 2001.
17. See Ringer 1978; Roberts 1980; Pyenson 1983.

18. Whewell 1847; Mach 1919; Hertz 1899; Duhem 1954; Poincaré 1913.
19. Arnold 1904, 249–56.
20. See Rocke 2000.
21. For physics in general, see Forman, Heilbron, and Weart 1975. For France, see Shinn 1979; Fox and Weisz 1980; Nye 1986. For Germany, see Manegold 1970; Borscheid 1976; Jungnickel and McCormmach 1986; Olesko 1991. For Great Britain, see Sanderson 1972; Divall 1994.
22. These included BASF (Badische und Soda-Fabrik), Höchst, Bayer, and AGFA (Aktiengesellschaft für Anilinfabrikation).
23. See Nye 1999, 2–5, 139.
24. For example, see Hettema 2000a, xxvii–xxix.
25. Gebhardt 1903.
26. Paletschek 2001a, 40–41.
27. Paletschek, 2001b; 2001a, 37–38.
28. Quoted in Stern 1999a, 68, from Einstein 1934.
29. Weber 1948a, 134. On Weber, see Ringer 2004.
30. Quoted in Barkan 1999, 226.
31. McCormmach 1983, 99, 100–101, 95 (quotation, 113). Max Planck's responsibilities, as described in John Heilbron's portrait of Planck in Berlin, are similarly daunting (Heilbron 1986, 1996). See also Hoffmann 2008a, 2008b.
32. McCormmach 1983, 112.
33. In Adams 1918, 77–78, quoted by Frank 2006, 6; 2009, 149.
34. Kessler 2000, 191–92. Also see Easton 2002.
35. In Déak, 1968, 13–15 quoted by Frank 2006, 8; 2009, 148. On Weimar Germany, see Gay 1968; Stern 2006, 51–88. Also see Görtemaker 2002; Hoffmann 2006; Goenner 2005.
36. Quoted in Ladd 1997, 110, from Baedeker 1923, 50.
37. Letter from Alfred Reis to Michael Polanyi, October 14, 1920 (MPP 1:11).
38. Weitz 2007, esp. 169–250 on architecture, photography, and cinema; Ladd 1997, 110.
39. Kessler 2000, 233.
40. Ibid., 364–67.
41. Frank 2006, 9; 2009, 150.
42. Kessler 2000, 325–26.
43. Ringer 1969; also Cassidy 1994, 158. More generally, see Proctor 1991.
44. On the manifesto of the ninety-three German intellectuals and its impact, see Schroeder-Gudehus 1978. The text of the manifesto appears in French and German in Schroeder-Gudehus 1966. On Planck, see Heilbron 1986, 69–79.
45. Rowe and Schulmann 2007.
46. Hahn 1970, 90. For biographical portraits of leading Berlin scientists, with emphasis on physical scientists, in the late nineteenth and earlier twentieth centuries, see Treue and Hildebrandt 1987. Portraits include Fischer, Planck, Haber, Nernst, Bodenstein, Einstein, Laue, Hahn, Meitner, Strassman, Volmer, Hertz, and others.
47. Wigner 1992, 71.
48. Ibid., 75.
49. Scott and Moleski 2005, 85, from a 1950 lecture in Chicago.
50. Quoted in Stoltzenberg 1994, 439–40, and in Stoltzenberg 2004, 226. See also Scott and Moleski 2005, 84; Hoffmann 2000. More generally, on Berlin science, see Laitko et al. 1987.

51. Stern 1999a, 109 (first quotation); Heilbron 1986, 69 (second quotation).
52. Heilbron 1986, 68; Henning and Kazemi 1998, 11–12.
53. Nye 1999, 11–12; but especially Cahan 1989.
54. Henning and Kazemi 1998, 71–74. See also Johnson 1990; Max-Planck-Gesellschaft 1999, 9–12.
55. On Einstein, among many other sources, see Clark 1972; Isaacson 2007.
56. Stern 1999a, 110; Ladd 1997, 112–13. Also see Macrakis 1989, 26; 1993.
57. Stern 1999a, 84–88; Stoltzenberg 2004, 77–105; Perutz 1996, 31–32.
58. Stoltzenberg 2004, 126–27. For Fischer's promise, see Willstätter 1965 (quotation, 212); Olby 1974, 28.
59. MacLeod 1998, 31, 34; Stern 1999a, 119; Henning and Kazemi 1998, 74.
60. Stern 1999a, 120–25; Stoltzenberg 2004, 132–48. Perutz 1996, 32, 34. Also see Von Leitner 1996.
61. Stoltzenberg 2004, 232–35.
62. By 1930, the number of research institutes in Dahlem was fourteen, and the total number in Germany was about thirty. The Max Planck Gesellschaft in 2009 numbered approximately twelve thousand staff members and nine thousand PhD students, postdoctoral researchers, guest scientists and others. See http://www.mpg.de/instituteProjekteEinrichtungen/index.html (accessed October 5, 2009).
63. Henning and Kazemi 1998, 74; Max-Planck-Gesellschaft 1999, 14–15; Heilbron 1986, 94–96; and Stoltzenberg 2004, 230–31. On Haber's gold project, see Hahn 1999.
64. On Sponer, see Vogt 1999, 134–35; Max-Planck-Gesellschaft 1999, 16.
65. Henning and Kazemi 1998, 52–53; Mark 1993, 20; Scott and Moleski 2005, 68.
66. Scott and Moleski 2005, 67, 74, 78, 91, 103; Marton 2006, 72.
67. See Wigner's interview by Kuhn (SHQP) and Chayut 2001, 59–63. See also Westfall 2008, 294.
68. Quoted in Scott and Moleski 2005, 85.
69. Letter from Michael Polanyi to Ernst Telschow, June 27, 1952, outlining Polanyi's history of employment at the KWG for the purposes of filing for his pension (MPG Abt. II, Rep. 1A, PA Polanyi 4, L:1952 B:07). Document of agreement between Herzog and Polanyi, January 1, 1922 (MPG Abt. II. Rep. 1A. PA Polanyi, Folder 1914–1932, Blattzahl 14, document of January 1, 1922). It was only in 1951 that a civil service act provided a means for those employees terminated, or "retired," from the KWG for racial reasons in the 1930s to apply for their rightful pensions or restitution. Approximately a hundred employees of the KWG were dismissed during National Socialism, of whom more than 90 percent were Jewish (defined as having at least one Jewish grandparent). Of these, thirty-six applied for compensation after 1945, and of these thirteen were granted some form of compensation. See Schüring 2006, 310–12.
70. See Polanyi's curriculum vitae (MPP 2:12); Wigner and Hodgkin 1977, 415.
71. Letter from Fritz Haber to Adolf von Harnack, June 9, 1923 (MPG Abt. I, Rep 0001, #1164 from the KWG Generalverwaltung. Akten zu KWI für physikalische Chemie und Elektrochemie. 7. Wissenschaftliche Mitglieder 1923 [Ladenburg, Polanyi]).
72. Henning and Kazemi 1998, 54–55; Scott and Moleski 2005, 85, 91–94; Scott 1983, 288; Mark 1962, 603; Wigner 1992, 78–80.
73. Keynes 2004, 77–82. Also see Stern 2006, 64.
74. Stern 2006, 64 on Einstein; Kessler 2000, 183.
75. Stern 2006, 65–66.

76. Clark 1972, 374.
77. Scott and Moleski 2005, 94.
78. Quoted in Heilbron 1986, 92. See Eckert and Märker 2004.
79. Palló 1998, 41–42.
80. Scott and Moleski 2005, 87.
81. Ibid., 87.
82. Ibid., 103.
83. Document of July 27, 1923, signed by Polanyi, Harnack, and Franz v. Mendelsohn (MPG Abt. II. Rep. 1A. PA Polanyi. Folder 1914–1932, Blattzahl 14). Scott and Moleski (2005, 103) mention as his clients Siemens Electric Works and the Osram Lamp Work in Berlin, the Philips Lamp Works in Eindhoven, and the United Lamp Works in Budapest.
84. Letter from Fritz Haber to Generalverwaltung, KWG, November 21, 1928; letter from Haber to Stefan Bogdandy, November 16, 1928, regarding the 30 percent policy (MPG Abt. I. Rep. 000 1A #1165).
85. Sachse 2009, 100–2; Kohler 1991, 157–58, 252. Also see Biedermann 2002.
86. Document 717D, November 14, 1930, on $7,000 grant. Interoffice correspondence from L. W. Jones's log, July 13, 1931, regarding Kallmann. Memo from W. E. Tisdale to Lauder W. Jones, January 20, 1932, regarding needs in Haber's Institute. Interoffice correspondence from Lauder W. Jones's log for trip to Germany April 7–April 22, 1932, dated April 11, 1932, on meeting with Freundlich, Polanyi, and Kallmann. Letter from Warren Weaver to Max Planck, October 16, 1932, regarding award of $13,200 (all in RFA RG 1.1, Series 717D, Box 13, F. 110).
87. Kohler 1991, 250–54.
88. Document of October 26, 1929, following a letter of July 4, 1929 from Haber to Polanyi regarding negotiations (MPG Hauptabteilung. II. Repositur 1A. PA Polanyi, Michael. Folder for 1914–1932, Blattzahl 14). Scott and Moleski (2005, 112–113, 116) write of offers from the German University in Prague in 1928 and from the University of Szeged in May 1929. Polanyi also had an ongoing offer from Izzo. Scott and Moleski also report that Polanyi received an offer from Harvard in 1929.
89. Wigner, 1991, 127, 131.
90. In Russell McCormmach's *Night Thoughts of a Classical Physicist*, the fictional Victor Jakob remembers the historical Drude: "Trim in your gymnastic suit with its dark sash, you looked like someone who got more than his share of satisfaction from life. When you took flight on the parallel bar or the wooden horse, you gave the impression that you could handle any problem with ease and confidence" (1983, 110–11). On Harnack House, see Henning and Kazemi 1998, 43–46; Macrakis 1989, 50–52. On Polanyi's activities, see Scott and Moleski 2005, 121–24.
91. Scott and Moleski 2005, 113, 119.
92. Crowther 1970, 63.
93. Ibid., 63–64.
94. Ibid., 65.
95. Stern 1999a, 108–9; Henning and Kazemi 1998, 7–12.
96. Polanyi also thought that faculty on American campuses, particularly at the land-grant public university of Minnesota, were expected to spend too much time on research that had practical applications (Scott and Moleski 2005, 115). See Polanyi's "Notes from a trip to America 1929" (MPP 44:3).
97. Polanyi, quoted in Palló 1998, 39.
98. Wigner 1992, 156–57.

99. Crowther 1970, 65, 66.
100. Kessler 2000, 399–400.
101. Undated letter from Cecile Polanyi to her sister-in-law, Irma Pollacsek, quoted in Scott and Moleski 2005, 182.
102. Letter from W. E. Tisdale to Lauder W. Jones, January 20, 1932 (RFA RG 1.1, Series 717D, Box 13, F. 110).
103. Stoltzenberg 2004, 231.
104. Schroeder-Gudehus, 1972, esp. 564–67; Stoltzenberg 2004, 231.
105. Polanyi 1930.
106. Carbon copy of letter from Polanyi to Arthur Lapworth, January 13, 1933 (MPP 44:4); also carbon copy of letter from Polanyi to F. G. Donnan, January 17, 1933 (MPP 2:11). Adolf Hitler was appointed chancellor of Germany on January 30, 1933.
107. Letter from Fritz Haber to Polanyi, June 27, 1932, in response to a note from Polanyi of June 26, 1932 (MPP 2:9).
108. Letter from Hugh S. Taylor to vice-chancellor Walter B. Moberly, January 8, 1932 (UMA Vice-Chancellor's Archives [VCA]/7/358 [Chemistry: Polanyi] 12/12).
109. On Allmand, see Anonymous 1949; Scott and Moleski 2005, 134–35.
110. Letter from Richard Willstätter to Vice-Chancellor Moberly, in English translation from the handwritten German, February 22, 1932 (UMA VCA/7/358 [Chemistry: Polanyi] 12/12); typescript, February 22, 1932, regarding Committee meeting chaired by the vice-chancellor on the chemistry chair (UMA VCA/7/358 [Chemistry: Heilbron] 10/10).
111. Letter from Arthur Lapworth to Michael Polanyi, March 1, 1932 (MPP 2:8).
112. Scott and Moleski 2005, 134.
113. Quoted from carbon copy of draft of letter from Michael Polanyi to Arthur Lapworth, March 15, 1932, in comparison with carbon copy of letter from Michael Polanyi to Arthur Lapworth, March 17, 1932, both written in German (MPP 2:8). Quotation with permission of John C. Polanyi.
114. Letter from Michael Polanyi to Herr Geheimrat [Fritz Haber], April 16, 1932 (MPP 2:8).
115. On Lenard and Stark, see Beyerchen 1977; Walker 1995; Hoffmann 2005, esp. 296–302.
116. Scott and Moleski 2005, 134. Also, regarding Polanyi's demands, see letter from Allmand to Michael Polanyi, May 17, 1932 (MPP 2:8). Regarding the Rockefeller Foundation initiative, see letter from F. G. Donnan to Michael Polanyi, May 19, 1932 (MPP 2:8).
117. Lauder W. Jones's log of June 1–16, 1932 (RFA RG 1.1, Series 717D, Box 13, F. 110); Lauder W. Jones's log of trip to England and Scotland, June 1–18, 1932; and letter from Lauder W. Jones to Vice-Chancellor Moberly, June 22, 1933 (RFA RG 1.1, Series 401D, Box 47, F. 616).
118. Letter from Fritz Haber to Polanyi, June 27, 1932, in response to Polanyi's letter of June 16, 1932 (MPP 2:9). Authorization for quotation from Special Collections Research Center, University of Chicago Library.
119. Lauder W. Jones's diary entry, July 14, 1932 (RFA RG 1.1, Series 717D, Box 13, F. 110).
120. Typewritten carbon copy of letter from Vice-Chancellor Moberly to Robert Robinson, November 1, 1932 (UMA VCA/7/358 [Chemistry: Polanyi] 12/12; original em-

phasis). Quotation reproduced by courtesy of the University Librarian and Director, The John Rylands University Library, The University of Manchester.

121. Letter from W. Pope to H. E. Armstrong, November 14, 1932; letter from H. E. Armstrong to Vice-Chancellor Moberly, November 19, 1932; letter from Robert Robinson to Moberly, November 8, 1932; letter from Robert Robinson to Moberly, November 14, 1932 (UMA VCA/7/358 [Chemistry: Polanyi] 12/12).

122. Scott and Moleski 2005, 135.

123. On the terms, see letter from Arthur Lapworth to Michael Polanyi, November 27, 1932 (MPP 2:10).

124. Letter from Vice-Chancellor Moberly to Michael Polanyi, December 15, 1932; typed addendum, December 23, 1932; letter from E. D. Simon to Michael Polanyi, December 22, 1932 (MPP 2:10).

125. Letter from Michael Polanyi to Arthur Lapworth, January 13, 1933 (MPP 2:11).

126. Letter from Michael Polanyi to F. A. Donnan, January 17, 1933 (MPP 2:11).

127. Quoted from Szilard 1969, 96.

128. Einstein 2007, 269.

129. Letter from Planck to Einstein, March 19, 1933, quoted in Heilbron 1986, 155.

130. Quoted in Stoltzenberg 2004, 200.

131. Scott and Moleski 2005, 137, based on Scott's interview with Magda Polanyi.

132. Moore 1989, 265.

133. In chemistry, the law applied to the institutes for physical chemistry, biochemistry, and medical research (Deichmann 1999, 19, 21).

134. See Sime 1996, 138.

135. Stoltzenberg 2004, 278–79.

136. Quoted in Deichman 1999, 24 (emphasis added).

137. Quoted in ibid.

138. In Szilard 1969, 95–96. Also see Szilard 1978, 13.

139. Frank 2005, 209, 213; Szilard 1969, 95–96.

140. Quoted in Sime 1996, 139.

141. See Beyerchen 1996, 75–76.

142. Quoted in Stern 1999b, 154–55.

143. Quoted in Scott and Moleski 2005, 138. On his Austrian passport, still held at this time, see Scott and Moleski 2005, 142.

144. On Freundlich, see Gortner and Sollner 1941, 415. Glum explicitly wrote Haber on April 23, 1933, that only the lower-level assistants were meant to resign (Goran 1967, 160).

145. Quoted in Stoltzenberg 2004, 280.

146. For a discussion of the interview with Hitler, see Heilbron 1996, 153–54, 210–14. Also see Albrecht 1993.

147. Quoted in Hahn 1970.

148. Stern 1999a, 157–58.

149. Barkan 1999, 244. For details of his retirement, see Bartel and Huebener 2007, 327–28. On courses of action to take among non-Jews, in addition to Deichmann (1999, 19–22; where Ute Deichmann discusses obedience, anti-Semitism, and self-benefit), see Alan Beyerchen's categories of professional opposition, opportunism, and ideological conviction (1994, 138).

150. Moore 1989, 273, 276–77.

151. Sime 1996, 146.

152. Memo to Lauder Jones from I. Marcovich, Paris, May 3, 1933 (RFA RG 1.1, Series 717D, Box 13, F. 110). On Bodenstein, see Cremer 1987. On his forced retirement, see Bartel and Huebener 2007, 329, 336.

153. Heilbron 1986, 165–74.

154. Scott and Moleski 2005, 139 (on Herzog's letter), 135 (on the November 1932 dinner). Also see Wigner and Hodgkin 1977, 415–16.

155. Henning and Kazemi 1998, 56.

156. Quoted in Sime 1996, 140.

157. Willstätter 1965, 424.

158. Quoted in Sime 1996, 356; also see 144, 184–85.

159. Scott and Moleski 2005, 141.

160. Acting as an intermediary, Leo Szilard telegrammed Moberly from Holburn, London, on May 1, 1933 with the message: "Polanyi thanks for telegram looks forward to letter will wire immediately would like to keep matter secret until finished preparations for leaving Germany" (UMA VCA/7/358 [Chemistry: Polanyi] 12/12). Quotation reproduced by courtesy of the University Librarian and Director, The John Rylands Library, The University of Manchester. On newspaper announcements, see Scott and Moleski 2005, 139.

161. Gortner and Sollner 1941, 415. For the most complete account of the careers, migrations, and choices made by Jewish and non-Jewish chemists and biochemists under National Socialism, see Deichmann 2001.

162. See Deichman 1999, 52 (on Farkas), 66–68 (on Mark). Also on Mark, see Furukawa 1998.

163. Deichmann 2001, 118–25.

164. Deichmann 2001, 120, 484; Scott and Moleski 2005, 204–5. Also see Morton 1980.

165. See Deichman 1999, 66–68; 1998, 197–98.

166. Cassidy 1992, 487.

167. Heilbron 1986, 162.

168. Max-Planck-Gesellschaft 1999, 18–19; Henning and Kazemi 1998, 178.

169. Quoted in Stern 1999b, 162; and see Stoltzenberg 2004, 286–91, 299–300.

170. Deichmann 2001, 63 and 63n29; see also Stoltzenberg 2004, 306; Becker, Becker, and Block 1987, 179–80. On Bosch, see Willstätter 1965, 290, 293. Neuberg, who was Jewish, was forced to retired in September 1934 as director of the Kaiser Wilhelm Institute for Biochemistry, and he fled from Germany in 1937. See Henning and Kazemi 1998, 107–12.

171. Hoffmann and Walker 2004, 53.

172. Polanyi 1928b. Polanyi later reflected on his difficult leave-taking from Haber in 1933 and his wish for Haber to take an even stronger stand for his Jewish colleagues than just resigning himself, which was the action Haber took at the same time that he was ordered to request Polanyi's resignation (see Scott and Moleski 2005, 139). If Polanyi knew of Haber's earlier May 1931 letter to Hermann Dietrich (minister of Finance) advising the Weimar government to "adopt a dictatorship and a planned economy as its own program," thereby temporarily abandoning the rule of the market economy, Polanyi surely disagreed. On Haber's letter, see Stern 1999a, 150.

173. Polanyi 1962a. Polanyi drafted an essay twenty years earlier, in 1942, called "The City of Science" (MPP 27:6). On the unpublished essay, see Scott and Moleski 2005, 188.

174. Quoted in Cassidy 1994, 159 (original emphasis).

175. Quoted from Erwin Chargaff, *Das Feuer des Heraklit: Skizzen aus einem Leben vor der Nature* (Munich, 1984) in Ash and Söllner 1996a, 5.

CHAPTER THREE

1. See Scott and Moleski 2005, 327–50, for a bibliography of Michael Polanyi's papers with English titles.

2. Polanyi 1913a, 1913b, discussed in Scott 1983, 282–83; Wigner and Hodgkin 1977, 416. On the work on surface chemistry, also see Nye 2000a, 2001, 2002.

3. Letter from Michael Polanyi to Alfred Reis, December 11, 1912; Georg Bredig to Polanyi, February 1, 1913; Georg Bredig to Polanyi, February 12, 1913 (MPP 1:2). More than thirty years later, Percy Bridgmann wrote Polanyi that he was surprised to have just learned of this work and thought it unfortunate that it had escaped general notice (letter from Percy Bridgman to Polanyi, December 19, 1946 [MPP 5:2]).

4. See Scott and Moleski 2005, 28.

5. Ibid., 34.

6. Letter from Walter Nernst to Michael Polanyi, August 22, 1913; Nernst to Polanyi, August 30, 1913; draft of letter from Polanyi to Nernst, September 3, 1913; and Nernst to Polanyi, October 15, 1913 (MPP 1:2). See Polanyi 1913b, 340–41. On Nernst, the heat theorem, and Einstein, see Barkan 1999, 164–80.

7. Polanyi 1917b.

8. Scott and Moleski 2005, 35, referring to Arnold Eucken's paper on the theory of adsorption.

9. Polanyi noted that Eucken introduced the term *adsorptionspotential* (adsorption potential) in 1914, a few months before Polanyi's first paper on the subject (Polanyi 1963, 1013n2).

10. Polanyi's phrasing, in Polanyi 1963, 1010.

11. On Polanyi's original adsorption potential theory, see Scott and Moleski 2005, 24–25, 31–36, 42–43. Also see Radovic 2002, 7.

12. Polanyi 1914, 1916. See also Scott 1983, 283; Wigner and Hodgkin 1977, 417; but especially Brunauer 1944, 95–99.

13. Lewis 1916.

14. See Suits and Martin 1974; Gaines 1993. Some of Langmuir's relevant papers are Langmuir, 1915, 1916, 1917. On the history of the General Electric Laboratory, see Reich 1985.

15. Langmuir 1919a, 1919b, 1919c.

16. Langmuir 1922.

17. Cited as *Festschrift der Kaiser-Wilhelm-Gesellschaft* 1921, 171, in Polanyi 1929a, 431.

18. Polanyi 1963, 1011. For an analysis in terms of the core-set of researchers in adsorption studies, see Palló 2010.

19. See Gortner and Sollner 1941.

20. See Glasstone 1947, 1196; Taylor and Glasstone 1951, 598.

21. Mentioned in Polanyi 1963, 1010.

22. Ibid.

23. Freundlich 1926, 122.

24. Ibid., 815 (emphasis added).

25. Scott 1983, 284; also see Mark 1962, 603.

26. Polanyi 1963, 1012.

27. See Scott and Moleski 2005, 107; Wigner and Hodgkin 1977, 418.

28. For example, see Laidler 1993, 347. The famous Heitler-London paper is Heitler and London 1927.
29. Goldmann and Polanyi 1928; Polanyi and Welke 1928; Heyne and Polanyi 1928.
30. See Scott and Moleski 2005, 107–8; Goldmann and Polanyi 1928; and Brunauer 1944, 116–19.
31. Haber 1929. The January 1929 issue of *Zeitschrift für Elektrochemie und angewandte physikalische Chemie* carried the initial announcement of the colloquium on page 1. Also see the April 1929 issue, pages 161–62, for the program.
32. See Michael Polanyi's diary entry of May 9, 1929 (MPP 44:4).
33. Wigner 1992, 78–80.
34. Wigner and Hodgkin 1977, 417. See also Polanyi 1929b.
35. Polanyi 1929a. This is in part a response to Zeise 1929. See also Brunauer 1944, 75–76.
36. See Brunauer 1944, 115–16, who faults McBain and Britton (1930) for deducing erroneous consequences from the potential theory and then claiming that the potential theory had been refuted by experimental data inconsistent with these (erroneous) consequences.
37. Schwab 1937, 194.
38. Gavroglu 1995, 50, 59.
39. Ibid., 67–68. See London and Eisenschitz 2000.
40. London 2000a.
41. London and Polanyi 1930.
42. Introductory remarks by Robert Mond, in Faraday Society 1932, 130.
43. Taylor, in Faraday Society 1932, 132. In his footnote to the published text, Taylor cites Polanyi's 1928 articles with F. Goldmann and K. Welke.
44. Taylor, in Faraday Society 1932, 138, citing Eyring and Polanyi 1931 and Eyring 1931. On Taylor, see also Radovic 2002, 18.
45. Freundlich, in "Diskussion," ZE 35 (1929): 585; Freundlich 1932, 198.
46. The same points are emphasized in Polanyi 1929a, 431. One of the matters at issue in Polanyi's critique of Langmuir's equation was the exact way in which constants in the equation are temperature dependent. A difficulty for Polanyi's theory was that his isotherm does not obey Henry's law at low temperatures. Langmuir's theory explains this behavior in terms of chemical forces that are different from the physical forces in Henry's law. The law, dating to 1803, requires the amount of gas dissolved in a liquid to increase as pressure increases and to decrease as temperature increases.
47. Polanyi 1932a, 321–22.
48. Langmuir 1966, 308, 321–22. Langmuir refers to Langmuir 1918, with special reference to pages 1399–1400.
49. On the nominations by Bohr, see Friedman 2001, 165–70.
50. Letter of nomination from Freundlich, October 6, 1928 (PKVA 1929: Letters Section, pp. 59–60).
51. Report of the Chemistry Committee to the Academy, September 27, 1928 (PKVA 1929: 9pp.), signed Henrik G. Söderbaum, Theodor Svedberg, Wilhelm Palmer, Oskar Widman, and Ludwig Ramberg. Söderbaum's field was agricultural chemistry; Svedberg, colloid chemistry and biophysical chemistry; Palmer, physical chemistry and electrochemistry; and Widman, organic chemistry; see Friedman 2001, 284–85. The 1928 nomination of Freundlich is discussed on p. 3 and the 1928 nomination of Langmuir on p. 4 of the 1928 Report of the Chemistry Committee

(PKVA 1929). Appendices to the Chemistry Committee's report include Bilagar 4, The Svedberg, March 20, 1928, "Kort Redogörelse för H. Freundlichs arbeten efter 1921," pp. 46–49; and Bilagar 5, C. W. Oseen and L. Ramberg, "Utredning rörande Dr. Irving Langmuirs till belonging med Nobelpris föreslagna arbeten," pp. 50–81. Freundlich's nomination of Langmuir is dated October 6, 1928, and is included in the file for discussion in 1929. Nominations for a year's award had to be received by February 1 of that year.

52. On the nominations, see Crawford et al. 1987.

53. Report of the Chemistry Committee, September 25, 1931 (PKVA 1931), signed by Söderbaum, Ramberg, Svedberg, Palmer, and Euler. Report of the Chemistry Committee, October 1, 1932 (PKVA 1932), signed by the same five men. Svedberg's nominating letter of January 1, 1931 nominates Langmuir for his work on chemical surfaces (PKVA 1931: Letters Section, p. 46).

54. Nominating letter from Hans von Euler, dated January 30, 1932 (PKVA 1932: Letters Section, p. 107). Euler nominated Langmuir to share the 1932 Nobel Prize in Chemistry with G. Embden and C. Neuberg or with Paul Karrer.

55. Report of the Chemistry Committee, October 1, 1932 (PKVA 1932), Appendices, Bilagar 10, The Svedberg, "Kompletterande utredning rörande I. Langmuirs arbeten," May 6, 1932, pp. 57–61. Svedberg also wrote a resume in 1932 on the work of G. N. Lewis, but Lewis never would receive a Nobel award. Report of the Chemistry Committee, October 1, 1932 (PKVA 1932), Appendices, Bilagar 11, The Svedberg, "Kort utredning rörande G. N. Lewis's teori för den kemiska valensen," April 25, 1932, pp. 62–65.

56. Nominating letter from Karl Landsteiner, dated January 9, 1931 (PKVA 1931: Letters Section, p. 45). See discussion in Friedman 2001, 197–198.

57. Söderbaum 1934.

58. Polanyi 1935a.

59. From Scott's interview with Erika Cremer, quoted in Scott and Moleski 2005, 112.

60. "Die Voraussetzungen 1. und 2. hat bereits vor mir A. Eucken eingeführt und verwertet; mein Beitrag bestand darin, die in der Voraussetzung 3. gelegene Approximation zu prüfen. In ihrer Verwendung liegt der grundsätzliche Unterschied gegenüber dem Ansatz von Langmuir, der die Wirkung der Kohäsionskräfte zwischen den adsorbierten Molekülen vernachlässigt." In Polanyi 1929a, 431.

61. Polanyi 1969a, 94.

62. Quoted in personal communication from Gunther Stent, April 13, 1998, from Macdougall 1943.

63. Quoted in personal communication from Gunther Stent, April 13, 1998, from Taylor 1931 (emphasis added).

64. See Brunauer 1944. Polanyi's name and work figure throughout the text to a considerably larger extent than is indicated by references in the name index.

65. Brunauer, Emmett, and Teller 1938. The BET theory modified Langmuir's theory in favor of multilayer adsorption, with reference, among others, to Goldman and Polanyi's paper of 1928 focusing on adsorption isotherms for charcoal.

66. Brunauer 1944, 60, 96. Also see Glasstone 1947, 1202–3.

67. See the discussion by Radovic 2002, 19–22; Gregg and Sing 1982, 220. In Dubinin 1955, there is no mention of Polanyi. On Dubinin, see Russian Chemical Bulletin 1993.

68. Radovic 2002, 1, 21.

69. Henning and Kazemi 1998, 53; Scott and Moleski 2005, 68.

70. Mark 1962, 603.
71. Letter from Albert Reis to Michael Polanyi, October 14, 1920 (MPP 1:11), quoted in Scott and Moleski 2005, 71. For Reis's expertise, see Reis 1920. See comment about Reis in Mark 1993, 80. On Polanyi's work in X-ray diffraction, see Nye 2000a, 2001.
72. On the discovery and confirmation of X-ray diffraction, see Forman 1969.
73. Olby 1974, 23–24.
74. See James 2007, 24.
75. Bragg 1975, 31–32. See design of the spectrometer in figure 2, p. 31. Also see Sands 1975, 89.
76. Scherrer 1962, 643.
77. Hull 1962, 584.
78. Olby 1974, 26.
79. Herzog, Jancke, and Polanyi 1920a.
80. Herzog, Jancke, and Polanyi 1920b; Olby 1974, 26, 29.
81. Scott and Moleski 2005, 75.
82. Furukawa 1998, 30–36; see also Furukawa 2003.
83. Letter from Reginald Herzog to Michael Polanyi, October 8, 1920 (MPP 1:11).
84. Mark 1976, 179. On Polanyi's remark about Freudenberg's paper in an interview with Robert Olby, Scott and Moleski 2005, 76.
85. See Furukawa 1982, 7–9.
86. Quoted from Staudinger's *Arbeitserrinerungen* in Furukawa 1998, 67.
87. Polanyi 1962b, 99.
88. Ibid.
89. Mark 1976, 179.
90. Polanyi 1962b, 99.
91. On Mark, see Morawetz 1995.
92. Sponsler and Dore 1926; see also Mark 1976, 179.
93. Polanyi 1928a.
94. Furukawa 1998, 67.
95. Quoted in Furukawa 1998, 72 from Mark 1981, 529.
96. See Furukawa 1998, 72–75.
97. Meyer and Mark 1928. By the 1950s, cellulose was estimated to have an average molecular weight of 400,000, corresponding to 2,500 glucose units (now estimated at 3,000–5,000 units). As Meyer and Mark proposed in 1928, the glucose chains form micelles, or sheets, of parallel chains held together by side-to-side hydrogen bonds. The diameter of a micellar unit has been found to correspond to 100–200 cellulose chains, rather than the 40–60 Meyer and Mark proposed, and the length of the chains is about 200 glucose units, rather than 30–50. The sheets are held together in staggered layers by Van der Waals forces. The mechanical strength of cellulose, as well as its chemical stability (in contrast to protein) is considered to result from the micellar structure. For the late 1950s, see Fieser and Fieser 1957, 280.
98. See James 2007, quoting from Perutz 1990, 192. Perutz especially mentions the impact on chemists of Linus Pauling's *The Nature of the Chemical Bond* (1939) in which Pauling marshaled evidence from X-ray crystallography for molecular structures.
99. Scott 1983, 288; Mark 1962, 603.
100. Wigner 1992, 78–80. The title of Wigner's thesis, defended at the Berlin Technische Hochschule in 1925, was "Bildung und Zerfall von Molekülen" (Formation and decay of molecules). On Weissenberg, see Chayut 2001, 63–69. On Gomperz, see Scott and Moleski 2005, 78.

101. Polanyi 1921a, 1921b, 1921c. See also Polanyi1962b, 99; Scott and Moleski 2005, 76–77.
102. Schiebold 1922; Polanyi and Weissenberg 1922; Polanyi, Schiebold, and Weissenberg 1924; Ettisch, Polanyi, and Weissenberg 1921. On Weissenberg, see Buerger 1973.
103. See Mark 1993, 24; Polanyi 1921b, eq. 2 and figs. 1 and 3; Polanyi, Schiebold, and Weissenberg 1924, 339. Also see Bragg 1975, 139; Sands 1975, 90–92, 95–97; Lonsdale 1949, 79.
104. Polanyi 1962b, 100; Mark, 1993, 24; Chayut, 2001, 65. See Polanyi, Schiebold, and Weissenberg 1924; Weissenberg 1924, discussed in Scott and Moleski 2005, 77.
105. See Scott and Moleski 2005, 77–78, 88–89.
106. Scott and Moleski 2005, 89; Hellner and Ewald 1962, 462.
107. Scott 1983; Polanyi 1934.
108. Polanyi 1934. See also Braun 1992.
109. See Hume-Rothery 1962; Lonsdale 1962.
110. Polanyi 1962b, 102. See also Taylor 1934; Orowan 1934. On more recent studies of dislocation, see Service 1996.
111. Braun 1987. Also see Hoch 1986, 1987.
112. See Polanyi 1962b, 100.
113. Mark 1993, 25.
114. Furukawa 1998, 89.
115. Mark 1993, 31. Kallmann and Mark 1926. On the Compton effect, see Stuewer 1975.
116. Mark 1993, 29.
117. Carbon copy of letter from Michael Polanyi to Wichard von Moellendorff, March 27, 1929 (MPP 2:5).
118. Eckert and Schubert 1990, 184–85.
119. Orowan 1965.
120. Weart 1992, 643. Also see Weaire and Windsor 1987, xiv, on Great Britain.
121. Interview of Michael Polanyi by Thomas Kuhn, February 15, 1962, Berkeley, CA (SHQP, pp. 9–10).
122. Dubinin 1960, 235; and Hansen and Smolders 1962, 167.
123. Quotation from Polanyi 1963, 1011. See Stent 1972, 436. Stent's notion of prematurity was the subject of a 1997 conference and book. See Hook 2002 and Stent 2002, 26–27, with reference to Polanyi and adsorption. Yves Gingras refers to the "death and resurrection of Michael Polanyi's potential theory of adsorption" as an illustration of "the crucial role of temporality in science" (1995, 145; Gingras, cited Polanyi 1963).
124. Quoted from Stent 1972, 436 (original emphasis).
125. Handwritten notes by a secretary or another assistant to Polanyi, 4 pages, on X-ray crystallography (MPP 43:4). Bragg refers, however, to the "classical rotation photograph method" "developed by Shiebold and Polanyi" (1975, 138–39).
126. Polanyi 1962b.
127. Polanyi 1963, 1012.
128. Russell 1952, 110–11, quoted in Polanyi, 1963, 1013.
129. Polanyi 1963, 1012.

CHAPTER FOUR
1. Pechukas 1982, 372.
2. Kaufman 1982, 362. Also see Ramsey 1997; and for a thorough discussion of Po-

lanyi's work, Nye 2007a. The term "semi-empirical" first appears in Eyring and Polanyi 1931, 300 as part of a section heading "Halbempirisches Verfahren zur schliesslichen Berechnung der Reaktion $H + H_2 \rightarrow H_2 + H$."

3. Polanyi 1920a. This paper was written in Karlsruhe.
4. See Kaufman 1982, 362.
5. Quoted in Ramsey 1999, 2, from Hinshelwood 1928, 340–41.
6. See Laidler 1985; Nye 1993, 105–38.
7. Arrhenius's formulation is based on the prior idea and equation of Van't Hoff 1884. See Arrhenius 1889; also see Laidler 1985, 55–57.
8. See Marcellin 1914; Eyring 1976, 90. See also Guéron and Magat 1971, 6.
9. Polanyi 1920b, 1920c, 1920d. Polanyi writes, "We want to show that a still unknown kind of energy source coming from empty space is possible" (1920d, 31–32).
10. Letter from Reginald Herzog to Michael Polanyi, October 8, 1920 (MPP 1:11). See Kragh 1989.
11. See Nye 1993, 121–29; as well as Laidler 1985; 1993, 263–65.
12. Letters from Hans Halban to Michael Polanyi, June 20, 1920, and July 28, 1920 (MPP 1:9 and 1:10, respectively). The Austrian (and later French) physicist Hans Halban was born in 1908, so it is unlikely that he is this correspondent.
13. Letter from Max Born to Michael Polanyi, June 13, 1921 (MPP 1:14).
14. Quoted in Scott 1983, 305–6n10; letter from Einstein to Born, December 30, 1921, in Born 1971.
15. Taylor 1962, 11.
16. See Laidler 1985, 69–73; and Hartley 1955, 180. Also Laidler 1993, 263–65; Hinshelwood 1926; Semenov 1935.
17. Stoltzenberg 1994, 475.
18. The discovery of chemiluminescence had been described by F. Haber and W. Zisch in 1922. Beutler and Polanyi 1925; and Beutler, Bogdandy, and Polanyi 1926.
19. Scott and Moleski 2005, 97, 305n5.
20. On Beutler and Hartel, see Scott and Moleski 2005, 94, 117–19. See discussion in Polanyi 1932b, 32–38.
21. Eyring 1982. Also see Dambrowitz and Kuznicki 2010.
22. Eyring 1982, 348–49; Kauzmann 1996.
23. Hirschfelder 1982, 351. On Dirac, see Farmelo 2009.
24. The act of chemical combination is to be explained "als den Erfolg stetig wirkender Kräfte aufzufassen und das Verhalten der Atome durch Angabe einer Funktion der Lage der Atome zueinander nach Art eines Kräftepotentials zu beschreiben" (London 1929, 552).
25. Also see Kallmann and London 1929, discussed in London 2000b.
26. By 1931, Polanyi was unhappy with Hartel's work, as indicated in a letter from Haber to Polanyi, dated August 5, 1931, in which Haber said he was bowing to Polanyi's judgment in terminating Hartel (MPP 2:7).
27. Chayut 1994.
28. Ibid., 243.
29. Ibid., 244.
30. Eyring and Polanyi 1930, 1931.
31. Eyring 1931.
32. Born and Oppenheimer 2000, 457.
33. In addition to the original papers by Eyring and Polanyi, I have relied upon Ramsey 1997a, esp. 630–38. Also see Glasstone, Laidler, and Eyring 1941, 1–13. Note that

the paper "On Simple Gas Reactions" uses the reaction designation Y + XZ → YX + Z, whereas *Theory of Rate Processes* uses X + YZ → XY + Z. I am using the system illustration in the original paper "On Simple Gas Reactions."

34. See the discussion in Glasstone, Laidler, and Eyring 1941, 2–5.
35. Hargittai 1997, 72.
36. Eyring and Polanyi 2000, 425. Also see the language and descriptions in Eyring 1931, 2540–42; Glasstone, Laidler, and Eyring 1941, 96–97.
37. Eyring and Polanyi 2000, 444.
38. For an example of the language, see Eyring and Polanyi 1931, 304. On credit to Polanyi, see Scott 1983, 298. On Eyring, see Scott and Moleski 2005, 117.
39. Heitler and London 1927; Sugiura 1927; Morse 1930; and Farkas 1930.
40. See Polanyi 1932b, 20; and Eyring 1930.
41. See Glasstone, Laidler, and Eyring 1941, 96.
42. Scott and Moleski 2005, 116–17.
43. Ramsey 1997a, 633–34.
44. Eyring and Polanyi 1931, 301; 2000, 442. Eyring mentions another possible result at the saddle point, namely something like what Wigner would write in 1932 about tunneling: "The quantum mechanics provides still another alternative. The point may pass through the energy mountain without bothering to go over the divide" (1931, 2540–41; Eyring cites Langer 1929).
45. Eyring 1931, 2540–41.
46. Ramsey 1997a.
47. Eyring and Polanyi 2000, 450.
48. Ibid., 426.
49. For a history, see Kritsman, Zaikov, and Emanuel 1995, 182–210.
50. Eyring 1931; Kauzmann 1996.
51. Letter from M. G. Evans to Michael Polanyi, February 20, 1933 (MPP 2.11); Milly Evans to Polanyi, April 14, 1933 (MPP 2.11); M. G. Evans to Polanyi, July 10, 1933 (MPP 2.12); Richard A. Ogg Jr. to Polanyi, May 5, 1933 (MPP 2:12).
52. Pelzer and Wigner 1932; see Eyring 1961, 28.
53. Wigner 1932; see also Laidler 1993, 248.
54. Scott and Moleski 2005, 128–29. Polanyi and Cremer coauthored a series of papers from 1931 to 1933.
55. Letter from Henry Eyring to Michael Polanyi, October 28, 1933 (MPP 2.13), noting that he and Wheeler were recalculating the activation energies of the hydrogen molecule with halogens, taking account of directed valence.
56. Letter from Henry Eyring to Michael Polanyi, October 28, 1933 (MPP 2:13).
57. Eyring 1935.
58. Eyring 1982, 349.
59. See Bunsen Gesellschaft 1982, 348–464.
60. Eyring 1935.
61. Evans and Polanyi 1935, 894, with acknowledgements to Wigner for helpful discussions.
62. Evans and Polanyi 1935, 877.
63. Ibid., 893.
64. Ibid., 1935, 879n10. Polanyi cites Polanyi 1920a, 1920c; Wigner and Polanyi 1928.
65. Urey is quoted in Kritsman 1995, 187, from Heath 1980, 89.
66. Hirschfelder 1982, 352.
67. Kritsman 1995, 187.

68. Coolidge and James 1934, 811, 816, 817; James and Coolidge 1933. For a general discussion of computation in the early history of the quantum theory using *ab inito* methods, see Park 2009, with reference to James and Coolidge 1933, 45–49. Both learned quantum mechanics from Edwin C. Kemble at Harvard. James was a physics graduate student and Coolidge was a professor of chemistry in 1933. In his historical article, Park noted that there were rumors that their work on the hydrogen problem took James and Coolidge three years, but Park suggested a figure of six months (2009, 61).

69. Kritsman 1995, 187. On Lindemann (later Lord Cherwell) as Churchill's scientific advisor, see Harrod 1959.

70. Hirschfelder 1941, 645–47. The method had proven successful in predicting the energy of resonance between various Lewis configurations for carbon molecules (646), but the method remained very approximate because of its assumptions that $s$, $p$, and $d$ electrons could be treated equally and because of its silence on the influence of neighboring groups on activation energy (647). Glasstone provided a theoretical demonstration that the rate of a bimolecular reaction in solution is higher than the rate of reaction in the gas phase by a value equal to the ratio of the activity coefficients of the starting molecules to the activity coefficient of the activated complex. Discussed in Kritsman 1995, 199.

71. Letter from Henry Eyring to Michael Polanyi, February 18, 1937 (MPP 3.8).

72. Faraday Society 1937, 27, 72.

73. Ibid., 77, 123.

74. Eyring 1937a, 3.

75. Dirac 1929, 714.

76. Eyring 1937b, 41–42.

77. Faraday Society 1937, 70–71.

78. Ibid., 28.

79. London 2000a, 419. Among the few other German researchers who were developing semi-empirical methods in the 1930s was Hans G. A. Hellmann, who studied in the late 1920s with Otto Hahn and Lise Meitner at the Kaiser Wilhelm Institute for Chemistry. After taking his doctorate in physics with Erich Regener at Stuttgart, Hellmann applied semi-empirical methods to chemical bonding in a paper of 1933. He left Germany in 1934 for the Karpov Institute in Moscow, where he was executed during the purges of 1938. He authored the first German textbook in quantum chemistry, *Einführung in die Quantenchemie*, which appeared in Russian and in German in 1937. See Schwarz et al. 1999; Karachalios 2000, esp. 504–6.

80. Hirschfelder 1941, 645.

81. G. Porter 1992, 117.

82. See Chemical Society 1962. On the semi-empirical method vs. reductionism, see Woody 2000; Ramsey 1997b; Scerri 1994. By this time, Henry Eyring, a third-generation Mormon, had long since returned to the University of Utah (in 1946).

83. Herschbach 1987.

84. Dagani 1999; G. Porter 1992. Porter was an undergraduate student of Evans's at Manchester.

85. Nye 1993, 164–65, 179–82. On Manchester science, see Pickstone 2007.

86. R. Norman Burkhardt, "Some Famous Manchester Scientists," January 4, 1983, typescript for Association for Science Education; typed memorandum by committee on chair of chemistry, May 19, 1938 (UMA VCA/7/40/1 [Chemistry Chairs: Todd]).

87. Letter from F. A. Donnan to Michael Polanyi, April 7, 1933 (MPP 2:11).

88. Letter from Ian Heilbron to John Moberly, vice-chancellor, May 3, 1933 (UMA VCA/7/358 [Chemistry: Heilbron] 10/10).
89. Newspaper clipping dated July 14, 1933, with headline "University Post for a Foreigner: Professor Critic of a City Appointment"; newspaper clipping from the *Guardian*, July 13, 1933 (UMA VCA/7/358 [Chemistry: Polanyi] 11/12).
90. See Scott and Moleski 2005, 142, 145, 147–48, 150. On Schmalz, see letter from Michael Polanyi to Norman Smith, registrar, July 17, 1933 (UMA VCA/7/358 [Chemistry: Polanyi] 11/12); on Evans and heavy water, letter from Meredith Evans to Michael Polanyi, July 10, 1933 (MPP 2:12).
91. Scott and Moleski 2005, 150–51.
92. Letter from Michael Polanyi to Arthur Lapworth, June 13, 1933 (UMA VCA/7/358 [Chemistry: Polanyi]) 11/12).
93. Letter from Ernest Simon to VC Walter Moberly, November 9, 1934; carbon copy of a letter from Ernest Simon to Christopher Needham, November 30, 1934 (UMA, VCA/7/358 [Chemistry] 2/12).
94. Letter from Lauder W. Jones to Vice-Chancellor Walter H. Moberly, June 22, 1933 (UMA VCA/7/358 [Chemistry: Rockefeller Grants] 9/12).
95. Unsigned memo, "Memorandum on Needs of the Chemistry Department," November 7, 1934; carbon copy of letter from Michael Polanyi to Major G. I. Taylor, FRS (UMA VCA/7/358 [Chemistry] 2/12). Quotation reproduced by courtesy of the University Librarian and Director, The John Rylands University Library, the University of Manchester.
96. Quote from signed memo, "On the Necessity of a New Building for Research in Physical Chemistry," from Michael Polanyi to Vice-Chancellor Moberly, March 8, 1935 (UMA VCA/7/358 [Chemistry: Heilbron] 10/10). Also see memo, "Teaching Staff in the Department of Chemistry," by Michael Polanyi, March 11, 1935, (UMA, VCA /7/358 [Chemistry] 2/12). Quotation reproduced by courtesy of the University Librarian and Director, The John Rylands University Library, the University of Manchester.
97. Calvin 1991–1992, 40–42; and Scott and Moleski 2005, 154–55.
98. See Kay 1993, 148–50; and Kohler 1991, 344. Linus Pauling, handwritten note, October 24, 1933 (from PP Science Box 14.037, Folder 4, by authorization of Linus and Ava Helen Pauling Papers, Special Collections, Valley Library, Oregon State University), quoted in Nye 2000b, 482. Weaver had been a member of the physics faculty at Caltech from 1917–1920.
99. Polanyi 1911a, 1911b.
100. Calvin 1991–1992, 41.
101. Scott and Moleski 2005, 156–57.
102. Grant-in-Aid #36160, June 26, 1936, to University of Manchester for Michael Polanyi; Grant-in-Aid #37070, n.d., to University of Manchester for Michael Polanyi (RFA RG 1.1, Series 401D, Box 47, Folder 616).
103. Scott and Moleski 2005, 157. See Calvin 1961.
104. See Nye 1993, 163–226.
105. See Scott and Moleski 2005, 157, on his collaborators in Manchester.
106. Discussed in Bergmann 1961, 37–38.
107. For example, see Ogg and Polanyi 1935a, 1935b; Evans and Polanyi 1935.
108. J. Polanyi 2003, 121.
109. See discussion in Nye 1993, 207–21. See also Ingold 1935; Branch and Calvin 1941.

110. Ogg and Polanyi 1935a, 620.
111. Carbon copy of letter from Michael Polanyi to Christopher Ingold, January 11, 1938; letter from Christopher Ingold to Michael Polanyi, January 14, 1938 (MPP 3:11); letter from Michael Polanyi to Cyril Hinshelwood, June 5, 1935 (MPP 3:3).
112. Warren Weaver diary excerpt, June 16, 1936 (RFA RG 1.1, Series 401D, Box 47, Folder 616).
113. Letter from Michael Polanyi to F. Glum, June 18, 1936; Von Cranach to Peter Debye, June 25, 1936; Von Cranach to Michael Polanyi, June 25, 1936; Michael Polanyi to Von Cranach, June 26, 1936 (MPG Abt. I, 1a/1077/Gruppe, 4), courtesy of Dieter Hoffmann.
114. Typed note for the vice-chancellor from Sir Ernest Simon, May 27, 1937; purple typescript from Michael Polanyi to the vice-chancellor, June 22, 1937; letter from the vice chancellor to Michael Polanyi, June 23, 1937 (UMA VCA/7/358 [Chemistry] 3/12).
115. Fellowship report of T. R. Hogness, University of Chicago, August 16, 1937 (RFA RG 1.1, Series 401D, Box 47, Folder 616); see also Calvin 1991–1992.
116. Todd 1983, 52.
117. Letter from Michael Polanyi to the vice-chancellor, December 21, 1938; carbon copy of letter from Alexander Todd to vice-chancellor, March 7, 1940; letter to vice-chancellor from F. S. Spring, March 12, 1940; letter from G. N. Burkhardt to vice-chancellor, March 16, 1940; letter from Michael Polanyi to vice-chancellor, July 2, 1941 (UMA VCA/7/358/[Chemistry] 5/12).
118. Scott and Moleski 2005, 191. Janos Plesch's patients also included Albert Einstein and John Maynard Keynes. See J. Bernstein 2004.
119. Plesch 2007–2008, 40–43; see also Plesch, 1946; Evans et al. 1946.
120. The first-year students used Pauling 1939. Third-year students were given materials from Pauling and Wilson (1935). Alan A. R. Battersby recalled the lectures on chemical bonding in a letter to William T. Scott, September 22, 1982 (information courtesy of Martin X. Moleski). Brian G. Gowenlock kept copies of Polanyi's typed lectures that were duplicated and distributed for the honors course, the chemical bonding course (titled "Chemical Links"), and the quantum mechanics course. Thanks to Martin X. Moleski and Brian Gowenlock for copies of the lectures and information about them.
121. I am grateful to Derry W. Jones, Emeritus Professor of Chemical and Forensic Sciences, University of Bradford, for discussion about Manchester physics and chemistry during 1945 to 1948. On 1943 to 1948, see Gowenlock 2008–2009, 34.
122. Letters from M. G. Evans to vice-chancellor, November 17, 1947, and M. G. Evans to "Sir John" (vice-chancellor), July 12, 1948 (UMA VCA/ 70/40/4 [Chemistry Chair: Evans]); letter from M. G. Evans to Sir John Stopford VC (UMA VCA/7/358/ [Chemistry] 8/12).
123. See Brush 1999; Simões 2003.
124. Van Vleck and Sherman 1935, 168–69.
125. Typed memo, "Teaching Staff in the Department of Chemistry," by Polanyi, March 11, 1935; typed memo "Organization of the Chemistry Department," by Polanyi, March 11, 1935 (UMA VCA/7/358 [Chemistry] 2/12).
126. Ruthenberg 1997, 2009; Laidler 1993, 265–66; and more extensively, Emeléus 1960. On Polanyi and Paneth's getting together for an evening in Berlin, see Polanyi's diary entry for April 27, 1929 (MPP 44:4).
127. Ramsey 2006.

128. Polanyi 1936a.
129. Ibid., 36.
130. J. Polanyi 2003, 120.
131. Quoted in Plesch 2007–2008, 44. Also see Polanyi 1962b, 1963.

CHAPTER FIVE

1. Musil 1996, 102. On Musil's interpretations of Viennese intellectual culture, see Luft 1980.
2. Quoted in Coen 2007, 247. Serafin Exner explained in his inaugural speech that the circulation of money depends on the self-interest of innumerable independent individuals and thus has the character of a chance phenomenon. The "most probable state" therefore is not an equal distribution of wealth—"that would instead be highly improbable." There should naturally emerge a large middle class and a smaller upper and lower class, with all deviations from the average distributed along a normal curve. The liberal state is the most probable state (249). For a minimal definition of Austrian Liberalism, see Coen 2007, 10.
3. On Adolf Polanyi's economic monograph, see L. Polányi 1995, 9.
4. John Stuart Mill wrote in his *Principles of Political Economy*, first published in 1848, that: "In so far as the economical condition of nations turns upon the state of physical knowledge, it is a subject for the physical sciences, and the arts founded on them. But in so far as the causes are moral or psychological, dependent on institutions and social relations, or on the principles of human nature, their investigation belongs not to physical, but to moral and social science, and is the object of what is called Political Economy" (quoted in Schabas 2005, 5).
5. Polanyi 1962a.
6. Letter from Karl Polanyi to Michael Polanyi, January 21, 1957, in English (MPP 12:12), quoted in Nagy 1994, 108. The question mark appears in parentheses in the original text.
7. Polanyi-Levitt 1990b, 119. And Duczynska, n.d., 13; Litván 1990, 34.
8. Scott and Moleski 2005, 42; Vezér 1990, 23. The quote is from Congdon 1997–1998, 7, quoting from Zeisel 1968, 172.
9. See Rabinbach 1983; Leonard 1998, esp. 7; and Cartwright et al. 1996, 57–60. On deprivation and misery in Vienna during the First World War, see Healy 2004.
10. Hacohen 2000, 117–18.
11. Mendell 1990, 68.
12. Scott and Moleski 2005, 15, 51.
13. Litván 1990, 34; 2006, 230.
14. Nagy 1994, 93, 85–87.
15. Polanyi-Levitt 1990b, 119.
16. Rosner 1990, 56. Caldwell 2006, 26–27. On Neurath, see Cartwright et al. 1996.
17. See Spiegel 1971, 426–427; McCraw 2007, 40–45; März 1991, 99–100; and Cartwright et al. 1996, 3, 20, 51–53.
18. Caldwell 2004, 114–16.
19. Ibid., 118.
20. Caldwell, 2006, 24–25.
21. Cat 2004, 3–4; Rosner 1990, 56.
22. Mendell 1990, 69, 71 (quotation, 71). Also see Morgan 2003, 278–79.
23. See Rosner's summary (1990, 57–59) of K. Polanyi 1922, 1925.
24. Rosner 1990, 62–63.

25. McCraw 2007, 427–428.
26. Caldwell 2004, 140. Also see Hayek 1994, 58; Mises 1969, 15–17.
27. Carnap, Hahn, and Neurath 1929, 29–30. More generally, see Stadler 1997; Johnston 1972; Richardson and Uebel 2008; Uebel 2008. Also see Hacohen 2000, 187–90; Cartwright et al. 1996, 77–78.
28. Watkins 1997; Popper 1976, 87; and Brudny 2002, 61–72. On Karl Polanyi's participation in activities of the Vienna Circle group, see Stadler 2003, 343, 345.
29. See Antiseri 2004, 120–21, 134; Menger 1883.
30. Krohn 1993, 15.
31. Ibid., 40.
32. Condon 1997–1998, 8, citing K. Polanyi 1932–1933, 301–3.
33. Polanyi-Levitt 1990b, 121; Vezér 1990, 23; Scott and Moleski 2005, 153–54. See also Duczynska 1978, 12. Toni Stolper assisted Duczynska in the translation of this book from the original (Duczynska 1975). On Kari Polanyi-Levitt's education and career, see Mendell 2005.
34. Cartwright et al. 1996, 83.
35. Quoted in Scott and Moleski 2005, 109.
36. Scott and Moleski 2005, 110. Scott and Moleski cite Robertson's book as one called *Production*, but there is no title by this name in the British Library.
37. On differences between Keynes and Robertson, see Moggridge 2004.
38. Interview of Michael Polanyi by Thomas S. Kuhn, February 15, 1962, transcript, pp. 10–12 (SHQP).
39. On Marschak, see Rosner 1990, 963; Krohn 1993, 173.
40. Scott and Moleski 2005, 121; on Haber, Joffe, and Moellendorff, see letter from Michael Polanyi to Toni Stolper, January 25, 1930 (MPP 2:6). On Moellendorff's collaboration with Rathenau, see Caldwell 2004, 118. Rathenau was the inspiration for the figure of the Berlin industrialist and writer Arnheim in Musil's *Man without Qualities*. See Luft 1980, 116, 242.
41. See Klausinger 2001, 241–42.
42. Ibid. Toni Stolper's biography of her husband does not mention either Michael or Karl Polanyi (Stolper 1960).
43. Polanyi 1930 (original emphasis)
44. Cassidy 2000, 46; Backhouse 2006, 47.
45. Klausinger 2001, 246–47; for quotation, see Keynes 2004, 41.
46. Klausinger, 2001, 252.
47. Letter from Michael Polanyi to Toni Stolper, January 25, 1930 (MPP 2:6).
48. Quoted in Scott and Moleski 2005, 121. According to Scott and Moleski, Söllner, who was not invited to join the group, "complained that Polanyi formed the study group in order to be 'surrounded by a society of adorers'" (121). Söllner became a faculty member in the field of physiological chemistry at the University of Minnesota.
49. Interview of Michael Polanyi by Thomas S. Kuhn, February 15, 1962, transcript, pp. 11–12 (SHQP).
50. Letter from N. Semenoff to Michael Polanyi, February 15, 1930 (MPP 2:6). Scott and Moleski 2005, 126, 134. For the offer from Semenoff, Scott and Moleski cite the letter from Michael Polanyi to Nikolai Semenoff, December 17, 1931 (MPP 2:7).
51. On Joffe, see Josephson 2005, 72–73, 77–79, 86; and Graham 1993, 175, 209–10.
52. Scott and Moleski 2005, 159, citing Scott's interview with John Jewkes; letter from Michael Polanyi to Toni Stolper, September 17, 1933 (MPP 2:13).

53. Chilvers 2003, 422–26; Brown, 2005, 105–6.
54. See Szapor 1997.
55. Polanyi 1935b.
56. See Scott and Moleski 2005, 160–61.
57. Interview of Michael Polanyi by Thomas S. Kuhn, February 15, 1962, transcript, p. 11 (SHQP); On Lippmann and Simon, see Scott and Moleski 2005, 161.
58. Scott and Moleski (2005, 163) write that Polanyi had been convinced since 1929 that a motion picture about economics would be a good educational project.
59. Scott and Moleski 2005, 143, 163.
60. Handwritten draft of letter from Michael Polanyi to Toni Stolper, November 4, 1935, pp. 1–2 (MPP 3:5).
61. Letter from Oscar Jászi to Michael Polanyi, November 24, 1935, mentioning the film (MPP 3:5).
62. Scott and Moleski 2005, 164, based on Scott's interviews with Wilfried Heller and Cecil Bawn.
63. Ibid.
64. Scott and Moleski 2005, 164, drawing upon Scott's interviews with Daniel Eley and Melvin Calvin.
65. Schabas 2005, 46–47.
66. Ibid., 59. Phillips's model had three transparent tanks corresponding to production, stocks, and consumer demand; and tubes through which water was pumped. On the way the model worked, see Wulwick 1989; Morgan and Boumans 2004.
67. Scott and Moleski 2005, 164.
68. Ibid., 165, citing letter from Toni Stolper to Michael Polanyi, November 21, 1936 (MPP 3:7).
69. Scott and Moleski 2005, 166–67. On Gaumont, see copy of letter from Michael Polanyi to Charles Vale, September 4, 1937 (MPP 3:9). On soundless, see the carbon copy of letter from Michael Polanyi to J. D. Bernal, September 13, 1938 (MPP 3:12).
70. "Money Is Star of This Film," *Evening News* (Manchester), March 10, 1938, newspaper clipping (MPP 47:3).
71. Michael Polanyi, "Memorandum on Economic Films," typescript, undated [1938], pp. 2, 3 (MPP 3:6); Scott and Moleski 2005, 168.
72. Letter from Association of Scientific Workers to Michael Polanyi, August 24, 1938; letter from J. D. Bernal to Michael Polanyi, September 10, 1938 (MPP 3:12).
73. Grant-in-Aid Application, February 7, 1939 (RFA RG 1.1, Series 401S, Box 82, F. 1084). On speed and jerkiness, see letter from R. S. Lambert at the British Film Institute, London, to Michael Polanyi, August 4, 1938 (MPP 3:12); on receipt of Rockefeller funds, see carbon copy of letter from Michael Polanyi to Charles Vale, April 28, 1939 (MPP 3:15); carbon copy of letter from John Jewkes to Robert Letort, Rockefeller Foundation, Paris, May 18, 1939 (MPP 3:15).
74. Scott and Moleski 2005, 179.
75. Letter from Michael Polanyi to Tracy B. Kittredge, Rockefeller Foundation, La Baule, Loire, March 7, 1940; Michael Polanyi to Tracy Kittredge, April 2, 1940; Michael Polanyi to Tracy Kittredge, New York City, May 24, 1940; radiogram, June 13, 1940; carbon copy of letter from Michael Polanyi to Prof. J. B. Condliffe, Dept. of Economics, UC Berkeley, July 15, 1940; memo from Joseph H. Willits, Rockefeller Foundation, to Michael Polanyi, August 30, 1940 (RFA RG 1.1, Series 401S, Box 83, Folder 1085).

76. List of attendees at Museum of Modern Art, November 7, 1940 (RFA, RG 1.1, Series 401S, Box 83, F. 1086).
77. Letter from John B. Condliffe to Tracy Kittredge, November 25, 1940 and July 3, 1941; Oskar Morgenstern to Tracy Kittredge, March 25, 1942; Kenneth M. Spang to Tracy Kittredge, June 15, 1942; memo from Joseph H. Willits to Walter Stewart, Institute for Advanced Study, November 19, 1940 (RFA RG 1.1, Series 401S, Box 83, Folder 1086).
78. "The Film in Economics Classes. A WEA [Workers' Educational Association] Experiment" (London: Hereford Times, Ltd, n.d.), 15 pages (Roper, pp. 5–6; Dawes, p. 7; Raybould, p. 15) (RFA RG 1.1, Series 401S, Box 83, Folder 1088).
79. Letter from Michael Polanyi to Joseph Willitts, December 12, 1944 (RFA RG 1.1, Series 401S, Box 83, Folder 1088).
80. Polanyi 1938, 19.
81. Polanyi 1945b, v for acknowledgements. On Beveridge and Robbins, see Caldwell, 2004, 173–74. Also on Robbins, see Howson 2004.
82. Polanyi 1945b, xvi. Keynes 1964, 378, 380. See Thorpe 2009, 76–82.
83. Polanyi 1945b, 146.
84. Keynes 1964, 150; also see Cassidy 2001, esp. 44. Schumpeter's *Theory of Economic Development*, published in German in 1911 but not in English until 1934, describes the psychological elements of the joy of creating, the impulse to fight, and so forth, developing the notions of innovation and entrepreneurship. Keynes does not refer to the book, however, nor pay much attention to German and Austrian economic traditions. See McCraw 2007, 70–74, 273.
85. Keynes 1964, 374.
86. Polanyi 1945b, 146.
87. Polanyi 1945b, 27–29, 31, 37–39. Also see Roberts 2005, esp. 129–30.
88. On similarities and differences between Keynes and Polanyi, see Manucci 2005, esp. 156–57 on the "principle of neutrality."
89. Polanyi 1945b, 67, 73, 79–80. See also, on this point, Polanyi 1946d, mentioned in Congdon 2001, 79.
90. Polanyi 1945b, 125, 131–32 (quotation, 134).
91. Keynes 1964, 33.
92. Ibid., 254.
93. Polanyi 1945b, 149.
94. Polanyi, 1945b, 88–89 (original emphasis).
95. Ibid., 88.
96. Ibid., 149–50.
97. Keynes, 1964, 60.
98. Polanyi, 1945b, 149.
99. Ibid., 103, 150.
100. Thorpe 2009, 29–30.
101. Johns 2006, 7, 154. Also Scott and Moleski 2005, 185.
102. Polanyi 1945b, 149–50.
103. Johns 2006, 153, 155–56. See Polanyi 1944a. Also, on the general topic of industrial research laboratories, scientists, and entrepreneurs, see Shapin 2008.
104. Scott and Moleski 2005, 185 (on Jewkes), 199 (on Haberler). Haberler wrote the foreword for Polanyi 1945b. Jewkes soon after published *Ordeal by Planning* (1948).

105. Letter from Michael Polanyi to Karl Mannheim, April 14, 1944 (MPP 4:1), quoted in Nagy 199–43, 87.
106. Scott and Moleski 2005, 196, from letter from Karl Polanyi to Michael Polanyi, November 1, 1945 (MPP 4:13). Also in Congdon 2001, 82.
107. Letter from Karl Polanyi to Irene Grant, October 13, 1933, on his intention to get in touch with Keynes, Tawney, Cole, and others, quoted in Congdon 2001, 19.
108. See Maucourant 2005, 31; Polanyi-Levitt 1990b, 121.
109. Caldwell 2004, 215; Congdon 1997–1998, 8.
110. Letter from Karl Mannheim to Walter Adams, November 19, 1933, quoted in Mannheim 2003, 90–91.
111. Quoted in Duczynska 1977, xvi.
112. Congdon 2001, 40, 53–54. For a contemporary review, see Edwin Ewart Aubry's article in *Journal of Religion* 17, no. 3 (July 1937): 343–45.
113. See Congdon 1997–1998, 8; quotation in Polanyi-Levitt 1990b, 121.
114. K. Polanyi 1937 (quotations, 56).
115. Quoted in Nagy 1994, 99, from manuscript in the papers of Karl Polanyi, Box 6, Folder: Bulletin No. 4, pp. 8, 19, Karl Polanyi Institute of Political Economy, Concordia University, Montréal, Quebec.
116. Letter from Michael Polanyi to Karl Polanyi (MPP: 1944, II), quoted in Nagy 1994, 100.
117. K. Polanyi 1944.
118. Ibid., in the acknowledgments. Peter Drucker (1979) has written about the Polanyis in an autobiographical book, but some of the information and anecdotes are unreliable.
119. See the Karl Polanyi Institute of Political Economy website: http://artsandscience1 .concordia.ca/polanyi/about/ (accessed July 22, 2010).
120. Quoted in Congdon 2001, 82, from letter from Karl Polanyi to Michael Polanyi, August 22, 1941 (MPP 17:11).
121. For Keynes 1964, 3.
122. K. Polanyi 1944, 68.
123. Schabas 2005, 99. Schabas notes that the phrase "invisible hand" appears only three times in Smith: once in *The Theory of Moral Sentiments*, once in *The Wealth of Nations*, and a third time in a posthumously published essay on the history of astronomy, where the hand is that of Jupiter (95, 99). On freedom of movement and the labor market, see Torpey 2000, 9, 59, 89.
124. K. Polanyi 1944, 40, 42, 68, 78, 80–83, 115 (quotations, 126).
125. K. Polanyi 1944, 46. Cf. Veblen 1994.
126. For Weber, see Weber 1923.
127. K. Polanyi, 1944, 46–51, 153; Malinowski 1922.
128. K. Polanyi, 1944, 53. As Schabas writes, Aristotle's *Politics* and his *Nicomachean Ethics* include analyses of money, market exchange, and household production. He used the terms "oikonomike" or "oeconomia" for the art of household management and "chrematistike" for commerce. The terms precede Aristotle, however, perhaps going back to Homer. See Schabas 2005, 3–4.
129. Tönnies 1887.
130. K. Polanyi, 1944, 3, 9–10, 23, 243 (quotation, 244).
131. Ibid., 1944, 256, 258A.
132. K. Polanyi 1944, 258. See Maucourant 2005, 41. See Talcott Parsons's similar ar-

guments in *Structure of Social Action* (1937), as discussed in Brick 2006, 122–28, 181–83.

133. Polanyi-Levitt 2007, xiv–xv. On the intellectual origins of American institutional economics, see Mayhew 1987.

134. See Neale 1990, 150n1. On institutionalism in the Economics Department at Columbia University, see Rutherford 2004.

135. Kindleberger 1973, 45.

136. Ibid., 47, 50–51.

137. Polanyi 1945b, 136.

138. Letter from Karl Polanyi to Michael Polanyi, November 1, 1945 (MPP 17:1; original emphasis), quoted in Condon 2001, 83.

139. Letter from Michael Polanyi to Karl Polanyi, in 1953 (MPP), quoted in Nagy 1994, 105.

140. Scott and Moleski 2005, 193, quoting from a letter from Michael Polanyi to Karl Polanyi, June 12, 1944, in the letters collected by Kari Levitt-Polanyi.

141. Litván 2006, 508. In the English translation of Litván's book, the German word *Kauderwelsch* appears in the text with the translation "Double Dutch" in brackets. Another possible translation for *Kauderwelsch* is the English word "gibberish."

142. Ibid.

143. Letter from Oscar Jászi to Karl Mannheim, December 19, 1936, quoted in Mannheim 2003, 183.

144. Litván 2006, 508.

145. See Letter from Karl Mannheim to Joseph Houldsworth Oldham, April 1941, quoted in Mannheim 2003, 261–62.

146. Bernstein 2001, 46, 123; Morgan and Rutherford 1999.

147. See Brick 2006, 237–38; Krugman, 2009, 38.

148. Polanyi 1962a. *Minerva* was launched by Edward Shils at the University of Chicago as "A Review of Science, Learning and Policy." On laissez-faire economics in the work of Polanyi, see, e.g., Thorpe 2009; Mirowski 1998–1999; Hollinger, 1996b.

149. Polanyi, 1945b, 149.

150. Quoted from Hayek in Caldwell 2004, 197. Stephen Turner points out that the idea of spontaneous coordination was a commonplace of late nineteenth-century continental liberalism and that it is a term used by Auguste Comte. Turner 2005.

151. Caldwell 2004, 238. See Jacobs 1999. Another example of an economist writing on the growth of knowledge as a market place is Machlup 1980–1984.

152. Hayek comments on their relationship in England and at the University of Chicago in an interview of Hayek by James Buchanan in 1978, under the auspices of the Oral History Program of the University Library at UCLA. See Mirowski 1998–1999, 30.

153. Hayek 1960, 160, quoted in Caldwell 2004, 294.

154. Polanyi 1948, 150, 153; Polanyi 1946e. See Polanyi's reiteration of polycentricity as an argument against central planning in Polanyi 1969b. This paper was an essay in honor of Hayek, in Stressler 1969, 165–69. On this theme, see Allen 1998, 154–56; Vinti 2005, 134.

155. Scott and Moleski 2005, 186, 188.

156. Scott and Moleski 2005, 188, citing "The City of Science" (MPP 27:6).

157. Polanyi 1943b, 19.

158. Polanyi 1943b, quoted in Hollinger 1996c, 143. Also see Turner 2005, 85.

159. Polanyi 1946e, 63.

160. Polanyi 1962a, 54–55. Thorpe notes Polanyi's use of the "invisible hand" metaphor in a speech at a July 1953 conference, "Science and Freedom," in Hamburg. See Thorpe 2009, 21.

161. Polanyi 1962b (quotation, 60).

162. Ziman 2000, 21. For a critical analysis of the political and moral implications of Polanyi's economic metaphor for science, see Mirowski 1997, esp. S135–S138.

163. Thanks to Catherine Herfeld and Beatrice Collina for clearly stating this insight at a 2009 summer school in Vienna.

164. Polanyi 1962a, 69, 72, for "Explorers." The comment by Jewkes is quoted in Wigner and Hodgkin (1977, 242, from Jewkes 1976).

165. Polanyi 1962a, 72. Jordi Cat notes Karl Polanyi's concern in *The Great Transformation* with guarantees in any socialist society of individual freedom and nonconformity within institutional structures. See Cat 2004, 16–17, citing K. Polanyi 1944, 255.

166. Polanyi 1962a, 64.

CHAPTER SIX

1. Scott and Moleski 2005, 143–44.

2. Ibid., 143, 152.

3. See Greenaway 1932, 12–13; Robinson 1976, 27.

4. See Allen 1998, 1–2. See Michael Polanyi, "The Liberal Conception of Freedom" (MPP 26:8).

5. Typescript dated December 20, 1946, marked "not sent" and "Blackett" (MPP 5:2). Quoted with permission of John C. Polanyi.

6. Adelman 1987, 2–4; Pugh 1993, 260, 264.

7. Pugh 1993, 260, 268. See also Labour Party 1934.

8. See Waters and Van Helden 1992.

9. Werskey 1978, 239n; Barberis, McHugh, and Tyldesley 2003, 355–56.

10. On Wells, see Rose and Rose 1970, 52–53.

11. In a letter to Karl Mannheim in 1944 Polanyi recalled that "as a boy and young man I was a materialist and eager disciple of H.G. Wells" (quoted in Nagy 1994, 87, from a letter from Michael Polanyi to Karl Polanyi, in English, April 14, 1944 [MPP 4:11]). For his letter to Fajans, see Scott and Moleski 2005, 41; Polanyi's diary entries for January 10, 1929, and April 15, 1929 (MPP 44:4).

12. Chilvers 2003, 426; Rose and Rose 1999, 143.

13. Crowther 1930.

14. Chilvers 2003, 421–22; Brown 2005, 109.

15. See Huxley 1932.

16. Quoted in Brown 2005, 110.

17. MacLeod and MacLeod 1976.

18. See Paul Forman's discussion of Bukharin's April 1931 address in Forman 2007, 23.

19. Werskey 1978, 170; Huxley 1934 (foreword by Hyman Levy, v, and author's preface, x).

20. Kevles 1992, 241, citing Clark 1968, 204; Baker 1976, 212.

21. Blackett 1934.

22. Ibid., 224.

23. Blackett 1935.

24. See Scott and Moleski 2005, 37. More broadly on Blackett, see Nye 2004.

25. Polanyi 1949.

26. See, for example, the history of the establishment of institutes and faculties of sciences in late-nineteenth-century France supported by local industries in Nye 1986.
27. See, for example, Pyenson 1983.
28. Quoted in Shapin 2008, 24, from Gregory 1928, 50.
29. Blackett's introduction to the first J. D. Bernal lecture, which was given by Dorothy Hodgkin, October 23, 1969, at Birkbeck College (BP: H.142, p. 4). See also Hessen 1932, 147–212. On Blackett's interest in the history of science and the Marxist and Bernalist impact on history of science in Great Britain in the 1930s, see Nye 2008.
30. See Werskey 1978, 138–40; McGucken 1984, 72–73. Werskey's volume focuses on Hyman Levy, J. B. S. Haldane, Lancelot Hogben, J. D. Bernal, and Joseph Needham as scientists of the British Left.
31. Chilvers (2003) notes the articles "Bukharin Shows His Colours: Propaganda and Hate," *Daily Mail* (London), July 6, 1931; "Soviet Delegates Pulled Up," *Morning Post* (London), July 6, 1931; and "Only a Few Minutes Each: How Soviet Scientists Were Treated at Congress," *Daily Worker* (London), July 6, 1931 (426nn 62, 65). On the *Guardian* and the *Spectator*, see Werskey 1978, 140.
32. Werskey 1978, 140, 140–41n.
33. Quoted in Werskey 1978, 145, from letter from Dorothy Singer to Joseph Needham, August 10, 1943.
34. Crowther 1970, 79.
35. Crowther 1935, 1937, 1941.
36. Quoted from Crowther 1941, 432, in Chilvers 2003, 428.
37. Crowther 1970, 86–87.
38. Blackett 1933 (quotation, 67).
39. P. M. S. Blackett, Highgate February 1936, 1-page ms notes (BP: F.1).
40. P. M. S. Blackett, Draft for article? 1938? (BP: F.3).
41. Crowther 1970, 178.
42. Hessen 1932, 1.
43. P. M. S. Blackett, History of Science. Faculty of Arts 1938–9? (BP: F.2).
44. Polanyi 1930.
45. Polanyi 1939, 63.
46. Polanyi 1939, 63–65.
47. Sarton 1931, 21, 177. Also, for example, in a specific riposte against Blackett's history of science, see Taylor 1945, 7, 10.
48. Quoted in Werskey 1978, 40, from "The A.S.W.-Twenty Years History," *Scientific Workers* 12 (1939): 68. On its formation, see Rose and Rose 1970, 52–53.
49. Rose and Rose 1970, 54; McGucken 1984, 13–20.
50. Brown 2005, 121; McGucken 1984, 19–35.
51. Bernal 1933–1934, discussed by Hobsbawm 1999, xii. Also see Rose and Rose 1999, 139. On the BAAS, see McGucken 1984, 36–46, 119, 132; Werskey 1978, 246.
52. Langevin's well-known remark is quoted in Montagu 1999, 212. Also, see Bensaude-Vincent 1987.
53. Montagu 1999, 215.
54. Calder 1999, 161.
55. Hobsbawm 1999, xii; and Rose and Rose 1999, 141.
56. Brown 2005, 112.
57. Mott 1986, 51.
58. Calder 1999, 162–63.
59. Quoted in Calder 1999, 165.

60. Webb and Webb 1935, 1941.
61. Scott and Moleski 2005, 155.
62. Polanyi 1997a (long quotation, 59). Sidney Webb wrote a letter to Polanyi dated March 17, 1937, in which he defended the value of providing a picture of a society rather than providing enumeration of facts and figures that do not form a pattern (MPP 3:8).
63. Graham 1993, 210. Also see Hall 1999.
64. Marton 2006, 95.
65. Koestler 1952, ix–x.
66. Gulick 2003–2004, 7–8.
67. Weissberg 1952, 13; on Blackett, see Koestler 2001, 69–70. Michael Polanyi also spoke with Blackett on behalf of Weissberg, as explained in Letter from Michael Polanyi to C. Lauriston (at Caltech), December 6, 1937 (MPP 3:10). On Fritz Houtermans, who worked with Heisenberg on the German uranium project, see Rhodes 1986, 370–71.
68. Gulick 2003–2004, 8. On Laura (Mausi) Polanyi Striker, see Szapor 1997.
69. Letter from H. Rabinovitch to Michael Polanyi, June 24, 1937, dated from Geneva (MPP 3:9).
70. Koestler 2001, 71.
71. Letter from Michael Polanyi to Karl Polanyi, quoted in Nagy 1994, 100.
72. Marton 2006, 139.
73. Koestler 1945.
74. Michael Polanyi, "Communist Revolts," *Manchester Guardian*, March 28, 1952.
75. Josephson 2005, 85. For names of many prominent scientists who were executed, see Graham 1998, 53–54.
76. Bernal 1967, esp. 310, 317.
77. Ibid., 223–231.
78. Quoted in Roll-Hansen 2005, 287.
79. Roll-Hansen 2005, 128–39.
80. Polanyi 1939, 75.
81. Letter from P. M. S. Blackett to Michael Polanyi, August 26, 1939 (MPP 4:1). Scott and Moleski (2005, 175) describe the Polanyi family's arrival in Cork on August 15, 1939, and Michael Polanyi's departure alone for Manchester on September 1 after hearing news on the radio of the German invasion of Poland. His family returned shortly thereafter to Manchester.
82. Letter from Michael Polanyi to the vice-chancellor of the University of Manchester, August 24, 1939 (UMA VCA/7/358 [Chemistry: Polanyi] 11/12).
83. On Polanyi's naturalization papers, see letter from Costanza Blackett to Michael Polanyi, September 5, 1939 (MPP: 4.1); letter from Michael Polanyi to the vice-chancellor of the University of Manchester, August 30, 1939, saying that he had returned from Ireland to Manchester where he had found a telegram saying that the naturalization certificate had been issued (UMA VCA/7/358 [Chemistry: Polanyi] 11/12).
84. Scott and Moleski 2005, 180.
85. Ibid., 178.
86. Polanyi 1940a, quotation, 139.
87. Ibid., 139–40.
88. Ibid., 130–31 (quotation, 132; emphasis added).
89. See Scott and Moleski 2005, 183, discussing Polanyi 1941.

90. Letter from Cyril N. Hinshelwood to Michael Polanyi, January 27, 1941 (MPP 4:6).
91. Baker 1939, 174.
92. Quoted in Scott and Moleski 2005, 184, from letter from Michael Polanyi to John Baker, June 4, 1962.
93. Jessica Reinisch (2000) has written an unpublished thesis.
94. Letter from Max Born to Blackett, July 22, 1941, from Edinburgh (BP: J. 9). Also see McGucken 1984, 266–75. See Baker 1939, 174–75.
95. Quoted from letter from A. V. Hill to Tansley, June 6, 1941, in McGucken 1984, 288. Rose and Rose 1970, 61, 63–64.
96. Society for Freedom in Science, List of Members (June 1947), photocopy from Museum of Comparative Zoology Library, Harvard University, accession date October 17, 1949. Thanks to Kristin Johnson for this photocopy. On the ecologists, see Cameron and Forrester 1999. Needham's views were expressed clearly in sermons as Master of Caius College at Cambridge. See Davies 1997, 99.
97. On Bridgman, see Schweber 2000, 6–8.
98. The anonymous editor of *Science in War* (Anonymous 1940) was Solly Zuckerman.
99. Crowther, Howarth, and Riley 1942. See Nall 2007, 5–6.
100. Hayek 2007, 12.
101. Crowther 1941, 1942; Haldane 1941; Waddington 1941.
102. Tansley 1942; Baker 1942. See the discussion in Nall 2007, 7–8.
103. Hayek 2007, 10.
104. Quoted in Mirowski 1998–1999, 32. Letter from F. A. von Hayek to Michael Polanyi, January 7, 1941 (MPP 4:5).
105. Hayek 2007.
106. Ibid., 202–3.
107. Stephen Turner (2005, 84–85) points out that the idea of spontaneous coordination was a commonplace of late-nineteenth-century continental liberalism and that it is a term used by Auguste Comte.
108. Caldwell 2004, 238. Also see Nall 2007, 29–31; Jacobs 1999.
109. On this point, see Nall 2007, 31.
110. For the BBC broadcasts, see Polanyi 1944b, 1949.
111. H. J. Muller published a report in the *Bulletin of the Atomic Scientists* in December 1948 on the repression of the Soviet geneticists. C. D. Darlington (1949) discussed British efforts to find out what had happened. See Dale et al. 1949; Jones 1988, 23–24.
112. Darlington 1947; and Zirkle 1959.
113. Quoted from Eric Ashby, "Science without Freedom," BBC Third Programme, January 1948, in Harman 2003, 325.
114. Polanyi 1951, 56–57.
115. Quoted from Rauschning 1939, 224–25 in Polanyi, 1951, 59.
116. Quoted from 1932 conference proceedings in Polanyi 1951, 63. As footnoted on p. 62, Polanyi had read extracts from the 1939 conference report.
117. Quoted in Polanyi, 1951, 67 (discussion of Vavilov, p. 59–65).
118. Polanyi 1964.
119. Quoted in Saunders 1999, 79.
120. Scott and Moleski 2005, 222.
121. Weaver 1945; for Weaver's letter to the *Times* see Warren Weaver, "Free Science Sought: Control, It Is Argued, Would Hamper Advances." *New York Times*, September 2, 1945.

122. Letter from Michael Polanyi to Warren Weaver, March 19, 1946 (RFA RG 1.1, Series 401D, Box 39, Folder 507).

123. Memo, February 10, 1953, from Warren Weaver to L. K., on Nicholas Nabokov's request for aid; letter from Michael Polanyi to Warren Weaver, February 23, 1953, supporting the request; memo from Warren Weaver to D. R., March 10, 1953; Warren Weaver, diary excerpts, and March 11 and 17, 1953; grant-in-aid, RF #52199 for $10,000 to the Congress for Cultural Freedom (RFA RG 1.2, Series 100D, Box 25, Folder 179). By 1960 the CCF's budget was about one million dollars; it was doubled by 1967. The money came from the CIA but also from the state department and foundations which included the Fairfield Foundation, the Hoblitzelle Foundation, and the Ford Foundation. See Krige 2006, 162. On the U.S. congressional hearings, see Krige 2006, 142–147 and 309–10n102.

124. Shils 1954; Polanyi 1953. See also Hollinger 1996c, 80–96, 155–74; and, more broadly, Jones 1988.

125. Memo from Warren Weaver to Dean Rusk, May 6, 1954; letter from Dean Rusk to the Secretary of State, January 21, 1955; letter from Robert Murphy, deputy undersecretary of state, to Dean Rusk, February 5, 1955 (RFA RG 1.2, Series 100D, Box 25, Folder 179); letter from Michael Polanyi to Warren Weaver, January 9, 1956; Warren Weaver diary excerpt, January 2, 1957; letter from Warren Weaver to Michael Polanyi, April 12, 1957; letter from Michael Polanyi to Warren Weaver, May 7, 1957; memo from R. S. M. to J. G. H., May 16, 1957; memo from J. C. B. to Warren Weaver, May 22, 1957; letter from Michael Polanyi to Warren Weaver, June 17, 1957 (RFA, RG 1.2, Series 100D, Box 25, Folder 180).

126. On CCF, Saunders 1999, 371; Krige 2006, 161–62. Quotation from Kennan in Saunders 1999, 408.

127. On this, see Scott and Moleski 2005, 267–68. On the American Committee for Cultural Freedom, see news clipping from the *New York Times* (RFA RG 02, OMR, Series D (Civic Interests), Box 10, Folder 56 [American Committee for Cultural Freedom], 61.14).

128. See McGucken 1984, 350–51.

129. Memos from 1945–1946 and undated typescript following the March 11, 1946, meeting, entitled "The Balanced Development of Science in the Universities of UK" (MPP 22.11).

130. Wilkie 1991, 47–53. Clement Atlee's postwar government also acted positively upon Blackett's recommendation for the establishment of a National Research Development corporation with right of first refusal for patenting results of research supported by public funds.

131. Garfield 1982, 513.

132. Hodgkin, 1980, 28, cited in Garfield 1982, 513–14.

133. Werskey 2007b, 313; see also Synge, 1999.

134. Wilson 1983, 565; Werskey 1978, 81.

135. Garfield 1982, 515, 518, drawing from Perutz 1981, and from a personal communication from W. Traub, March 29, 1982.

136. Werskey 1978, 81.

137. Garfield 1982, 514; Brown 2005, 46, 78. On Bernal and his women collaborators, see Abir-Am 1987. On mutual research interests with Polanyi, Letter from Michael Polanyi to J. D. Bernal, April 29, 1935 (MPP 3:2).

138. Bernal 1967, 4, 5 (quoting from Sarton 1931, 68).

139. Ibid., 13.

140. Ibid., 11.

141. Bernal 1967, 20, 25, 13; P. M. S. Blackett, Science Today (BP:F.3).
142. On Singer, Werskey 1978, 145.
143. Quoted in Werskey 1978, 145, from J. D. Bernal, "Science and Society," *Spectator*, July 11, 1931, reprinted in Bernal 1949, 338. Bernal's approach was used in the work of Samuel Lilley, who studied for his doctoral degree in mathematics at Cambridge during 1936 to 1938 and became an important Marxist historian of science by the late 1940s. Lilley's notion of the impact of the dual roles of the scholar and the craftsman in the development of modern science owed debts to Edgar Zilsel. Lilley edited a special issue of *Centaurus* on the "social relations of science" in 1953 (see Enebakk 2009, 581–82).
144. Letter from Michael Polanyi to J. D. Bernal, September 20, 1938 (MPP 3:12).
145. Bernal 1967, 85, 310, 317, 441.
146. Ibid., 223–31.
147. Ibid., 65, 310, 321–23 (quotation, 323).
148. A. V. Hill 1933, 952, quoted in Bernal 1967, 395.
149. Bernal 1967, 395, 381–82.
150. Ibid., 95, 321–323.
151. Huxley 1928, quoted in Bernal 1967, 97.
152. Werskey 1978, 155; Bernal 1967, 97–98.
153. Graham 1998.
154. Polanyi 1940b, 13, 14.
155. Henry Dale 1951, 7.
156. Polanyi 1997b, 76.
157. Fine 1996, 7–11, 109–11; see Darling 2003.
158. Blackett 1933, 74; Blackett, Draft for article? 1938? (BP: F3).
159. Bernal 1967, 36, 97, 113.
160. Werskey 1978, 189. See Merton 1942. A recent analysis of Merton's norms is found in Turner 2007.
161. Bernal 1967, 416 (emphasis added).
162. Waddington 1941, 77.
163. See Polanyi 1943b; 1946b, 48–52; 1946c, 63–64; 1962a, 54.
164. Thorpe 2009; Hollinger 1996b; Mirowski 1997 (quotation, 62).
165. Sachse and Walker 2005a, 17–19.

CHAPTER SEVEN

1. Heilbron 1998, 505. For a briefer discussion of the political foundations of the philosophy of science of Popper, Kuhn, and Polanyi, see Nye 2010.
2. Wilson 1968. The geophysicist Anthony Hallam (1973) also tied the claim of scientific revolution to Thomas Kuhn's historical and philosophical analysis.
3. Cox 1973, 4–5.
4. Godfrey-Smith 2003, 57.
5. For example, among other sites, see http://www.takingchildrenseriously.com/the_education_of_karl_popper (accessed December 12, 2009).
6. Irving quoted Popper to the effect that "a theory may be true even though nobody believes it, and even though we have no reason for accepting it, or for believing it is true; and another theory may be false, although we have comparatively good reasons for accepting it" (Irving, 1964, frontispiece). Also, private communications with Edward Irving, May 1, 2001, and October 26, 2001.
7. See Golinski 1998, 17: "Kuhn adopted the phrase 'tacit knowledge' from Michael

Polanyi to characterize a large part of what the scientist learns." For some standard references to Polanyi in science studies literature, see Thorpe 2001, 19–20; Collins 1974, 1981b.

8. Hollinger 1995; Reisch 2006; Fuller 2000b.
9. Biographies of Popper include Hacohen 2000; Brudny 2002; Magee 1973; and Watkins 1997. See also Popper's autobiography: Popper 1976. On Kuhn, see Andersen 2001; Fuller 2000b. On Polanyi, see Scott and Moleski 2005; Wigner and Hodgkin 1977. On Kuhn and Popper comparatively, see Fuller 2004.
10. Andresen 1999.
11. Brudny 2002, 97.
12. Brudny 2002, 62–36; Scott and Moleski 2005, 74.
13. Popper 1976, 9; Brudny 2002, 52; Watkins 1997, section on Vienna.
14. Recalled in Popper 1976, 19.
15. Popper 1976, 33–35. On Popper in Vienna, see Gattei 2001.
16. Popper 1976, 37–39. Gerald Holton notes that Einstein made this statement about the redshift in the 1920 edition of *Relativity, the Special and General Theory*, but not in earlier editions and printings of 1917, 1918, and 1919 in which Einstein stated that the predicted bending of starlight and the predicted red shift were too small to be observed. In these earlier editions, Einstein expressed confidence that his theory would be confirmed (See Holton 1986, 8–9).
17. Brudny, 2002, 59.
18. Popper 1976, 7–8; 39–41.
19. Brudny 2002, 62, 69.
20. Watkins 1997, section on Vienna; Popper 1976, 72–78.
21. Popper 1976, 20.
22. Carnap 1929.
23. Popper 1976, 84–85.
24. See Nemeth 2003; Zilsel 1926. For Zilsel and his writings, see Zilsel 2000.
25. Reichenbach 1929, 249, 205, cited in Richardson 2008, 94. Also see Giere and Richardson 1996.
26. Feigl 1969b, 653, 657. Among the many histories and documents for the Vienna Circle, see especially Stadler 1997, 2003; Stöltzner and Uebel 2006.
27. Popper 2002, 6, 7, 10, 13.
28. Hacohen 2000, 198.
29. Popper 1976, 80–83.
30. Hacohen 2000, 215.
31. Popper 2002, 30–31.
32. Popper 2002, 96, 126. Herbert Feigl notes that Carnap used the notion of testability ("Prüfkarbeit" or "Nachprüfbarkeit") in 1928 (see Feigl 1969a, 6). In the 1993 legal case of *Daubert v. Merrell Dow Pharmaceuticals*, which was argued before the Supreme Court, Popper's criterion of falsifiability was applied in a judge's decision that expert witness testimony is the product of valid scientific knowledge. Judge William Overton ruled in an earlier case, in 1982 in Arkansas, that creation "science" is not science because it is not falsifiable.
33. Popper 2002, 210, 278. On the frequency interpretation, see Feigl 1969a, 9.
34. Watkins 1997, section on Vienna; Popper 1976, 87; Hacohen 2000, 279.
35. Popper 2004, xi; Watkins 1997, section on Vienna.
36. Weibel and Stadler 1995; Fleming and Bailyn 1969; see also Richardson 2002.
37. Popper 2004, ix; Popper 1976, 115; and Hacohen 2000, 383.

38. Popper 1962.
39. Popper 2004, 3, 140.
40. See Agassi 1963; Donagan 1964; Dray 1964, 1973; Danto 1965; White 1973; Gardner 1974.
41. Popper 2004, 83n45, 90, 120; 120–21n3. Also see Popper 2002, 57–61 against conventionalists.
42. Popper 2004, 80 (original emphasis).
43. Hacohen 2000, 396–97; Brudny 2002, 129–30.
44. Popper 1976, 36.
45. Popper 2004, 43, 58, 62–63, 69. Karl Mannheim's *Mensch und Gesellschaft im Zeitalter der Umbaus* (1935) appeared in English in 1940. See also Cat 2004. Fuller (2004, 61) remarks that Henri Bergson used the phrases of "closed society" and "open society" in *The Two Sources of Religion and Morality* (1932).
46. Popper 2004, 143–44 (original emphasis).
47. Popper 2003, 240–41.
48. Jarvie 2001b. On Popper's "social democratic" republic of science, in contrast to Polanyi's more conservative republic, see Jarvie, 2001a, 545–46.
49. Popper 2002, 22–26 (quotation, 24).
50. Hayek 2007, 50, 67, 220.
51. Hayek 2007 (introduction on p. 2).
52. On Hayek's help with Popper's book, see Watkins 1997.
53. Feyerabend 1995, 88, 108–9. Alan Musgrave compares the role that Watkins played for Popper to Thomas Henry Huxley's role as "Darwin's bulldog" (Alan Musgrave, "Obituary: Professor John Watkins," *Independent*, August 5, 1999, http://www.independent .co.uk/arts-entertainment/obituary-professor-john-watkins-1110722.html [accessed July 30, 2009]).
54. Polanyi 1950, 29, 39, 32, 36; Polanyi 1952b.
55. Watkins 1997, section on early days at LSE.
56. Shils 1995–996, 13. Antinomianism is the resistance of members of a religious group to obedience to religious law.
57. Popper 1976, 128.
58. Baltas et al. 2000, 285–86. For an analysis of differences between Popper and Kuhn, with an argument for the more revolutionary character of Popper's "critical rationalism," see Gattei 2009.
59. Heilbron 1998, 510.
60. Popper 1974, 1145, quoted in Hacohen 2000, 532.
61. Heilbron 1998, 506; Hufbauer 1998, 7.
62. Kuhn 1949.
63. Heilbron 1998, 506.
64. Baltas et al. 2000, 272–73.
65. Hufbauer 1998, 6.
66. Harvard University 1945, 39, 43, 47, 50, quoted in Hollinger 1995, 445.
67. I am grateful to Gerald Holton for an account of the Faculty meeting (personal communication, September 23, 2009).
68. Conant 1947. On Conant, see Hershberg 1993.
69. Conant 1947, 41, 16, quoted in Hufbauer 1998, 10, 11. Also see Hershberg 1993, 409–10.
70. Nash 1944; Cohen 1947; Holton 1947.
71. Hufbauer 1998, 9; Baltas et al. 2000, 287.

72. Conant, 1957; Nash 1963.
73. Holton 1952, 1973a; Holton and Roller 1958; Holton and Brush 2001; Cohen 1952, 1954, 1960. See also Holton 1999, S97.
74. Cohen and Watson 1952; Frank 1952; see also Frank, 1947, 1949.
75. See Reisch 2006, 230–31, 295–99; Frank 1951. On Frank, see Holton 1992.
76. Marcum 2005, 10.
77. Kuhn 1962, v.
78. Baltas et al. 2000, 293.
79. Hufbauer 1998, 14. See also Frank 1949, 288, 302; Collingwood 1945.
80. Baltas et al. 2000, 285, 297.
81. Kuhn 1962, vi–vii. On the lead from Reichenbach's book, see Baltas et al. 2000, 283. On Kuhn's reading Fleck, see Kuhn's foreword to Fleck 1979 (vii–xi). In referring to Fleck, Reichenbach mentions only Fleck's discussion of the difference between ancient and modern drawings of the human skeleton as a means of leading into Reichenbach's remark that there are different perspectives of the world from which we seek an intellectual integration of views (Reichenbach 1938, 224–25, 224n6).
82. See Harwood 1986, 174, 176.
83. Douglas 1987, 11–14. See also Fleck 1979, 46.
84. Fleck 1979, 103, 104, 142, 92.
85. Ibid., 46.
86. For an analysis of Fleck's participation in philosophical debates of the 1920s and 1930s, see Hedfors 2007, esp. 51–53, 73, 78–79 (quotation, 68). See also Fleck 1929, 1932.
87. Kuhn's foreword in Fleck 1979, ix–x.
88. Baltas et al. 2000, 296.
89. Conant 1951, 120–21, 304–5, 315–24, 339–40, 346–48, cited in Hershberg 1993, 588. See also Polanyi 1945a, 1946a, 1947, 1964.
90. See letter from Thomas S. Kuhn to William H. Poteat, February 28, 1967, cited in Scott and Moleski 2005, 246.
91. Hufbauer 1998, 14–16. For a discussion of mid-eighteenth-century use of the term "paradeigmata" to describe patterns of explanation in physical science, as well as its subsequent use, see Toulmin 1972, 106–7. Toulmin (1961) and N. R. Hanson (1958) employed the term "paradigm" in their books. Cf. Watson 1938.
92. Guerlac 1963, quoted in Mercer 2005, 8.
93. Quoted in Hufbauer 1998, 20–21 (original emphasis; pages 42–51 include the text of Kuhn's Guggenheim Fellowship application).
94. Ibid., 64n86.
95. Marcum 2005, 13–14. On the History of Science Department at Harvard, see Cohen 1999.
96. Hufbauer 1998, 21–22; Kuhn 1957.
97. Kuhn (1962, notes on 44–45) refers to Wittgenstein 1953 in regard to the word "paradigm"; he also refers to Bruner and Postman 1949 (Kuhn 1962, 63).
98. Marcum 2005, 14.
99. Heilbron 1998, 509; Marcum 2005, 15. On Quine, see Zammito 2004, 63.
100. Baltas et al. 2000, 296. Kuhn said that he regarded Hanson 1958 as more important to his thinking than was Polanyi or Toulmin (Baltas et al. 2000, 311). Hanson died in 1967 at the age of 43.
101. Baltas et al. 2000, 296 (original emphasis). Cf. Toulmin 1961, esp. 16–17, 57, 61, 109–15.

102. For the footnote referring to Polanyi, see Kuhn 1962, 44n1, with reference particularly to chapters 5 and 6 of Polanyi 1958.
103. See Holton 1964.
104. Discussed in Hufbauer 1998, 21.
105. Kuhn revised "The Role of Measurement in the Development of Natural Science," first given to the Social Sciences Colloquium in Berkeley (Kuhn 1961a). See Kuhn 1977, 188, 192 for quotations.
106. Kuhn 1959, 229, 236, 227.
107. On Polanyi's concerns, see Moleski 2006-2007, esp. 14; and other articles in this issue of *Tradition and Discovery*. Also see Jacobs 2009.
108. For information about the draft and for Feyerabend's critiques, see Hoyningen-Huene 1995, 2006. Also see Kuhn's acknowledgments (1962, xii).
109. Letters from Conant to Kuhn, June 5, 1961, and December 19, 1962, quoted in Hershberg 1993, 860n84.
110. Polanyi 1961, 375; Kuhn 1961a.
111. Crombie 1963, 394-95 (original emphasis).
112. Quoted in Scott and Moleski 2005, 183, from the essay "The Social Message of Science."
113. Polanyi 1964, 21, 25-28, 28; Michael Polanyi, "Reflections on John Dalton." *Manchester Guardian*, July 22, 1944.
114. Polanyi 1964, 28-31 (long quotation, 31).
115. Ibid., 43, 47-48, 49-51, 45, 43, 44.
116. Ibid., 49. These three categories also appear in Polanyi 1958, 135-36.
117. Polanyi 1962a, 57-58; with similar language in Polanyi 1964, 49-50. Also see Polanyi 1964, 91-94. The Terry Lectures were published as Polanyi 1966 (see esp. pp. 63-75).
118. Polanyi 1962a, 57-58.
119. Ibid., 58-59.
120. Polanyi 1964, 51-52, 49, 67, 63.
121. Ibid., 34-35.
122. Benda 1969, 3-4, 11-12. On Polanyi's reading Benda, see Scott and Moleski 2005, 109.
123. Benda 1969, xi, 120-21, 27-28. Benda condemns Marx, Charles Maurras, and Houston Chamberlain, among others.
124. Ibid., 43 (original emphasis).
125. Ibid., 45, 161, 166-167.
126. Ibid., xi, for Tolstoy, frontispiece for Renouvier.
127. Weber 1948a, 134.
128. Weber 1948a, 138-39, 150.
129. Polanyi 1964, 64, 73, 82.
130. See Baltas et al. 2000, 296, on extrasensory perception. See Kuhn in Crombie 1963, 394-95 on social implications.
131. Scott and Moleski 2005, 184, citing letter of June 4, 1962.
132. Ibid., 212, 216.
133. Ibid., 220.
134. Scott and Moleski 2005, 221; Grene 2002, 4-9; Mullins 2002, 34, 37, 45-46. Also see memos of November 13, 1951, and November 30, 1952; letter from John Nef to Joseph Willits, October 18, 1951; memo from Joseph Willits to C. I. B., November 9, 1951 (RFA, RG 1.1, Series 200S, Box 412, Folder 4880).

135. Polanyi 1958, 268, 9–15 (on the Michelson-Morley experiment as a textbook falsehood).
136. Hollinger 1995, 1996b. More broadly, see Jones 1988; Saunders 1999. See Merton 1942. The most famous argument for the autonomy of the scientific community within a system of governmental support is Bush 1945. On scientists' political actions in the United States, see Smith 1965.
137. Mayer 2000, 676, 679–83.
138. Fuller 2000b, ch. 5; 2004, 121.
139. Kuhn 1959, 239.
140. Kuhn 1962, 164, 167–68.
141. Kuhn 1959, 236; 1962, 146–47.
142. Kuhn 1962, 144–47.
143. Ibid., 146.
144. Ibid., 169–70.
145. Franklin 2000.
146. On this, see Heilbron 1998, 511.
147. See Nickles 1998.
148. See Hempel 1959.
149. Feyerabend, in letter published by Hoyningen-Huene 1995, 355, 360, 371, 373.
150. Kuhn 1962, 92–94 (quotation, 94).
151. Feyerabend, in Hoyningen-Huene 1995, 374 (original emphasis).
152. Ibid., 357.
153. Ibid., 358 (original emphasis).
154. Ibid., 361; and in Hoyningen-Huene 2006, 625–26.
155. Feyerabend 1978, 98–99 (quotation, 8–9). Also see Feyerabend 1975.
156. Baltas et al. 2000, 308.
157. See the discussion in Vinti 2005, 141.
158. See Lakatos and Feyerabend 1999, 27–31.
159. Polanyi 1951, vi (original emphasis).
160. Quoted in Moleski 2006–2007, 11.
161. Kuhn 1962, 172–73.
162. Kuhn 1992, 14, 19–20. Thanks to Gerald Holton for comments on the audience's reaction to the lecture (personal communication, September 23, 2009.)
163. Toulmin 1959, 189.
164. Quoted in Scott and Moleski 2005, 202. Among those also receiving honorary degrees on this occasion were Niels Bohr, Linus Pauling, and Reinhold Niebuhr.
165. Richardson 2006, 4.
166. Richardson 2009. Also see Reisch 2006, 233, regarding Kuhn.
167. Quoted in Marcum 2005, 162; Hesse (1963) forcefully argued the importance of the kind of scientific reasoning that is not formal or mathematical. For a discussion of some connections of Kuhn's ideas with the views of Hanson, Toulmin, Polanyi, and others, see Gattei 2008.

CHAPTER EIGHT
1. On the Gifford Lectures, see http://www.giffordlectures.org/aberdeen.asp (accessed December 19, 2009).
2. Scott and Moleski 2005, 203–4, 211–21.
3. Polanyi 1958, ix.
4. Magda Polanyi 1956, 1967.

5. Polanyi 1966, vi. Acknowledgments include the Terry Lectures at Yale University in 1962. The purpose of the Terry Lectureship is discussion of ways in which science and philosophy inform religion and the application of religion to human welfare. http://www.yale.edu/terrylecture/ (accessed December 19, 2009).

6. Mead 2008, 5–6.

7. Scott and Moleski 2005, 187–90 (quotation, 190). Also see letter from Michael Polanyi to Joseph Willits, December 12, 1944 (RFA, RG 1.1, Series 401S, Box 83, Folder 1088).

8. Polanyi 1958, viii, 18.

9. Ibid., 375 (original emphasis).

10. Grene 1958.

11. Polanyi 1958, 269–74, with reference to Kant's 1781 *Critique of Pure Reason*, which Polanyi read for the first time while on holiday in Wales in August 1947; Scott and Moleski 2005, 206. On Kantian influences in Polanyi, see Jha, 20002, esp. 3, 95, 222, 276–77n21.

12. Polanyi 1958, 139–42, and, on satisfaction, 106.

13. Ibid., 139–42, 151–53; on "totalitarianism which tries to fulfill the Laplacian program," see 213. Polanyi specifically mentions Bernal on page 237 of *Personal Knowledge* in a discussion of socialism, scientism, and nihilism, 225–44.

14. Polanyi 1958, 122–23, 134–35.

15. Ibid., 143.

16. Ibid., 151.

17. Ibid., 49–63. He returns to connoisseurship on p. 351.

18. Ibid., 55.

19. Ibid., 59, 64, 195, on "indwelling."

20. Ibid., 394.

21. Ibid., 12–13, 13n1.

22. Ibid., 214.

23. Ibid., 265.

24. Ibid., 266–67.

25. Ibid., 105–6.

26. Ibid., 257, 262–63, 263n1 on Turing 1950; quotation in Polanyi 1958, 387. Also see Polanyi 1951–1952.

27. Polanyi 1958, 381.

28. Ibid., 324.

29. Ibid., 405.

30. See Toulmin 1971.

31. On Toulmin, see William Grimes, "Stephen Toulmin, a Philosopher and Educator, Dies at 87," *New York Times*, December 11, 2009, http://www.nytimes.com/2009/12/11/education/11toulmin.html (accessed December 29, 2009). On the trip to Leeds, see Elizabeth Sewell, "Memoir of Michael Polanyi," typescript, p. 18 (MPP 46:12). Toulmin 1958, 1972.

32. Toulmin, 1959b, 212, 214. Thanks to Martin X. Moleski for a copy of this review.

33. Buchdahl 1965, 59–60.

34. Ibid., 64.

35. Anonymous 1959.

36. Earle 1959.

37. Oakeshott 1958, 79. See discussion of Oakeshott in Mitchell 2006, 141–44.

38. Brodbeck 1960, 583.

39. Popper 2002, xxvi.
40. See Curtis 1991, 125.
41. See Jha 2006, 328. For Fuller, see for example Fuller 2000a, esp. 26, 29.
42. Jha 2006, 318–19, 326.
43. Ibid., 329.
44. Lakatos and Musgrave, 1970, vi, 25.
45. Quoted in Palló 1996b, para. 5, trans. roughly from Hungarian for me by Susanne Kovacs and Gabor Kovacs as "Parallels and Intersections: Imre Lakatos and Mihaly Polanyi."
46. Lakatos 1970, 104. Also see Jha 2006, 334.
47. Lakatos 1970, 115–16, 119, 121, 155 (quotation, 116).
48. Jha 2006, 335; Palló 1996b, para. 6, 8. Polanyi 1970, 1972.
49. Lakatos 1999, 30.
50. Taylor 1959; Rawlins 1958. Taylor's reference is to Polanyi 1957.
51. See, e.g., essays in Holton 1973b, 2000, 2005.
52. Holton 1992–1993, 25–26 (original emphasis); 1969.
53. Mullins 2009–2010; Auxier and Hahn 2002.
54. Mullins 2009–2010, 65n9.
55. Quoted in Mullins 2002, 37.
56. Grene 1977, 165–67.
57. Ibid., 44.
58. Cohen 2005, 7.
59. Grene 1999, 110, 116.
60. Mullins 2009–2010, 58.
61. Mullins 2009–2010, 45–46.
62. Grene, 1977, 168.
63. Cohen 2005, 10 (original emphasis).
64. Mead 2008, 5–6. On theologians, see Mitchell 2006, 119.
65. Holton 1992–1993, 22.
66. Elizabeth Sewell, "Memoir of Michael Polanyi," typescript, p. 6 (MPP 46:12).
67. Quotation in Mead 2008, 10. On the Polanyi Society, see http://www.missouriwestern .edu/orgs/polanyi/psmbrshp.htm (accessed December 30, 2009). Other organizations focused on Michael Polanyi and his work are the Michael Polanyi Liberal Philosophical Association in Budapest, which was founded in 1990 and has published the journal *Polanyiana* since autumn 1991; and the Society for Post-Critical and Personalist Studies in Leicester, which took over the publication of *Appraisal* in 2004. See http://www.kfki.hu/~cheminfo/polanyi/ and http://www.spcps.org.uk/ (accessed December 30, 2009).
68. Polanyi 1958, 195–202, 279–96.
69. Ibid., 279–80.
70. Ibid., 198–99.
71. Ibid., 284–85.
72. See Mitchell 2006, 117–22; Moleski 2005–2006; Congdon 2005–2006, for their views on the question and on the views of others, such as Father Terence Kennedy (Polanyi was not religiously committed), Thomas Torrance (certainly a Christian), and Lady Drusilla Scott (who aimed to establish a foundation for religious faith; cited in Mitchell 2006, 120). It was Mitchell's view that Polanyi was critical of Catholicism and found Paul Tillich's Protestantism congenial (Mitchell 2006, 121); Moleski believed that he identified himself with Christianity (Moleski 2005–2006,

37); and Condon found him to be an unbeliever in dogmas of historic Christianity (2005–2006, 14).

73. Letter, March 10, 1948, from Michael Polanyi to the Reverend Dr. S. James Knox, quoted in Moleski 2007–2008, 42–43.

74. Scott and Moleski 2005, 196–97.

75. Quotations from Mitchell 2006, 121; Mullins 1997, 183.

76. An excellent starting place is Brooke 1991.

77. Shapin 2010a, 3.

78. Scott and Moleski 2005, 231. On Coulson, see Simões 2004.

79. Quotations from Einstein's 1930 credo "What I Believe" and from a 1927 letter to M. Shayer, a Colorado banker, in 1927, in Isaacson 2007, 387–88. "What I Believe" was published as "The World as I See It," in Einstein 1954, 8–11.

80. Quoted in Heilbron 1986, 185.

81. On the underdetermination thesis and its roots in Duhem's scientific practice, see Darling 2002. See also Quine 1951; Duhem 1991.

82. Polanyi 1958, 146n1.

83. On Duhem's scientific career and historical and philosophical views, Nye 1986, 208–23. Also see Jaki 1984; Martin 1991.

84. Quotations in Darling 2003, 1128–31, from Duhem 1991, 25–27.

85. Polanyi 1958, 395–96.

86. Darling 2003, 1134, from Duhem 1991, 334–35 (Duhem's emphasis).

87. Darling 2003, 1134–35, quoting from Fine's reading of Einstein, in Fine 1996, 110, 109 (original emphasis).

88. Cooper 2000. See also Attridge 2009.

89. See Gelwick 2005a, b. Also see Apczynski 2005; Gulick 2005.

90. Polanyi 1958, 40. Thanks to Martin X. Moleski, SJ, for this insight.

91. Woodger 1937; Hobsbawm 2009.

92. Quoted in Hobsbawm 2009. See also Needham 1936.

93. Polanyi 1958, 355–56.

94. Ibid., 382–85, 395 (quotations, 395).

95. Karl Popper, who occasionally attended meetings of the Biotheoretical Gathering in 1935 and the 1940s, later attempted to modify Darwinian ideas with a dose of Lamarckism. See Aronova 2007, 40–44.

96. Woodger 1960, 70.

97. Ibid., 68.

98. Polanyi, 1966, 70, 74, 85. Although Polanyi does not mention Max Weber in *Personal Knowledge*, Raymond Aron (1961) discusses similarities and differences in their thought. Klaus Allerbeck (2002) has emphasized Polanyi's rejection of Weber's view that the sciences can and should be value free or ethically neutral.

99. Polanyi 1958, 234, 243.

100. Ibid., 219.

101. Ibid., 264.

102. Quoted in Mannheim 2003, 183. Also see Scott and Moleski 2005, 194.

103. Quoted in ibid., 183.

104. Mannheim 1936, 264, 5. On Mannheim, see Loader 1985.

105. Mannheim 1940.

106. Shils 1973, 84–87.

107. Mannheim 1936, 117–46.

108. Ibid., 271, 300–1.
109. Ibid., 276.
110. Ibid., 296.
111. Ibid., 297. Max Horkheimer and other members of the neo-Marxist Critical Theory movement at the Institute for Social Research in Frankfurt criticized Mannheim for undermining the Marxist distinction between true and false consciousness and for attempting to justify objective truth through the "relationism" of partial truths (see Jay 1973, 64).
112. Mannheim 1936, 271, 283, 306 (emphasis added).
113. Scott and Moleski 2005, 194–96; 315n112. Also see Mannheim 2003, 309.
114. Mannheim 2003, 314; also see Scott and Moleski 2005, 194–95.
115. Mannheim 1936, 267, 268.
116. Quoted in Scott and Moleski 2005, 195, from letter from Karl Mannheim to Michael Polanyi, April 26, 1944 (MPP 4:11).
117. Quoted in Mannheim 2003, 317–19.
118. Michael Polanyi, "Planning for Freedom," *Manchester Guardian*, July 3, 1951; Michael Polanyi, "Mannheim's Historicism," *Manchester Guardian*, December 9, 1952.
119. On the largely negative reaction of American sociologists' to Mannheim's work, see Kettler and Meja 1995.
120. On Talcott Parsons, see Brick 2006, 131–33, 177–83. Also on structural functionalism and the academy as a model for society, see Cohen-Cole 2009, 246–260.
121. Kaiser 1998, 69. See also Merton 1937, 1941, 1945, 1973a.
122. Quoted in Mannheim 2003, 260, in reference to Merton 1941.
123. Merton 1973a, 17.
124. Merton 1941, quoted in Kaiser 1998, 70.
125. Merton,1945, quoted in Kaiser, 1998, 71.
126. Merton 1973a, 21.
127. Ibid., 40, 34–35.
128. Michael Polanyi, "Applied Sociology," *Manchester Guardian*, October 27, 1953.
129. Ben-David 1971, 3–4; 1972, 377, mentioned in Storer 1973, xvi.
130. Merton with Zuckerman 1973, 491n53.
131. Merton 1972, 100–1.
132. Offprint of Robert K. Merton, 1972, *American Journal of Sociology* (MPP 57: 1). *Meaning* was published after Polanyi's death, as revised and completed by Harry Prosch (Polanyi and Prosch 1975). On the collaboration between Polanyi and Prosch, which proved unsatisfactory to Polanyi, see Scott and Moleski 2005, 281–84.
133. See Cole and Zuckerman 1975.
134. Price 1963, 19.
135. On new formats and policies at *Science*, see Golec 2009. On Greenberg's hiring, see Greenberg 1999, xxv.
136. Eisenhower 1961.
137. Weinberg 1961, 164.
138. In Greenberg 1999, 35, from the commentaries in *Science* in 1965 and 1966, respectively.
139. Ibid., xxv–xxvi, 35.
140. Ibid., 5, from Polanyi 1951, 89.
141. Ibid., 4.

142. Ibid., 43–46. For Barber, see Barber 1961. For Polanyi's article, see the discussion in chapter 3.
143. Quoted in Greenberg 1999, 292, from John F. Kennedy's speech on the occasion of the Centennial of the National Academy of Sciences.
144. Quoted in Greenberg 1999, 301.
145. Snow 1968, 34. See Shapin 1999, xix.
146. See Shapin 1999, xv–xvi.
147. Snow 1993, 4–5, 10–11, 41–51. See Ortolano 2009, esp. pp. 34–35, 52–58, 76–85, 96; and the "Two Cultures?" issue of *History of Science* 43 (2005), including Edgerton 2005.
148. One-page carbon copy of review of Paul Freedman, *The Principles of Scientific Research* (London: Macdonald, 1949), for the *Manchester Guardian*, August 3, 1949 (MPP 32:7).
149. Polanyi 1959a, 44–46.
150. Ortolano 2009, 187–89.
151. Ibid., 221, 246–47. Hollinger 1996c, 166.
152. Ziman 1968, ix.
153. Ibid., v, xi, 8, 8n, 37–38.
154. Ibid., 8, 116, 83–85.
155. Ibid., 137–38; Ziman 2000, 24. Diana Crane (1972) refers to Polanyi's notion of entanglement of scientific fields in "The Republic of Science" as a "honeycomb structure" (104).
156. Ziman 2000, 21–22, 23, 24.
157. Ziman 1968, epigraph (i).
158. Ravetz 1971, 75n2; also see viii–ix.
159. Ibid., 422–29 (quotation, 422–23).
160. Ziman 2000, 23.

EPILOGUE

1. Edge 1995, 6.
2. See Records of the Science Studies Oral History Project, University of Edinburgh, at http://www.nahste.ac.uk (accessed June 20, 2010); Henry 2008; MacKenzie 2003.
3. Edge 1995, 6–8, 10; also see Fuller 2000b, 324–31; Kaiser 1998, 75. See Bijker 2004, 131, 135.
4. Quoted in Friedman 1998, 240, 243, from Collins 1992; Barnes and Bloor 1982, 27–28, 23. The aim to establish a scientific sociology of knowledge in order to supersede traditional philosophy of science is paradoxical in its aping of the aim of the Vienna Circle's logical empiricists to establish a scientific philosophy (see Richardson 2008, 88–96).
5. Bloor 1973, 174; 1991, 7.
6. Shapin 1982, 159, quoted in Friedman 1998, 244–45.
7. Collins 1981a, 3.
8. Mannheim 1952, 135, quoted in Laudan 1977, 220.
9. On some of these points, see R. Porter 1992, 41; Shinn and Ragouet 2005, 14, 25, 35. For an overview of SSK and constructivism, see Zammito 2004, 123–82; Bucchi 2004. The argument that substantial scientific achievements took place in totalitarian regimes is made in Graham 1998 and in essays in Sachse and Walker 2005b.
10. For example, see Latour and Woolgar 1986. The Parex (Paris-Sussex) project for

studying the social relations of science began in 1970. See the preface by the editors in Fox and Weisz 1980.

11. See Seguin 2000. See Bowker and Latour 1987, 725–26 (for Latour on Foucault), 723–724 (on Gaston Bachelard), 731 (on Michel Serres).

12. Bowker and Latour 1987, 724, 727, 729, 740. For Mill on the authoritarian character of Comte's positivism, see Mill 1865.

13. Latour 1988, 1990.

14. Bowker and Latour 1987, 726 on Foucault (quotation on Bourdieu, 717). See Latour and Woolgar 1986, 128–29; Latour 1987. Latour's points of reference include Foucault 1967; Bourdieu 1975, 1976, 1984. For an overview of different approaches to a theory of science in this period, see Giere 1988, 22–61, and on constructivism, 56–58.

15. Bourdieu, 2004, vii–viii. On Bourdieu and the classic sociology of knowledge, see Ringer 1990, esp. 270–74, 279.

16. Bourdieu 2004, 38, 39, 73, 82.

17. Ibid., 1.

18. Ibid., 72, 84, 77.

19. Gross and Levitt 1994; Holton 1998; Gross et al. 1997; Labinger and Collins 2001; Parsons 2003.

20. Shapin 1999, xv; and more broadly, Shapin 2008, xiv, 5, 16, 33–34, 67–68, 233, 309, 312–13ff.

21. Shapin 2008, xv, 2, 5, 292.

22. Shapin 1992, 357, 359.

23. Werskey 2007a, 443.

24. See Zammito 2004, 213–221; Solomon 2008, 241; Haraway 1991, 191; Longino 1992, 199. In her introduction to Harding 1986, Ruth Bleier wrote that SSK was oblivious to feminist scholarship and to the question of gender as a significant social category for understanding the construction of scientific knowledge. On the larger context for social production of scientific knowledge, see Jasanoff 2004.

25. Edge 1995, 12. See Fuller 2000b, xiv–xvi, 335, 380, 422. On the division within science studies, Sismondo 2008, 18, 20.

26. Latour 2004, 227, 230 (original emphasis).

27. Collins and Evans 2002, 236, 241. In the chronological framework of Collins and Evans, the "First Wave" ran from the 1950s through the 1960s (culminating in Kuhn) and the "Second Wave" (SSK) lasted from the early 1970s to the present.

28. Latour 2004, 231–35; 2007, 812.

29. Polanyi 1944b.

30. Koestler 1959. On Koestler and Polanyi see, Scott and Moleski 2005, 200, 205–6, 217; Caesarani 1998, 450, 452, 470. According to Caesarini, on the basis of correspondence in the Koestler Papers at the University of Edinburgh, Magda Polanyi bitterly reproached Koestler in a letter of August 1980 for exploiting Polanyi's ideas. Elena Aronova has noted Koestler's concern with the split between reason and belief in Koestler 1949 (personal communication from Elena Aronova to the author, March 29, 2010).

31. Ascherson 2010.

32. Quoted in Scott and Moleski 2005, 208.

REFERENCES

Aaserud, Finn. 1990. *Redirecting Science: Niels Bohr, Philanthropy and the Rise of Nuclear Physics*. Cambridge: Cambridge University Press.

Abir-Am, Pnina. 1987. "The Biotheoretical Gathering, Trans-disciplinary Authority and the Incipient Legitimation of Molecular Biology in the 1930s: New Perspective on the Historical Sociology of Science." *History of Science* 25: 1–70.

Adams, Henry. 1918. *The Education of Henry Adams*. Boston: Houghton Mifflin.

Adelman, Paul. 1987. *British Politics in the 1930s and 1940s*. Cambridge: Cambridge University Press.

Agassi, Joseph. 1963. *Towards an Historiography of Science*. Vol. 2, *History and Theory*. The Hague: Mouton.

Albrecht, Helmuth. 1993. "Max Planck: 'Mein Besuch bei Adolph Hitler.' Anmerkungen zum Wert einer historischen Quelle." In *Naturwissenschaft und Technik in der Geschichte*, 41–63. Stuttgart: GNT.

Allen, R. T. 1998. *Beyond Liberalism: The Political Thought of F. A. Hayek and Michael Polanyi*. New Brunswick, NJ: Transaction Publishers.

Allerbeck, Klaus. 2002. "The Republic of Science Revisited." *Appraisal* 4 (1): 3–6.

Andersen, Hanne. 2001. *On Kuhn*. London: Wadsworth.

Andresen, Jensine. 1999. "Crisis and Kuhn." *Isis* 90:S43–S67.

Anonymous. 1940. *Science in War*. Harmondsworth, UK: Penguin.

——. 1949. "Chemistry at King's College, London: Prof. A. J. Allmand, F.R.S." *Nature* 164:989.

——. 1959. "Short Reviews: *Personal Knowledge*, by Michael Polanyi." *Scientific American* 201 (1): 164–65.

——. 1961. *The Logic of Personal Knowledge: Essays Presented to Michel Polanyi on his Seventieth Birthday, 11th March 1961*. Glencoe, IL: Free Press.

——.1966. "Nuclear Pioneer Lecturer: Kasimir Fajans." *Journal of Nuclear Medicine* 7:402–4.

Antiseri, Dario. 2004. *La Vienne de Popper*. Translated by Nathalie Janson, with revision by Alban Bouvier. Paris: Presses Universitaires de France.

Apczynski, John V. 2005. "The Discovery of Meaning through Scientific and Religious Forms of Indwelling." *Zygon: Journal of Religion and Science* 40:77–88.

Arnold, Matthew. 1904. *A French Eton: Higher Schools and Universities in France. Higher Schools and Universities in Germany*. Vol. 12, *Works*. London: Macmillan.

Aron, Raymond. 1961. "Max Weber and Michael Polanyi." In Anonymous 1961, 99–115.

Aronova, Elena. 2007. "Karl Popper and Lamarckism." *Biological Theory* 2 (1): 37–51.

Arrhenius, Svante. 1889. "Über die Reaktionsgeschwindigkeit bei der Inversion von Rohrzucker durch Säuren." *ZpC* 4: 226–48.

Ascherson, Neal. 2010. "Raging towards Utopia." *London Review of Books*, April 22, 3–8.

Ash, Mitchell G., and Alfons Söllner. 1996a. Introduction to Ash and Söllner 1996b, 1–19.

———, eds. 1996b. *Forced Migration and Scientific Change: Emigré German-Speaking Scientists and Scholars after 1933*. Cambridge: Cambridge University Press.

Aston, Francis W., Gregory P. Baxter, Bohuslav Brauner, A. Debierne, A. Leduc, T. W. Richards, Frederick Soddy, and G. Urbain. 1923. "Report of the International Committee on Chemical Elements." *JACS* 45: 867–74.

Attridge, Harold W., ed. 2009. *The Religion and Science Debate: Why Does It Continue?* New Haven, CT: Yale University Press.

Auxier, Randall E., and Lewis Edwin Hahn, eds. 2002. *The Philosophy of Marjorie Grene*. The Library of Living Philosophers 29. Chicago: Open Court.

Backhouse, Roger E. 2006. "Hayek on Money and the Business Cycle." In Feser 2006, 34–50.

Badash, Lawrence. 1995. *Scientists and the Development of Nuclear Weapons: From Fission to the Limited Test Ban Treaty 1939–1963*. Atlantic Highlands, NJ: Humanities Press.

Baedeker. 1923. *Berlin and Its Environs: Handbook for Travellers*. 6th ed. Leipzig: Baedeker.

Baker, J. R. 1939. "Counter-Blast to Bernalism." *New Statesman and Nation*, July 29, 174–75.

———. 1942. *The Scientific Life*. London: G. Allen and Unwin.

———. 1976. "Julian Sorell Huxley." *Biographical Memoirs of Fellows of the Royal Society* 22: 212.

Baltas, Aristides, Kostas Gavroglu, and Vassiliki Kindi. 2000. "A Discussion with Thomas S. Kuhn." In *The Road since Structure: Thomas S. Kuhn*, edited by James Conant and John Haugeland, 255–323. Chicago: University of Chicago Press.

Barber, Bernard. 1961. "Resistance by Scientists to Scientific Discovery." *Science*, September 1, 596–602.

Barberis, Peter, John McHugh, and Mike Tyldesley. 2003. "Next Five Years Group." In *Encyclopedia of British and Irish Political Organizations: Parties, Groups and Movements of the Twentieth Century*, entry no. 1262, 355–356. London: Continuum International Publishing Group.

Barkan, Diana Kormos. 1999. *Walther Nernst and the Transition to Modern Physical Science*. Cambridge: Cambridge University Press.

Barnes, Barry, and David Bloor. 1982. "Relativism, Rationalism and the Sociology of Knowledge." In *Rationality and Relativism*, edited by Martin Hollis and Steven Lukes, 21–47. Oxford: Blackwell.

Barnes, Barry, and Steven Shapin. 1977. "Where Is the Edge of Objectivity?" *BJHS* 10:61–66.

Bartel, Hans-Georg, and Rudolf P. Huebener. 2007. *Walther Nernst: Pioneer of Physics and of Chemistry*. Singapore: World Scientific.

Becker, Anne, Kurt A. Becker, and Jochen H. Block. 1987. "Fritz Haber." In Treue and Hildebrandt 1987, 167–82.

Benda, Julien. 1927. *La Trahison des Clercs*. Paris: Grasset.

———. 1969. *The Treason of the Intellectuals*. Translated by Richard Aldington. New York: W. W. Norton.

Ben-David, Joseph. 1971. *The Scientist's Role in Society*. Englewood Cliffs, NJ: Prentice Hall.

———. 1972. "The Profession of Science and Its Powers." *Minerva* 10:362–83.

Bensaude-Vincent, Bernadette. 1987. *Langevin: Science et Vigilance*. Paris: Belin.

Bergmann, Ernst D. 1961. "The Size and Shape of Molecules as a Factor in their Biological Activity." In Anonymous 1961, 37–46.

Bernal, J. D. 1933–1934. "The Scientist and the World Today: The End of a Political Delusion." *Cambridge Left*, Winter, 36–45.

———. 1949. *The Freedom of Necessity*. London: Routledge and Kegan Paul.

———. 1967. *The Social Function of Science*. Cambridge, MA: MIT Press.

Bernstein, Jeremy. 2004. "Janos Plesch: Brief Life of an Unconventional Doctor 1878–1957." *Harvard Magazine*, January–February. Accessed December 24, 2010. http://harvardmagazine.com/2004/01/janos-plesch.html.

Bernstein, Michael A. 2001. *A Perilous Progress: Economists and Public Purpose in Twentieth-Century America*. Princeton, NJ: Princeton University Press.

Beutler, H., and Michael Polanyi. 1925. "Reaktionsleuchten und Reaktionsgeschwindigkeit." *Die Naturwissenschaften* 13:711–13.

Beutler, H., St. v. Bogdandy, and M. Polanyi. 1926. "Über Luminescenz hochverdünnter Flammen." *Die Naturwissenschaften* 14:164–65.

Beyerchen, Alan D. 1977. *Scientists under Hitler: Politics and the Physics Community in the Third Reich*. New Haven, CT: Yale University Press.

———. 1994. "What We Now Know about Nazism and Science." In Jacob 1994, 128–55.

———. 1996. "Emigration from Country and Discipline: The Journey of a German Physicist into American Photosynthesis Research." In Ash and Söllner 1996b, 71–85.

Biedermann, Wolfgang. 2002. "Zur Finanzierung der Institut der Kaiser-Wilhelm-Gesellschaft zur Förderung der Wissenschaften Mitte der 20er bis zur Mitte der 40er Jahre des 20. Jahrhunderts." In *Wissenschaft und Innovation, Wissenschaftsforschung Jahrbuch 2001*, edited by Heinrich Parthey and Günter Spur, 143–72. Berlin: Gellsellschaft für Wissenschaftsforschung.

Bijker, Wiebe E. 2004. "In Memoriam: Robert K. Merton, Dorothy Nelkin, and David Edge." *Science, Technology, and Human Values* 29 (Spring): 131–38.

Blackett, P. M. S. 1933. "The Craft of Experimental Physics." In *University Studies: Cambridge 1933*, edited by Harold Wright, 67–96. London: Ivor Nicolson and Watson.

———. 1934. "Pure Science: Discussion with Professor P.M.S. Blackett." In *Scientific Research and Social Needs*, edited by Julian Huxley, 203–24. London: Watts.

———. 1935. "The Frustration of Science." In *The Frustration of Science*, edited by Sir Daniel Hall, J. G. Crowther, and J. D. Bernal, 129–44. London: Allen and Unwin.

Bloor, David. 1973. "Wittgenstein and Mannheim on the Sociology of Mathematics." *Studies in History and Philosophy of Science* 4:171–91.

———. 1991. *Knowledge and Social Imagery*. 2nd ed. Chicago: University of Chicago Press.

Born, Max. 1971. *The Born-Einstein Letters: Correspondence between Albert Einstein and Max and Hedwig Born from 1916 to 1955*. New York: Walker.

Born, Max, and Robert Oppenheimer. 2000. "On the Quantum Theory of Molecules." In Hettema 2000b, 1–14.

Borscheid, Peter. 1976. *Naturwissenschaft, Staat und Industrie in Baden (1848–1914)*. Stuttgart: Klett.

Bourdieu, Pierre. 1975. "The Specificity of the Scientific Field and the Social Conditions of the Progress of Reason." *Social Science Information* 14:19–47.

——. 1976. "La production de la croyance: contribution à une économie des biens symboliques." *Actes de la Recherche en Sciences Sociales* 13:3–43.

——. 1984. *Homo academicus*. Paris: Editions du Minuit.

——. 2004. *Science of Science and Reflexivity*. Translated by Richard Nice. Chicago: University of Chicago Press.

Bowker, Geof, and Bruno Latour. 1987. "A Blooming Discipline Short of Discipline: (Social) Studies of Science in France." *Social Studies of Science* 17:715–47.

Bragg, Lawrence. 1975. *The Development of X-Ray Analysis*, edited by D. C. Phillips and H. F. Lipson. New York: Hafner Press.

Branch, Gerald E. K., and Melvin Calvin. 1941. *The Theory of Organic Chemistry: An Advanced Course*. New York: Prentice-Hall.

Braun, E. 1987. Introduction to *Solid State Science: Past, Present and Predicted*, edited by D. L. Weaire and C. G. Windsor 1–9. Bristol, UK: Adam Hilger.

——. 1992. "Mechanical Properties of Solids." In Hoddeson et al. 1992, 317–58.

Brick, Howard. 2006. *Transcending Capitalism: Visions of a New Society in Modern American Thought*. Ithaca, NY: Cornell University Press.

Brodbeck, May. 1960. "Review." *American Sociological Review* 25:582–83.

Brooke, John Hedley. 1991. *Science and Religion: Some Historical Perspectives*. Cambridge: Cambridge University Press.

Brown, Andrew. 2005. *J. D. Bernal: The Sage of Science*. Oxford: Oxford University Press.

Brudny, Michelle-Irène. 2002. *Karl Popper: un philosophe heureux*. Paris: Grasset.

Brunauer, Stephen. 1944. *The Adsorption of Gases and Vapours*. Vol. 1, *Physical Adsorption*. Oxford: Oxford University Press.

Brunauer, Stephen, Paul H. Emmett, and Edward Teller. 1938. "Adsorption Gases in Multimolecular Layers." *JACS* 60:309–19.

Bruner, J. S., and Leo Postman. 1949. "On the Perception of Incongruity: A Paradigm." *Journal of Personality* 18:206–23.

Brush, Stephen G. 1999. "Dynamics of Theory Change in Chemistry. Part 1. The Benzene Problem, 1865–1945. Part 2. Benzene and Molecular Orbitals, 1945–1980." *Studies in History and Philosophy of Science* 30:21–79, 263–302.

Bucchi, Massimiano. 2004. *Science in Society: An Introduction to Social Studies of Science*. Translated by Adrian Bolton. London: Routledge.

Buchdahl, Gerd. 1965. "A Revolution in Historiography of Science." *History of Science* 4:55–69.

Buerger, Martin J. 1973. "Karl Weissenberg and the Development of X-Ray Crystallography." Accessed July 5, 2010. http://weissenberg.bsr.org.uk/2/x-ray%20crystallography.htm.

Bunsen Gesellschaft. 1982. *Fifty Years of Chemical Dynamics. Berlin, 12–15 October 1981, at the Fritz Haber Institut, Berlin-Dahlem: In Memorial Henry Eyring 20 June 1901–26 December 1981*. In *Berichte der Bunsen Gesellschaft* 86:348–464.

Bush, Vannevar. 1945. *Science—The Endless Frontier*. Washington DC: U.S. Government Printing Office.

Caesarani, David. 1998. *Arthur Koestler: The Homeless Mind*. New York: The Free Press.

Cahan, David. 1989. *An Institute for an Empire: The Physikalisch-Technische Reichsanstalt 1871–1918*. New York: Cambridge University Press.

Calder, Ritchie. 1999. "Bernal at War." In Swann and Aprahamian 1999, 161–90.

Caldwell, Bruce. 2004. *Hayek's Challenge: An Intellectual Biography of F. A. Hayek*. Chicago: University of Chicago Press.

——. 2006. "Hayek and the Austrian Tradition." In Feser 2006, 13–33.

Calvin, Melvin. 1961. "The Path of Carbon in Photosynthesis." Nobel Lecture, December 11. Accessed August 27, 2009. http://nobelprize.org/nobel_prizes/chemistry/laureates/1961/calvin-lecture.pdf.

———. 1991–1992. "Memories of Michael Polanyi in Manchester." *Tradition and Discovery* 18 (2): 40–42.

Cameron, Laura, and John Forrester. 1999. "'A Nice Type of the English Scientist': Tansley and Freud." *History Workshop Journal* 48:65–100.

Carnap, R., H. Hahn, and O. Neurath. 1929. *Wissenschaftliche Weltauffasung: Der Wiener Kreis*. Vienna: A. Wolf Verlag.

Carson, Cathryn. 2010. *Heisenberg in the Atomic Age: Science and the Public Sphere*. Cambridge: Cambridge University Press.

Cartwright, Nancy, Jordi Cat, Lola Fleck, and Thomas E. Uebel. 1996. *Otto Neurath: Philosophy between Science and Politics*. Cambridge: Cambridge University Press.

Cash, John M. 1977. "Guide to the Papers of Michael Polanyi." Special Collections, Regenstein Library, University of Chicago.

Cassidy, David C. 1992. *Uncertainty: The Life and Science of Werner Heisenberg*. New York: W. H. Freeman.

———. 1994. "Heisenberg, German Science, and the Third Reich." In Jacob 1994, 158–75.

Cassidy, John. 2000. "The Price Prophet." *New Yorker*, February 7, 44–51.

———. 2001. "When Will It End? Recession and the Bin Laden Effect." *New Yorker*, December 17, 40–46.

Cat, Jordi. 2004. "The Philosophical Adventures of Robinson Crusoe: Social Rationality, Objectivity and Unity in Philosophy of Science and Social Thought in Neurath, Polanyi and Popper, and the Alleged Tension between the Private Language Argument and the Central Planning Model." Manuscript copy, author's personal collection.

Chayut, Michael. 1994. "From Berlin to Jerusalem: Ladislaus Farkas and the Founding of Physical Chemistry in Israel." *HSPS* 24:237–64.

———. 2001. "From the Periphery: The Genesis of Eugene P. Wigner's Application of Group Theory to Quantum Mechanics." *Foundations of Chemistry* 3: 55–78.

Chemical Society. 1962. *The Transition State*. London: Chemical Society.

Chilvers, C. A. J. 2003. "The Dilemmas of Seditious Men: The Crowther-Hessen Correspondence in the 1930s." *BJHS* 36:417–35.

Clark, Ronald W. 1968. *The Huxleys*. 1968. New York: McGraw-Hill.

———. 1972. *Einstein: The Life and Times*. New York: Avon.

Coen, Deborah. 2007. *Vienna in the Age of Uncertainty: Science, Liberalism, and Private Life*. Chicago: University of Chicago Press.

Cohen, Benjamin. 2005. "Interview of Marjorie Grene with Benjamin Cohen." *Believer*, March. Accessed December 29, 2009. http://www.believermag.com/issues/200503/?read=interview_grene.

Cohen, I. Bernard. 1947. "Benjamin Franklin's Experiments: A New Edition of Franklin's Experiments and Observations on Electricity." PhD diss., Harvard University.

———. 1952. "The Nature and Growth of the Physical Sciences: An Outline of the Topics Covered in Natural Science 3 Together with a Selected Group of Readings Illustrative of the Development of Science." Mimeograph, Harvard University Archives. Cambridge, MA.

———. 1954. *The Nature and Growth of the Physical Sciences: An Introduction to Some of the Major Principles of Physical Science and Their Historical Background*. New York: Wiley.

———. 1960. *Birth of a New Physics*. New York: Doubleday Anchor.

———. 1999. "The Coming of Age of the History of Science Society." *Isis* 90:S28–S42.

Cohen, I. Bernard, and Fletcher G. Watson, eds. 1952. *General Education in Science. Papers Presented at the Workshop in Science in General Education held at the Harvard Summer School July 1950*. Cambridge, MA: Harvard University Press.

Cohen-Cole, Jamie. 2009. "The Creative American: Cold War Salons, Social Science, and the Cure for Modern Society." *Isis* 100:219–62.

Cole, Jonathan R., and Harriet Zuckerman. 1975. "The Emergence of a Scientific Specialty: The Self-Exemplifying Case of the Sociology of Science." In *The Idea of Social Structure: Papers in Honor of Robert K. Merton*, edited by Lewis A. Coser, 139–74. New York: Harcourt Brace Jovanovich.

Collingwood, R. G. 1945. *The Idea of Nature*. Oxford: Clarendon Press.

Collins, Harry M. 1974. "The TEA Set: Tacit Knowledge and Scientific Networks." *Science Studies* 4:165–86.

———. 1981a. "Stages in the Empirical Programme of Relativism." *Social Studies of Science* 11:3–10.

———. 1981b. "The Place of the 'Core-Set' in Modern Science: Social Contingency with Methodological Propriety in Science." *History of Science* 19:6–19.

———. 1992. *Changing Order: Replication and Induction in Scientific Practice*. 2nd ed. Chicago: University of Chicago Press.

Collins, Harry M., and Richard Evans. 2002. "The Third Wave of Science Studies: Studies of Expertise and Experience." *Social Studies of Science* 32:235–96.

Conant, James Bryant. 1947. *On Understanding Science: An Historical Approach*. New Haven, CT: Yale University Press.

———. 1951. *Science and Common Sense*. New Haven, CT: Yale University Press.

Conant, James Bryant, ed. 1957. *Harvard Case Histories in the Experimental Sciences*. 2 vols. Cambridge, MA: Harvard University Press.

Congdon, Lee. 1991. *Exile and Social Thought: Hungarian Intellectuals in Germany and Austria, 1919–1933*. Princeton, NJ: Princeton University Press.

———. 1997–1998. "Between Brothers: Karl and Michael Polanyi on Fascism and Communism." *Tradition and Discovery* 24 (2): 7–28.

———. 2001. *Seeing Red: Hungarian Intellectuals in Exile and the Challenge of Communism*. DeKalb: Southern Illinois University Press.

———. 2005–2006. "Polanyi and the Sadness of Unbelief." *Tradition and Discovery* 32 (3): 12–14.

Coolidge, Albert Sprague, and Hubert M. James. 1934. "The Approximations Involved in Calculations of Atomic Interaction and Activation Energies." *Journal of Chemical Physics* 2:811–17.

Cooper, William F. 2000. "The External Review Committee Report: Baylor University." Waco: Baylor University. Accessed January 1, 2010. http://www.texscience.org/pdf/001017polanyi.pdf.

Coser, Lewis A. 1984. *Refugee Scholars in America: Their Impact and Their Experiences*. New Haven, CT: Yale University Press.

Cox, Allan, ed. 1973. *Plate Tectonics and Geomagnetic Reversals: Readings, Selected, Edited and with Introductions*. San Francisco: W. H. Freeman.

Crane, Diana. 1972. *Invisible Colleges: Diffusion of Knowledge in Scientific Communities*. Chicago: University of Chicago Press.

Crawford, Elisabeth, J.L. Heilbron, and Rebecca Ullrich. 1987. *The Nobel Population. 1901–1937*. Berkeley, CA: Office for History of Science and Technology.

Cremer, Erika. 1987. "Walther Nernst und Max Bodenstein." In Treue and Hildebrandt 1987, 183–202.

Crombie, A. C., ed. 1963. *Scientific Change: Historical Studies in the Intellectual, Social, and Technical Conditions for Scientific Discovery and Technical Invention, from Antiquity to the Present. University of Oxford, 9–15 July 1961.* London: Heinemann.

Crowther, J. G. 1930. *Science in Soviet Russia.* London: Williams and Norgate.

———. 1935. *British Scientists of the Nineteenth Century.* London: K. Paul, Trench, Trubner & Co.

———. 1937. *American Men of Science.* New York: W. W. Norton.

———. 1941. *The Social Relations of Science.* New York: Macmillan.

———. 1942. *Soviet Science.* Harmondsworth, UK: Penguin.

———. 1970. *Fifty Years with Science.* London: Barrie and Jenkins.

Crowther, J. G., O. J. Howarth, and D. P. Riley, eds. 1942. *Science and World Order.* Harmondsworth, UK: Penguin.

Curtis, R.C. 1991. "Popularizing Science: Polanyi or Popper?" *Minerva* 29:116–30.

Dagani, Ron. 1999. "Zewail: Glimpses of the Ultrafast." *Chemical and Engineering News* 77 (42): 12–13.

Dale, Henry. 1951. "Speeches Made at the Dinner Held to Celebrate the Tenth Anniversary of the Foundation of the Society." *SFS Occasional Pamphlet* 11:4–11.

Dale, Henry, R.A. Fisher, and J. Baker. 1949. "Papers on the Soviet Genetics Controversy." *SFS Occasional Pamphlet* 9, January.

Dambrowitz, K. A. and S. M. Kuznicki. 2010. "Henry Eyring: A Model Life." *Bulletin for the History of Chemistry* 35 (1): 46–52.

Dannen, Gene. 1997. "The Einstein-Szilard Refrigerators." *Scientific American*, January, 90–95.

Danto, Arthur C. 1965. *Analytical Philosophy of History.* Cambridge: Cambridge University Press.

Darling, Karen Merikangas. 2002. "The Complete Duhemian Underdetermination Argument: Science Language and Practice." *Studies in the History and Philosophy of Science* 33:511–33.

———. 2003. "Motivational Realism: The Natural Classification for Pierre Duhem." *Philosophy of Science* 70:1125–36.

Darlington, C. D. 1947. "Retreat from Science in Soviet Russia." *Nineteenth Century and After* 142:157–68.

———. 1949. "Letter to the Editor: The Lysenko Controversy." *New Statesman and the Nation*, January 22, 81–82.

Davies, Mansel. 1997. "Joseph Needham." *British Journal for the History of Science* 30: 95–100.

Déak, István. 1968. *Weimar Germany's Left-Wing Intellectuals: A Political History of the Weltbühne and Its Circle.* Berkeley: University of California Press.

Deichmann, Ute. 1999. "The Expulsion of Jewish Chemists and Biochemists from Academia in Nazi Germany." *Perspectives on Science* 7:1–86.

———. 2001. *Flüchten, Mitmachen, Vergessen: Chemiker und Biochemiker in der NS-Zeit.* Weinheim: Wiley-VCH.

Dirac, P. A. M. 1929. "Quantum Mechanics of Many-Electron Systems." *Proceedings of the Royal Society of London* A123:714–33.

Divall, Colin. 1994. "Education for Design and Production: Professional Organization, Employers, and the Study of Chemical Engineering in British Universities, 1922–1976." *Technology and Culture* 35:258–88.

Donagan, Alan. 1964. "Historical Explanation: The Popper-Hempel Theory Reconsidered." *History and Theory* 4:3–26.

Douglas, Mary. 1987. *How Institutions Think*. London: Routledge and Kegan Paul.

Dray, William. 1964. *Philosophy of History*. Englewood Cliffs, NJ: Prentice-Hall.

———. 1973. "The Politics of Contemporary Philosophy of History: A Reply to Hayden White." *Clio* 3:54–76.

Drucker, Peter F. 1979. *Adventures of a Bystander*. New York: Harper and Row.

Dubinin, M. M. 1955. "A Study of the Porous Structure of Active Carbons using a Variety of Methods." *Quarterly Reviews, Chemical Society* 9:101–14.

———. 1960. "The Potential Theory of Adsorption of Gases and Vapors for Adsorbents with Energetically Non-Uniform Surfaces." *Chemical Reviews* 60:235–41.

Duczynska, Ilona. n.d. "Karl Polanyi (1886–1964): A Family Chronicle and a Short Account of His Life." Typescript, MPP 17:16.

———. 1975. *Der demokratische Bolschewik*. Munich: List.

———. 1977. "Karl Polanyi: Notes on His Life." In *The Livelihood of Man*, by Karl Polanyi, edited by Harry W. Pearson, xi–xx. New York: Academic Press.

———. 1978. *Workers in Arms: The Austrian Schutzbund and the Civil War of 1934*. With an introduction by E. J. Hobsbawm. New York: Monthly Review Press.

Duhem, Pierre. 1991. *The Aim and Structure of Physical Theory*. Translated by Philip P. Wiener. 2nd ed. Princeton, NJ: Princeton University Press.

Earle, William. 1959. "Personal Knowledge." *Science* 129:831–32.

Easton, Laird M. 2002. *The Red Count: The Life and Times of Harry Kessler*. Berkeley: University of California Press.

Eckert, Michael, and Karl Märker, eds. 2004. *Arnold Sommerfeld—Wissenschaftlicher Briefwechsel*. Vol. 2, *1919–1951*. Diepholz Munich: GNT-Verlag.

Eckert, Michael, and Helmut Schubert. 1990. *Crystals, Electrons, Transistors: From Scholar's Study to Industrial Research*. Translated by Thomas Hughes. New York: American Institute of Physics.

Edge, David. 1995. "Reinventing the Wheel." In *Handbook of Science and Technology Studies*, edited by Sheila Jasanoff et al. 3–24. London: Sage.

Edgerton, David. 2005. "C. P. Snow as Anti-Historian of British Science: Revisiting the Technocratic Moment, 1959–1964," *History of Science* 43:187–208.

Egressy, Gergely. 2001. "A Statistical Overview of the Hungarian Numerus Clausus Law of 1920—A Historical Necessity or the First Step toward the Holocaust?" *East European Quarterly* 24:447–64.

Einstein, Albert. 1934. *Mein Weltbild*. Amsterdam: Querido Verlag.

———. 1954. *Ideas and Opinions*. New York: Random House.

———. 2007. "Political Manifesto, 11 March 1933." In *Einstein on Politics:His Private Thoughts and Public Stands on Nationalism, Zionism, War, Peace, and the Bomb*, edited by David E. Rowe and Robert Schulmann, 269–73. Princeton, NJ: Princeton University Press.

Eisenhower, Dwight D. 1961. "Farewell Speech of 17 January 1961." Accessed January 1, 2011. http://www.americanrhetoric.com/speeches/dwightdeisenhowerfarewell.html.

Emeléus, H. J. 1960. "Friedrich Adolf Paneth. 1887–1958." *Biographical Memoirs of Fellows of the Royal Society* 6:226–46.

Enebakk, Vidar. 2009. "Lilley Revisited: Or Science and Society in the Twentieth Century." *British Journal for the History of Science* 42:563–93.

Ettisch, M., M. Polanyi, and K. Weissenberg. 1921. "Über Faserstruktur bei Metallen." *ZP* 7:181–84.

Evans, A. G., D. Holden, P. H. Plesch, Michael Polanyi, H. A. Skinner, and M. A. Weinberger. 1946. "Friedel-Crafts Catalysts and Polymerization," *Nature* 157:102.

Evans, M. G., and M. Polanyi. 1935. "Some Applications of the Transition State Method to the Calculation of Reaction Velocities, Especially in Solution." *TFS* 31:875–94.

Ewald, P. P., ed. 1962. *Fifty Years of X-Ray Diffraction*. Utrecht: Oosthoek.

Eyring, Henry. 1930. "Verwendung optischer Daten zur Berechnung der Aktivierungswärme." *Die Naturwissenschaften* 18:915.

———. 1931. "The Energy of Activation for Bimolecular Reactions involving Hydrogen and the Halogens, according to the Quantum Mechanics." *JACS* 53: 2537–49.

———. 1935. "The Activated Complex in Chemical Reactions." *Journal of Chemical Physics* 3:107–15.

———. 1937a. "The Calculation of Activation Energies." *TFS* 33:3–11.

———. 1937b. "The Theory of Absolute Reaction Rates." *TFS* 33:41–48.

———. 1961. "Rates of Reaction." In Anonymous 1961, 25–35.

———. 1976. "Physical Chemistry: The Past 100 Years." *Chemical and Engineering News* 54:88–104.

———. 1982. "Reminiscences on My Stay in Berlin (1929–1930) and on the Events Leading to the Paper 'Über einfache Gasreaktionen.'" In Bunsen Gesellschaft 1982, 348–49.

Eyring, Henry, and Michael Polanyi. 1930."Zur Berechnung der Aktivierungs-wärme." *Die Naturwissenschaften* 18:914–15.

———. 1931. "Über einfache Gasreaktionen." *ZpC* B12: 279–311.

———. 2000. "On Simple Gas Reactions." In Hettema 2000b, 423–51.

Fajans, Kasimir. 1914. "Zur Frage der isotopen Elemente." *Physikalische Zeitschrift* 15:935–40.

Faraday Society. 1932. "The Adsorption of Gases: A General Discussion (12–13 January 1932)." *TFS* 28:129–447.

———. 1937. "Reaction Kinetics. Papers and Discussions." *TFS* 33:1–123.

Farkas, A. 1930. "Über die thermische Parawasserstoffumwandlung." *ZpC* B10:419–33.

Farmelo, Graham. 2009. *The Strangest Man: The Hidden Life of Paul Dirac, Mystic of the Atom*. New York: Basic Books.

Feigl, Herbert. 1969a. "The Origin and Spirit of Positivism." In *The Legacy of Logical Positivism: Studies in the Philosophy of Science*, edited by Peter Achinstein and Stephen F. Barker, 3–24. Baltimore, MD: The Johns Hopkins University Press.

———. 1969b. "The Wiener Kreis in America." In Fleming and Bailyn 1969, 630–73.

Fermi, Laura. 1968. *Illustrious Immigrants: The Intellectual Migration from Europe 1930–1941*. Chicago: University of Chicago Press.

Feser, Edward, ed. 2006. *The Cambridge Companion to Hayek*. Cambridge: Cambridge University Press.

Feyerabend, Paul. 1975. *Against Method: Outline of an Anarchistic Theory of Knowledge*. London: NLB.

———. 1978. *Science in a Free Society*. London: NLB.

———. 1995. *Killing Time: The Autobiography of Paul Feyerabend*. Chicago: University of Chicago Press.

Fieser, Louis F., and Mary Fieser. 1957. *Introduction to Organic Chemistry*. Boston: D. C. Heath.

Fine, Arthur. 1996. *The Shaky Game: Einstein, Realism and the Quantum Theory*. 2nd ed. Chicago: University of Chicago Press.

Fleck, Ludwik. 1929. "Zur Krise der Wirklichkeit." *Naturwissenschaften* 18:425–30.

——. 1932. "Besprechung: Mie, G. *Naturwissenschaft und Theologie.*" *Naturwissenschaften* 20:566.

——. 1935. *Enstehung und Entwicklung einer wissenschaftlichen Tatsache: Einführung in die Lehre von Denkstil und Denkkollektiv.* Basel: Benno Schwabe.

——. 1979. *Genesis and Development of a Scientific Fact.* Translated by Fred Bradley and Thaddeus J. Trenn. Edited by Thaddeus J. Trenn and Robert K. Merton. Chicago: University of Chicago Press.

Fleming, Donald and Bernard Bailyn, eds. 1969. *The Intellectual Migration: Europe and America, 1930–1960.* Cambridge, MA: Harvard University Press,

Forman, Paul. 1967. "The Environment and Practice of Atomic Physics in Weimar Germany: A Study in the History of Science." PhD diss., University of California, Berkeley.

——. 1969. "The Discovery of the Diffraction of X-Rays by Crystals: A Critique of the Myths." *Archive for History of Exact Sciences* 6:38–71.

——. 1971. "Weimar Culture, Causality, and Quantum Theory 1918–27: Adaptation by German Physicists and Mathematicians to a Hostile Intellectual Environment," *HSPS* 3:1–115.

——. 2007. "The Primacy of Science in Modernity, of Technology in Postmodernity, and of Ideology in the History of Technology." *History and Technology* 23:1–152.

——. 2010. *Quantum Mechanics and Weimar Culture: Revisiting the Forman Thesis, with Selected Papers by Paul Forman.* Edited by Cathryn Carson and Alexei Kojevnikov. London: World Scientific Books.

Forman, Paul, John L. Heilbron, and Spencer Weart. 1975. *Physics 'circa' 1900: Personnel, Funding, and Productivity of the Academic Establishments. HSPS* 5:1–185.

Foucault, Michel. 1967. *Les mots et les choses: une archéologie* du savoir. Paris: Gallimard.

Fox, Robert, and George Weisz, eds. 1980. *The Organization of Science and Technology in France, 1808–1914.* Cambridge: Cambridge University Press.

Frank, Philipp. 1947. *Einstein, His Life and Times.* Translated by George Rosen. Edited and revised by Shuichi Kusaka. New York: Knopf.

——. 1949. *Modern Science and Its Philosophy.* Cambridge, MA: Harvard University Press.

——. 1951. "The Logical and Sociological Aspects of Science." *Proceedings of the American Academy of Arts and Sciences* 80:16–30.

——. 1952. "What Teachers in General Education Courses in Science Should Know about Philosophy." In *General Education in Science: Papers Presented at the Workshop in Science in General Education held at the Harvard Summer School July 1950,* edited by I. Bernard Cohen and Fletcher G. Watson, 59–68. Cambridge, MA: Harvard University Press.

Frank, Tibor. 2001–2002. "Cohorting, Networking, Bonding: Michael Polanyi in Exile." *Tradition and Discovery* 28:5–19.

——. 2005. "Ever Ready to Go: The Multiple Exiles of Leo Szilard." *Physics in Perspective* 7:204–52.

——. 2006. "Berlin Junction: Patterns of Hungarian Intellectual Migrations, 1919–1933, Part I." *Storicamente Studi e Ricerche* 2. Accessed January 1, 2011. http://www.storicamente.org/05_studi_ricerche/02frank_print.htm.

——. 2009. *Double Exile: Migrations of Jewish-Hungarian Professionals through Germany to the United States, 1919–1945.* London: Peter Lang.

Franklin, James. 2000. "Thomas Kuhn's Irrationalism." *The New Criterion* 18 (10): 29–35.

French, Steven. 2010. "Notice." *Metascience* 19:157–58.

Freundlich, Herbert. 1926. *Colloid and Capillary Chemistry*. Translated by J. Stafford Hatfield. New York: Dutton.

———. 1932. "Introductory Paper to Section II." In "The Adsorption of Gases: A General Discussion (12–13 January 1932)." *TFS* 28: 195–201.

Friedman, Michael. 1998. "On the Sociology of Scientific Knowledge and Its Philosophical Agenda." *Studies in the History and Philosophy of Science* 29:239–71.

Friedman, Robert Marc. 2001. *The Politics of Excellence: Behind the Nobel Prize in Science*. New York: W. H. Freeman.

Fuller, Steve. 2000a. "Commentary II." *Minerva* 38:26–31.

———. 2000b. *Thomas Kuhn: A Philosophical History for Our Times*. Chicago: University of Chicago Press.

———. 2002. *Social Epistemology*, 2nd ed. Bloomington: Indiana University Press.

———. 2004. *Kuhn vs. Popper: The Struggle for the Soul of Science*. New York: Columbia University Press.

———. 2009. *The Sociology of Intellectual Life: The Career of the Mind in and around the Academy*. Los Angeles: Sage.

Furukawa, Yasu. 1982. "Hermann Staudinger and the Emergence of the Macromolecular Concept." *Historia Scientiarum* 22:1–18.

———. 1998. *Inventing Polymer Science: Staudinger, Carothers, and the Emergence of Macromolecular Chemistry*. Philadelphia: University of Pennsylvania Press.

———. 2003. "Macromolecules: Their Structures and Functions." In Nye 2003, 429–45.

Gaines, George L. Jr. 1993. "Irving Langmuir." In *Nobel Laureates in Chemistry 1901–1992*, edited by Laylin K. James, 205–210. Philadelphia: American Chemical Society and Chemical Heritage Foundation.

Galison, Peter, and Lorraine Daston. 2007. *Objectivity*. New York: Zone Books.

Gardner, Patrick, ed. 1974. *The Philosophy of History*. Oxford: Oxford University Press.

Garfield, Eugene. 1982. "J. D. Bernal—The Sage of Cambridge. 4S Award Memorializes His Contributions to the Social Studies of Science." In *Essays of an Information Scientist*. 15 volumes, 5:511–23. Philadelphia: Institute for Scientific Information.

Gattei, Stefano. 2001. "Review." *British Journal for the Philosophy of Science* 52:815–25.

———. 2008. *Thomas Kuhn's "Linguistic Turn" and the Legacy of Logical Positivism*. Farnham, UK: Ashgate.

———. 2009. *Karl Popper's Philosophy of Science: Rationality without Foundations*. London: Routledge.

Gavroglu, Kostas. 1995. *Fritz London: A Scientific Biography*. Cambridge: Cambridge University Press.

Gay, Peter. 1968. *Weimar Culture: The Outsider as Insider*. New York: Harper Torchbook.

Gebhardt, Bruno, ed. 1903. *Wilhelm von Humboldts politischen Denkschriften*. Vol. 1, *1802–1810*. Berlin: Behr.

Gelwick, Richard. 2005a. "Polanyi Scholarship and the Former Baylor Polanyi Center." Accessed December 30, 2009. http://www.creationismstrojanhorse.com/Gelwick_on_Polanyi.html.

———. 2005b. "Michael Polanyi's Daring Epistemology and the Hunger for Teleology." *Zygon: Journal of Religion and Science*. 40: 63–76.

Giere, Ronald N. 1988. *Explaining Science: A Cognitive Approach*. Chicago: University of Chicago Press.

Giere, Ronald N., and Alan Richardson, eds. 1996. *Origins of Logical Empiricism*. Minneapolis: University of Minnesota Press.

Gingras, Yves. 1995. "Following Scientists through Society? Yes, but at Arm's Length." In *Scientific Practice: Theories and Stories of Doing Physics*, edited by Jed Z. Buchwald, 123–50. Chicago: University of Chicago Press.

Glasstone, Samuel. 1947. *Textbook of Physical Chemistry*. 2nd ed. New York: D. Van Nostrand.

Glasstone, Samuel, Keith J. Laidler, and Henry Eyring. 1941. *The Theory of Rate Processes*. New York: McGraw-Hill.

Godfrey-Smith, Peter. 2003. *Theory and Reality: An Introduction to the Philosophy of Science*. Chicago: University of Chicago Press.

Goenner, Hubert. 2005. *Einstein in Berlin*. Munich: C. H. Beck.

Goldman, Alvin. 2006. "Social Epistemology." *Stanford Encyclopedia of Philosophy*. Accessed January 24, 2010. http://plato.stanford.edu/entries/epistemology-social.

Goldman, F., and Michael Polanyi. 1928. "Adsorption von Dämpfen an Kohle und die Wärmeausdehnung der Benetzungsschicht." *ZpC* 132:321–70.

Golec, Michael J. 2009. "*Science*'s 'New Garb': Aesthetic and Cultural Implications of Redesign in a Cold War Context." *Design Issues* 25:29–45.

Golinski, Jan. 1998. *Making Natural Knowledge: Constructivism and the History of Science*. Cambridge: Cambridge University Press.

Goran, Morris. 1967. *The Story of Fritz Haber*. Norman: University of Oklahoma Press.

Gormley, Melinda. 2006. "Geneticist L. C. Dunn: Politics, Activism, and Community." PhD diss., Oregon State University.

Görtemaker, Manfred. 2002. *Weimar in Berlin: Porträt einer Epoche*. Berlin-Brandenburg: Be.bra. Verlag.

Gortner, Ross Aikan, and Karl Sollner. 1941. "Obituary: Herbert Freundlich 1880–1941." *Science* 93 (2481): 414–16.

Gowenlock, Brian G. 2008–2009. "Michael Polanyi: Scientist and Philosopher: Some General Points and Personal Reminiscences." *Tradition and Discovery* 35 (1): 34–36.

Graham, Loren R. 1993. *Science in Russia and the Soviet Union: A Short History*. Cambridge: Cambridge University Press.

———. 1998. *What Have We Learned about Science and Technology from the Russian Experience?* Stanford, CA: Stanford University Press.

Greenaway, John. 1932. "Memorial Notice, 'William Henry Perkin.'" In *The Life and Work of Professor William Henry Perkin*, 7–38. London: The Chemical Society.

Greenberg, Daniel S. 1999. *The Politics of Pure Science*. New ed. Chicago: University of Chicago Press.

Gregg, S. J. and K. S. W. Sing. 1982. *Adsorption, Surface Area and Porosity*. 2nd ed. New York: Academic Press.

Gregory, Richard. 1928. *Discovery, or the Spirit and Service of Science*. New York: Macmillan.

Grene, Marjorie. 1958. "Personal Knowledge." *Encounter* 11 (4): 67–68.

———. 1977. "Tacit Knowing: Grounds for a Revolution in Philosophy." *Journal of the British Society for Phenomenology* 8 (3): 164–71.

———. 1999. *A Philosophical Testament*. Chicago: Open Court.

———. 2002. "Autobiography." In Auxier and Hahn 2002, 3–28.

Gross, Paul R., and Norman Levitt. 1994. *Higher Superstition: The Academic Left and Its Quarrels with Science*. Baltimore, MD: The Johns Hopkins University Press.

Gross, Paul R., Norman Levitt, and Martin W. Lewis, eds. 1997. *The Flight from Science and Reason*. New York: Annals of the New York Academy of Sciences.

Guerlac, Henry. 1963. "Some Historical Assumptions about the History of Science." In Crombie 1963, 797–812.

Guéron, Jules, and Michel Magat. 1971. "A History of Physical Chemistry in France." *Annual Review of Physical Chemistry* 22:1–25.

Gulick, Walter. 2003–2004. "Letters about Polanyi, Koestler, and Eva Zeisel" *Tradition and Discovery* 2 (2003–2004): 6–10.

———. 2005. "Polanyi on Teleology: A Response to John Apczynski and Richard Gelwick." *Zygon: Journal of Religion and Science*. 40:89–96.

———. 2008. "Michael and Karl Polanyi: Conflict and Convergence." *The Political Science Reviewer* 37:13–43.

Haber, Fritz. 1929. "Einleitung." 34th Hauptversammlung der Deutschen Bunsen-Gesellschaft für angewandte physikalische Chemie, 8–12 May 1929 in Berlin. "Die heterogene Katalyse," 10 May 1929. *ZE* 35: 533–34.

Hackett, Edward J., Olga Amsterdamska, Michael Lynch, and Judy Wajcman, eds. 2008. *The Handbook of Science and Technology Studies*. 3rd ed. Cambridge, MA: MIT Press.

Hacohen, Malachi. 1999. "Dilemmas of Cosmopolitanism: Karl Popper, Jewish Identity, and 'Central European Culture.'" *Journal of Modern History* 71:105–49.

———. 2000. *Karl Popper: The Formative Years, 1902–1945*. Cambridge: Cambridge University Press.

Hager, Thomas. 1995. *Force of Nature: The Life of Linus Pauling*. New York: Simon and Schuster.

Hahn, Otto. 1970. *My Life: The Autobiography of a Scientist*. Translated by Ernst Kaiser and Eithne Wilkins. New York: Herder and Herder.

Hahn, Ralf. 1999. *Gold aus dem Meer: Die Forschungen des Nobel preisträgers Fritz Haber in den Jahren 1922–1927*. Berlin: GNT-Verlag.

Haldane, J. B. S. 1941. *Science and Everyday Life*. Harmondsworth, UK: Penguin.

Hall, Karl P. 1999. "Purely Practical Revolutionaries: A History of Stalinist Theoretical Physicists." PhD diss., Harvard University.

Hallam, Anthony. 1973. *A Revolution in the Earth Sciences: From Continental Drift to Plate Tectonics*. Oxford: Clarendon Press.

Hansen, R. S., and C. A. Smolders. 1962. "Colloid and Surface Chemistry in the Mainstream of Modern Chemistry." *Journal of Chemical Education* 30:167.

Hanson, Norwood Russell. 1958. *Patterns of Scientific Discovery: An Inquiry into the Conceptual Foundations of Science*. Cambridge: Cambridge University Press.

Haraway, Donna. 1991. "Situated Knowledges: The Science Question in Feminism and the Privilege of the Partial Perspective." In *Simians, Cyborgs and Women: The Reinvention of Nature*, 183–201. New York: Routledge.

Harding, Sandra. 1986. *The Science Question in Feminism*. Ithaca, NY: Cornell University Press.

Hargittai, István.1997. "John C. Polanyi." *Polanyiana* 6 (2): 72–84.

———. 2006. *The Martians of Science: Five Scientists Who Changed the Twentieth Century*. Oxford: Oxford University Press.

Harman, Oren Solomon. 2003. "C. D. Darlington and the British and American Reaction to Lysenko and the Soviet Conception of Science." *Journal of the History of Biology* 36:309–52.

Harrod, Roy Forbes. 1959. *The Prof: A Personal Memoir of Lord Cherwell*. London: Macmillan.

Hartley, Harold. 1955. "Schools of Chemistry in Great Britain and Ireland. XVI. The University of Oxford." *Journal of the Royal Institute of Chemistry* 79:118–27, 176–84.

Harvard University. Committee on the Objectives of a General Education in a Free Society. 1945. *General Education in a Free Society: Report of the Harvard Committee.* Cambridge, MA: Harvard University Press.

Harwood, Jonathan. 1986. "Ludwik Fleck and the Sociology of Knowledge." *Social Studies of Science* 16:173–87.

Hayek, F. A. 1960. *The Constitution of Liberty.* Chicago: University of Chicago Press.

———. 1994. *Hayek on Hayek: An Autobiographical Dialogue,* edited by Stephen Kresge and Leif Wenar. Chicago: University of Chicago Press.

———. 2007. *The Road to Serfdom: Text and Documents—The Definitive Edition.* Edited by Bruce Caldwell. Chicago: University of Chicago Press.

Healy, Maureen. 2004. *Vienna and the Fall of the Hapsburg Empire: Total War and Everyday Life in World War I.* Cambridge: Cambridge University Press.

Heath, S. H. 1980. "Henry Eyring, Mormon Scientist." MA thesis, University of Utah.

Hedfors, Eva. 2007. "Fleck in Context." *Perspectives on Science* 15 (1): 49–86.

Heilbron, John L. 1986. *The Dilemmas of an Upright Man: Max Planck as Spokesman for German Science.* Berkeley: University of California Press.

———. 1996. *The Dilemmas of an Upright Man: Max Planck and the Fortunes of German Science.* Berkeley: University of California Press.

———. 1998. "Thomas Samuel Kuhn. 18 July 1922–17 June 1996." *Isis* 89:505–15.

Heitler, Walther, and Fritz London. 1927. "Wechselwirkung neutraler Atome und homopolare Bindung nach der Quantenmechanik." *ZP* 44:455–72.

Hellner, E. E., and P. P. Ewald. 1962. "Schools and Regional Development: Germany." In Ewald 1962, 456–68.

Hempel, Carl. 1959. "The Function of General Laws in History." In *Theories of History,* edited by Patrick Gardner, 345–56. Glencoe, IL: Free Press.

Henning, Eckart and Marion Kazemi. 1998. *Dahlem—Domain of Science: A Walking Tour of the Berlin Institutes of the Kaiser Wilhelm/Max Planck Society in the "German Oxford."* *Max-Planck-Gesellschaft Berichte und Mitteilungen* 1/98. Munich: Max Planck Society.

Henry, John. 2008. "Historical and Other Studies of Science, Technology and Medicine in the University of Edinburgh." *Notes and Records of the Royal Society* 62:223–35.

Herschbach, Dudley. 1987. "Autobiography." In *Les Prix Nobel. The Nobel Prizes 1986,* edited by Wilhelm Odelberg. Stockholm: Nobel Foundation. Accessed September 8, 2009. http://www.nobel.se/chemistry/laureates/1986/herschbach-autobio.html.

Hershberg, James. 1993. *James B. Conant: Harvard to Hiroshima and the Making of the Nuclear Age.* New York: Knopf.

Hertz, Heinrich. 1899. *Principles of Mechanics.* Translated by D. E. Jones and F. T. Walley. London: Macmillan.

Herzog, R. O., Willi Jancke, and Michael Polanyi. 1920a. "Röntgenspektrographische Beobachtungen an Zellulose I." *ZP* 3:196–98.

———. 1920b. "Roentgenspektrographische Beobactungen an Zellulose II." *ZP* 3: 343–48.

Hesse, Mary. 1963. *Model and Analogies in Science.* London: Sheed and Ward.

Hessen, Boris. 1932. "The Social and Economic Roots of Newton's *Principia.*" In *Science at the Crossroads,* 149–212. London: Kniga.

Hettema, Hinne. 2000a. "Philosophical and Historical Introduction." In Hettema 2000b, xvii–xxxix.

———. 2000b. *Quantum Chemistry: Classic Scientific Papers*. Edited and translated by Hinne Hettema. Singapore: World Scientific Publishing.

Heyne, W. and Michael Polanyi. 1928. "Adsorption aus Lösungen," *ZpC* 132:384–98.

Hill, Archibald Vivian. 1933. "International Status and Obligations of Science." *Nature* 132:952–54.

Hinshelwood, Cyril N. 1926. *The Kinetics of Chemical Change in Gaseous Systems*. Oxford: Clarendon Press.

———. 1928. "Chemical Kinetics." *Annual Reports of the Progress of Chemistry (for 1927)* 24:314–41.

Hirschfelder, Joseph O. 1941. "Semi-Empirical Calculations of Activation Energies." *Journal of Chemical Physics* 9:645–53.

———. 1982. "My 50 Years of Theoretical Chemistry, I: Chemical Kinetics." In Bunsen Gesellschaft 1982, 349–55.

Hobsbawm, Eric. 1999. Preface to Swann and Aprahamian 1999, ix–xx.

———. 2009. "Era of Wonders: Bomb, Book and Compass: Joseph Needham and the Great Secrets of China by Simon Winchester." *London Review of Books*, February 26, 2009, 19–20.

Hoch, Paul. 1986. "Formation of a Research School: Theoretical Solid State Physics at Bristol, 1930–1954." *BJHS* 19:19–44.

———. 1987. "Institutional versus Intellectual Migrations in the Nucleation of New Scientific Specialties." *Studies in the History and Philosophy of Science* 18:481–500.

Hoddeson, Lillian, Ernst Braun, Jurgen Teichmann, and Spencer Weart, eds. 1992. *Out of the Crystal Maze: Chapters from the History of Solid-State Physics*. New York: Oxford University Press.

Hodgkin, Dorothy C. 1980. "John Desmond Bernal." *Biographical Memoirs of Fellows of the Royal Society* 26:17–84.

Hoffmann, Dieter. 2000. "Physics in Berlin: Walking Tours in Charlottenburg and Dahlem and Excursions in the Vicinity of Berlin." *Physics in Perspective* 2:426–45.

———. 2005. "Between Autonomy and Accommodation: The German Physical Society during the Third Reich." *Physics in Perspective* 7:293–329.

———. 2006. *Einsteins Berlin: Auf den Spuren eines Genies*. Weinheim: Wiley-VCH.

———, ed. 2008a. *Max Planck: Annalen Papers*. Weinheim: Wiley-VCH.

———. 2008b. *Max Planck: Die Enstehung der modernen Physik*. Munich: C. H. Beck.

Hoffmann, Dieter, and Mark Walker. 2004. "The German Physical Society under National Socialism." *Physics Today* 57 (52): 52–58.

Hollinger, David A. 1995. "Science as a Weapon in Kulturkämpfe in the United States during and after World War II." *Isis* 86:440–54.

———. 1996a. "The Defense of Democracy and Robert K. Merton's Formulation of the Scientific Ethos," In Hollinger 1996c, 80–96.

———. 1996b. "Free Enterprise and Free Inquiry: The Emergence of Laissez-Faire Communitarianism in the Ideology of Science in the United States." In Hollinger 1996c, 97–120.

———. 1996c. *Science, Jews, and Secular Culture: Studies in Mid-Twentieth Century American Intellectual History*. Princeton, NJ: Princeton University Press.

Holton, Gerald. 1947. "Ultrasonic Propagation in Liquids at High Pressures." PhD diss., Harvard University.

———. 1952. *Introduction to Concepts and Theories in Physical Science*. Cambridge, MA: Addison-Wesley.

———. 1964. "On the Thematic Analysis of Science: The Case of Poincaré and Relativity." In *Mélanges Alexandre Koyré à l'occasion de son soixante-dixième anniversaire*, 2:257–68. Paris: Hermann.

———. 1969. "Einstein, Michelson, and the 'Crucial' Experiment." *Isis* 60: 132–97.

———. 1973a. *Introduction to Concepts and Theories in Physical Science*. With Stephen Brush. 2nd rev. ed. Reading, MA: Addison-Wesley.

———. 1973b. *Thematic Origins of Scientific Thought: Kepler to Einstein*. Cambridge, MA: Harvard University Press.

———. 1986. *The Advancement of Science, and Its Burdens: The Jefferson Lecture and Other Essays*. Cambridge: Cambridge University Press.

———. 1992. "Ernst Mach and the Fortunes of Positivism in America." *Isis* 83:27–60.

———. 1992–1993. "Michael Polanyi and the History of Science." *Tradition and Discovery* 19 (1): 16–30.

———. 1998. *Science and Anti-Science*. Cambridge, MA: Harvard University Press.

———. 1999. "Some Lessons from Living in the History of Science." *Isis* 90:S95–S116.

———. 2000. *Einstein, History and Other Passions: The Rebellion against Science at the End of the Twentieth Century*. Cambridge, MA: Harvard University Press.

———. 2005. *Victory and Vexation in Science: Einstein, Bohr, Heisenberg, and Others*. Cambridge, MA: Harvard University Press.

Holton, Gerald, and Stephen Brush. 2001. *Physics, the Human Adventure: From Copernicus to Einstein and Beyond*. 3rd ed. New Brunswick, NJ: Rutgers University Press.

Holton, Gerald, and Duane H. D. Roller. 1958. *Foundations of Modern Physical Science*. Under the editorship of Duane Roller. Reading, MA: Addison-Wesley.

Hook, Ernest B., ed. 2002. *Prematurity in Scientific Discovery: On Resistance and Neglect*. Berkeley: University of California Press.

Howson, Susan. 2004. "Robbins, Lionel Charles, Baron Robbins (1898–1984)." In *Oxford Dictionary of National Biography*, edited by C. G. Matthew and B. Harrison. Vol. 47, 72–76. Oxford; Oxford University Press.

Hoyningen-Huene, Paul. 1995. "Two Letters of Paul Feyerabend to Thomas S. Kuhn on a Draft of *The Structure of Scientific Revolutions*." *Studies in History and Philosophy of Science* 26:353–87.

———. 2006. "More Letters by Paul Feyerabend to Thomas S. Kuhn on *Proto-Structure*." *Studies in History and Philosophy of Science* 37:610–32.

Hufbauer, Karl. 1998. "Kuhn's Discovery of History." Manuscript copy, author's personal collection.

Hull, Albert W. 1962. "Autobiography." In Ewald 1962, 582–87.

Hume-Rothery, W. 1962. "Applications of X-Ray Diffraction to Metallurgical Science." In Ewald 1962, 190–211.

Huxley, Aldous. 1928. *Point Counter Point*. London: Chatto and Windus.

Huxley, Julian. 1932. *A Scientist among the Soviets*. London: Harper.

———. 1934. *Scientific Research and Social Needs*. London: Watts.

Ignotus, Paul. 1961. "The Hungary of Michael Polanyi."In Anonymous 1961, 3–12.

Ingold, Christopher. 1935. *Structure and Mechanism in Organic Chemistry*. Ithaca, NY: Cornell University.

Irving, Edward. 1964. *Paleomagnetism and Its Application to Geological and Geophysical Problems*. New York: John Wiley.

Isaacson, Walter. 2007. *Einstein: His Life and Universe*. New York: Simon and Schuster.

Jacob, Margaret C., ed. 1994. *The Politics of Western Science 1640–1990*. Atlantic Highlands, NJ: Humanities Press.

Jacobs, Struan. 1999. "Michael Polanyi's Theory of Spontaneous Orders." *Review of Austrian Economics* 11 (1–2): 111–27.

———. 2009. "Thomas Kuhn's Memory." *Intellectual History Review* 19: 83–101.

Jacobs, Struan, and R. T. Allen, eds. 2005. *Emotion, Reason and Tradition: Essays on the Social, Political and Economic Thought of Michael Polanyi*. Aldershot, UK: Ashgate.

Jaki, Stanley L. 1984. *Uneasy Genius: The Life and Work of Pierre Duhem*. The Hague: M. Nijhoff.

James, Hubert M., and A. S. Coolidge. 1933. "The Ground State of the Hydrogen Molecule." *Journal of Chemical Physics* 1:825–34.

James, Jeremiah. 2007. "Naturalizing the Chemical Bond: Discipline and Creativity in the Pauling Program, 1927–1942." PhD diss., Harvard University.

Jarvie, Ian. C. 2001a. "Science in a Democratic Republic." *Philosophy of Science* 68: 545–64.

———. 2001b. *The Republic of Science: The Emergence of Popper's Social View of Science*. Amsterdam: Rodopi.

Jasanoff, Sheila, ed. 2004. *States of Knowledge: The Co-Production of Science and the Social Order*. London: Routledge.

Jasanoff, Sheila, Gerald E. Markle, James C. Peterson, and Trevor Pinch, eds. 1995. *Handbook of Science and Technology Studies*. London: Sage.

Jay, Martin. 1973. *The Dialectical Imagination: A History of the Frankfurt School and the Institute of Social Research, 1923–1950*. Boston: Little, Brown and Co.

Jewkes, John. 1948. *Ordeal by Planning*. New York: Macmillan.

———. 1976. "Obituary Notice." *Nature*, May 26, 1976, 242.

Jha, Stefania Ruzsits. 2002. *Reconsidering Michael Polanyi's Philosophy*. Pittsburgh, PA: University of Pittsburgh Press.

———. 2006. "The Bid to Transcend Popper, and the Lakatos-Polanyi Connection." *Perspectives on Science* 14 (3): 318–46.

Johns, Adrian. 2006. "Intellectual Property and the Nature of Science." *Cultural Studies* 20:145–64.

Johnson, Jeffrey A. 1990. *The Kaiser's Chemists: Science and Modernization in Imperial Germany*. Chapel Hill: University of North Carolina Press.

Johnston, William M. 1972. *The Austrian Mind: An Intellectual and Social History 1848–1938*. Berkeley: University of California Press.

Jones, Greta. 1988. *Science, Politics and the Cold War*. London: Routledge.

Josephson, Paul R. 2005. *Totalitarian Science and Technology*. 2nd ed. Amherst, NY: Humanity Books.

Jungnickel, Christa, and Russell McCormmach. 1986. *Intellectual Mastery of Nature: Theoretical Physics from Ohm to Einstein*. Vol. 1, *The Torch of Mathematics 1800–1870*. Vol. 2, *The Now Mighty Theoretical Physics 1870–1925*. Chicago: University of Chicago Press.

Kaiser, David. 1998. "A Mannheim for All Seasons: Bloor, Merton, and the Roots of the Sociology of Scientific Knowledge." *Science in Context* 11:51–87.

———. 2005. *Drawing Theories Apart: The Dispersion of Feynman Diagrams in Postwar Physics*. Chicago: University of Chicago Press.

Kallmann, Harmut, and Fritz London. 1929. "Über quantenmechanische Energieübertragung zwischen atomaren Systemen." *ZpC* B2:207–43.

Kallmann, Harmut, and Hermann Mark. 1926. "Über einige Eigenschaften der Comptonstrahlung." *ZP* 36:120–42.

Karachalios, Andreas. 2000. "On the Making of Quantum Chemistry in Germany." *Studies in the History and Philosophy of Modern Physics* 31:493–510.

Kaufman, Frederick. 1982. "Progress and Prospect in Elementary Reaction Kinetics." In Bunsen Gesellschaft 1982, 362–67.

Kauzmann, Walter. 1996. "Henry Eyring. February 20, 1901–December 26, 1981." *National Academy of Sciences Biographical Memoirs* 70:45–57.

Kay, Lily E. 1993. *The Molecular Vision of Life: Caltech, the Rockefeller Foundation, and the Rise of the New Biology.* Oxford: Oxford University Press.

Kessler, Harry. 2000. *Berlin in Lights: The Diaries of Count Harry Kessler (1918–1937).* Edited and translated by Charles Kessler. New York: Grove Press.

Kettler, David, and Volker Meja. 1995. *Karl Mannheim and the Crisis of Liberalism: The Secret of These New Times.* New Brunswick, NJ: Transaction Publishers.

Kevles, Daniel J. 1987. *The Physicists: The History of a Scientific Community in Modern America.* Cambridge, MA: Harvard University Press.

———. 1992. "Huxley and the Popularization of Science." In *Julian Huxley: Biologist and Statesman of Science,* edited by C. Kenneth Waters and Albert van Helden, 238–51. Houston, TX: Rice University Press.

Keynes, John Maynard. 1964. *The General Theory of Employment, Interest, and Money.* San Diego: Harcourt, Inc.

———. 2004. *The End of Laissez-Faire. The Economic Consequences of the Peace.* Amherst, NY: Prometheus Books, 2004.

Kindleberger, Charles P. 1973. "*The Great Transformation.*" *Daedalus* 103:45–52.

Klausinger, Hansjörg. 2001. "Gustav Stolper, *Der deutsche Volkswirt,* and the Controversy on Economic Policy at the End of the Weimar Republic." *History of Political Economy* 33:241–67.

Knepper, Paul. 2005. "Michael Polanyi and Jewish Identity." *Philosophy of the Social Sciences* 35:263–93.

———. 2005–2006. "Polanyi, 'Jewish Problems' and Zionism." *Tradition and Discovery* 32 (1): 6–19.

Koestler, Arthur. 1945. *The Yogi and the Commissar, and Other Essays.* New York: Macmillan.

———. 1949. *Insight and Outlook: An Inquiry into the Common Foundations of Science, Art and Social Ethics.* London: Macmillan.

———. 1952. Preface to *Conspiracy of Science,* by Alex Weissberg. Translated by Edward Fitzgerald, i–xii. London: Hamish Hamilton.

———. 1959. *The Sleepwalkers: A History of Man's Changing Vision of the Universe.* London: Hutchinson.

———. 2001. "Arthur Koestler." In *The God That Failed,* edited by Richard Crossman, 15–74. New York: Columbia University Press.

Kohler, Robert E. 1991. *Partners in Science: Foundations and Natural Scientists, 1900–1945.* Chicago: University of Chicago Press.

Kovács, László Sr. 2003. "The Physical Tourist. Budapest: A Random Walk in Science and Culture." *Physics in Perspective* 5:310–48.

Kragh, Helge. 1989. "The Aether in Late 19th-Century Chemistry." *Ambix* 36:49–65.

Krige, John. 2006. *American Hegemony and the Postwar Reconstruction of Science in Europe.* Cambridge, MA: MIT Press.

Kritsman, V. A., G.E. Zaikov, and N.M. Emanuel. 1995. *Chemical Kinetics and Chain Reactions: Historical Aspects.* Translated by B. L. Kozhushin. Commack, NY: Nova Science Publishers.

Krohn, Claus-Dieter. 1993. *Intellectuals in Exile: Refugee Scholars and the New School for*

*Social Research*. Translated by Rita and Robert Kimber. Amherst: University of Massachusetts Press.

———. 1996. "Dismissal and Emigration of German-Speaking Economists after 1933." In Ash and Söllner 1996b, 175–197.

Krugman, Paul. 2009. "How Did Economists Get It So Wrong?" *New York Times Magazine*, September 6, 36–43.

Kuhn, Thomas S. 1949. "The Cohesive Energy of Monovalent Metals as a Function of Their Atomic Quantum Defects." Ph.D. diss., Harvard University.

———. 1957. *The Copernican Revolution: Planetary Astronomy in the Development of Western Thought*. Cambridge, MA: Harvard University Press.

———. 1959. "The Essential Tension: Tradition and Innovation in Scientific Research." In Kuhn 1977, 225–39.

———. 1961a. "The Function of Measurement in Modern Physical Science." In Kuhn 1977, 178–224.

———. 1961b. "The Function of Dogma in Scientific Research." In Crombie 1963, 347–69.

———. 1962. *The Structure of Scientific Revolutions*. Chicago: University of Chicago Press.

———. 1977. *The Essential Tension: Selected Studies in Scientific Tradition and Change*. Chicago: University of Chicago Press.

———. 1992. "The Trouble with the Historical Philosophy of Science: Robert and Maurine Rothschild Distinguished Lecture, 19 November 1991." Department of the History of Science, Harvard University, Cambridge, MA. Author's personal collection.

Labinger, Jay, and Harry Collins, eds. 2001. *The One Culture? A Conversation about Science*. Chicago: University of Chicago Press.

Labour Party. 1934. *For Socialism and Peace: the Labour Party's Programme of Action*. London: The Labour Party.

Ladd, Brian. 1997. *The Ghosts of Berlin: Confronting German History in the Urban Landscape*. Chicago: University of Chicago Press.

Laidler, Keith J. 1985. "Chemical Kinetics and the Origins of Physical Chemistry." *Archive for History of Exact Sciences* 32:43–75.

———. 1993. *The World of Physical Chemistry*. Oxford: Oxford University Press.

Laitko, Hubert et al. 1987. *Wissenschaft in Berlin: Von den Anfängen bis zum Neubeginn nach 1945*. Berlin: Dietz Verlag.

Lakatos, Imre. 1970. "Falsification and the Methodology of Scientific Research Programs." In *Criticism and the Growth of Knowledge*, edited by Imre Lakatos and Alan Musgrave, 91–196. Cambridge: Cambridge University Press.

———. 1999. "Lectures on Scientific Method (1974)." In *For and Against Method: Including Lakatos's Lectures on Scientific Method and the Lakatos-Feyerabend Correspondence*, by Imre Lakatos and Paul Feyerabend. Edited by Matteo Motterlini, 19–109. Chicago: University of Chicago Press.

Lakatos, Imre, and Paul Feyerabend. 1999. *For and Against Method: Including Lakatos's Lectures on Scientific Method and the Lakatos-Feyerabend Correspondence*, ed. Matteo Motterlini. Chicago: University of Chicago Press.

Lakatos, Imre, and Alan Musgrave, eds. 1970. *Criticism and the Growth of Knowledge*. Cambridge: Cambridge University Press.

Langer, R. M. 1929. "The Quantum Mechanics of Chemical Reaction." *Physical Review*, 34:92–108.

Langmuir, Irving. 1915. "Chemical Reactions at Low Pressures." *JACS* 37:1139–67.

———. 1916. "The Constitution and Fundamental Properties of Solids and Liquids. I. Solids." *JACS* 38:2221–95.

———. 1917. "The Constitution and Fundamental Properties of Solids and Liquids. II. Liquids," *JACS* 39:1848–906.

———. 1918. "The Adsorption of Gases on Plane Surfaces of Glass, Mica, and Platinum." *JACS* 40:1361–403.

———. 1919a. "Isomorphism, Isoterism and Covalence." *JACS* 41:1543–59.

———. 1919b. "The Arrangement of Electrons in Atoms and Molecules." *JACS* 41:868–934.

———. 1919c. "The Structure of Atoms and the Octet Theory of Valence." *Proceedings of the National Academy of Sciences* 5:252–59.

———. 1922. "The Structure of Molecules." In *Report of the Annual Meeting. British Association for the Advancement of Science, 1921, Edinburgh*, 468–69. London: British Association for the Advancement of Science.

———. 1966. "Surface Chemistry: Nobel Lecture in Chemistry in 1932." In *Nobel Lectures, Chemistry 1922–1941*, edited by Nobelstiftelsen [Nobel Foundation], 287–325. Amsterdam: Elsevier.

Laqueur, Walter. 1991. "Zionism and its Liberal Critics, 1896–1948." *Journal of Contemporary History* 26:255–76.

———. 1993. "The Right Chemistry." *New Republic*, February 15, 40–41.

Latour, Bruno. 1987. *Science in Action: How to Follow Scientists and Engineers through Society*. Milton Keynes, UK: Open University Press.

———. 1988. "Mixing Humans and Nonhumans Together: The Sociology of a Door-Closer." *Social Problems* 35:298–310.

———. 1990. "Postmodern? No, Simply AModern! Steps towards an Anthropology of Science. An Essay Review." *Studies in History and Philosophy of Science* 21:145–71.

———. 2004. "Why Has Critique Run Out of Steam? From Matters of Fact to Matters of Concern." *Critical Inquiry* 39:225–48.

———. 2007. "Turning around Politics: A Note on Gerard de Vries' Paper." *Social Studies of Science* 37 (5): 811–20.

Latour, Bruno, and Steve Woolgar. 1986. *Laboratory Life: The Construction of Scientific Facts*. Princeton, NJ: Princeton University Press.

Laudan, Larry. 1977. *Progress and Its Problems: Towards a Theory of Scientific Growth*. Berkeley: University of California Press.

Lendvai, Paul. 2003. *The Hungarians: A Thousand Years of Victory in Defeat*. Translated by Ann Major. Princeton, NJ: Princeton University Press.

Leonard, Robert J. 1998. "Ethics and the Excluded Middle: Karl Menger and Social Science in Interwar Vienna." *Isis* 89:1–26.

Lewis, Gilbert N. 1916. "The Atom and the Molecule." *JACS* 38:762–85.

Litván, György. 1990. "Karl Polanyi in Hungarian Politics (1914–64)." In Polanyi-Levitt 1990a, 30–37.

———. 2006. *A Twentieth-Century Prophet: Oscar Jászi 1875–1957*. Translated by Tim Wilkinson. Budapest: Central European University Press.

Loader, Colin. 1985. *The Intellectual Development of Karl Mannheim: Culture, Politics, and Planning*. Cambridge: Cambridge University Press.

London, Fritz. 1929. "Quantenmechanische Deutung des Vorgangs der Aktivierung," *ZE* 35:552–55.

———. 2000a. "On Some Properties and Applications of Molecular Forces." In Hettema 2000b, 400–22.

———. 2000b. "On the Theory of Non-Adiabatic Chemical Reactions." In Hettema 2000b, 32–60.

London, Fritz, and R. Eisenschitz. 2000. "On the Ratio of the Van der Waals Forces and the Homo-Polar Binding Forces." In Hettema 2000b, 336–68.

London, Fritz, and Michael Polanyi. 1930. "Über die atomtheoretische Deutung der Adsorptionskräfte." *Die Naturwissenschaften* 18:1099–100.

Longino, Helen. 1992. "Essential Tensions—Phase Two: Feminist, Philosophical, and Social Studies of Science." In *The Social Dimensions of Science*, edited by Ernan McMullin, 198–216. South Bend, IN: University of Notre Dame Press.

Lonsdale, Kathleen. 1949. *Crystals and X-Rays*. New York: D. Van Nostrand.

———. 1962. "X-Ray Diffraction and Its Impact on Physics." In Ewald 1962, 221–47.

Luft, David S. 1980. *Robert Musil and the Crisis of European Culture 1880–1942*. Berkeley: University of California Press.

Lukacs, John. 1993. *Budapest 1900: A Historical Portrait of a City and Its Culture*. London: Weidenfeld.

Macdougall, Frank Henry. 1943. *Physical Chemistry*. Rev ed. New York: Macmillan.

Mach, Ernst. 1919. *The Science of Mechanics: A Critical and Historical Account of Its Development*. 4th ed. Translated by Thomas J. McCormmach. Chicago: Open Court.

Machlup, Fritz. 1980–1984. *Knowledge: Its Creation, Distribution, and Economic Significance*. 3 vols. Princeton, NJ: Princeton University Press.

MacKenzie, Donald. 2003. "David Owen Edge 1932–2003." *Isis* 94:498–99.

MacLeod, Roy M. 1998. "Chemistry for King and Kaiser: Revisiting Chemical Enterprise and the European War." In *Determinants in the Evolution of the European Chemical Industry, 1900–1939*, edited by Anthony S. Travis, 25–49. Dordrecht: Kluwer.

MacLeod, Roy and Kay MacLeod. 1976. "The Social Relations of Science and Technology, 1914–1939." In *The Twentieth Century*, edited by Carlo M. Cipolla, 301–63. London: Collins/Fontana.

Macrakis, Kristie. 1989. "Scientific Research in National Socialist Germany: The Survival of the Kaiser Wilhelm Gesellschaft." PhD diss., Harvard University.

———. 1993. *Surviving the Swastika: Scientific Research in Nazi Germany*. Oxford: Oxford University Press.

Magee, Bryan. 1973. *Karl Popper*. Harmondsworth, UK: Penguin.

Malinowski, Bronislaw. 1922. *Argonauts of the Western Pacific: An Account of Native Enterprise and Adventure in the Archipelagoes of Melanesian New Guinea*. London: Routledge.

Manegold, Karl-Heinz. 1970. *Universität, Technische Hochschule und Industrie*. Berlin: Dunker und Humblot.

Mannheim, Karl. 1929. *Ideologie und Utopie*. Bonn: F. Cohen.

———. 1936. *Ideology and Utopia: An Introduction to the Sociology of Knowledge*. Translated by Louis Wirth and Edward Shils. New York: Harcourt, Brace and World.

———. 1940. *Man and Society in an Age of Reconstruction*. Translated by Edward Shils. London: Kegan Paul.

———. 1952. *Essays on the Sociology of Knowledge*. Edited by Paul Kecskemeti. London: Routledge and Kegan Paul.

———. 2003. *Selected Correspondence (1911–1946) of Karl Mannheim, Scientist, Philosopher, and Sociologist*. Edited by Eva Gábor, with assistance of Dézsö and R. T. Allen. Lewiston, NY: The Edwin Mellen Press.

Manucci, Monia. 2005. "Observations on Michael Polanyi's Keynesianism." In Jacobs and Allen 2005, 149–63.

Marcellin, René. 1914. *Contribution à la cinétique physico-chimique*. Paris: Gauthier-Villars.

Marcum, James A. 2005. *Thomas Kuhn's Revolution: An Historical Philosophy of Science*. London: Continuum International Publishing.

Mark, Hermann. 1962. "Recollections of Dahlem and Ludwigshafen." In Ewald 1962, 603–7.

———. 1976. "Polymer Chemistry: the Past 100 Years." *Chemical & Engineering News* 54:176–89.

———. 1981. "Polymer Chemistry in Europe and America—How It All Began." *Journal of Chemical Education* 58:527–34.

———.1993. *From Small Organic Molecules to Large: A Century of Progress*. Washington DC: American Chemical Society.

Martin, R. N. D. 1991. *Pierre Duhem: Philosophy and History in the Work of a Believing Physicist*. Chicago: Open Court.

Marton, Kati. 2006. *The Great Escape: Nine Jews Who Fled Hitler and Changed the World*. New York: Simon and Schuster.

März, Eduard. 1991. *Joseph Schumpeter: Scholar, Teacher and Politician*. New Haven, CT: Yale University Press.

Maucourant, Jérôme. 2005. *Avez-vous lu Polanyi?* Paris: La Dispute.

Max-Planck-Gesellschaft. 1999. *Fritz-Haber-Institut der Max-Planck-Gesellschaft. Max-Planck-Gesellschaft Berichte und Mitteilungen*. 1/99. Munich: Max-Planck-Gesellschaft.

Mayer, Anna-K. 2000. "Setting Up a Discipline: Conflicting Agendas of the Cambridge History of Science Committee, 1936–1950." *Studies in the History and Philosophy of Science* 31:665–89.

Mayhew, Anne. 1987. "The Beginnings of Institutionalism." *Journal of Economic Issues* 3:971–78.

McBain, J. W., and G. T. Britton. 1930. "The Nature of the Sorption by Charcoal of Gases and Vapors under Great Pressure." *JACS* 52:2198–222.

McCormmach, Russell. 1983. *Night Thoughts of a Classical Physicist: A Novel*. New York: Avon.

McCraw, Thomas K. 2007. *Prophet of Innovation: Joseph Schumpeter and Creative Destruction*. Cambridge, MA: Harvard University Press.

McGucken, William. 1984. *Scientists, Society and State: The Social Relations of Science Movement in Great Britain 1931–1947*. Columbus: Ohio State University Press.

Mead, Walter B. 2008. "Michael Polanyi (1891–1976): Introduction to an Unfinished Revolution." *Political Science Reviewer* 37:1–12.

Medawar, Jean, and David Pyke. 2001. *Hitler's Gift: Scientists Who Fled Nazi Germany*. London: Piatkus.

Meinel, Christoph. 1983. "Theory or Practice? The Eighteenth-Century Debate on the Scientific Status of Chemistry." *Ambix* 30:121–32.

Mendell, Marguerite. 1990. "Karl Polanyi and Feasible Socialism." In Polanyi-Levitt 1990a, 66–77.

———, ed. 2005. *Reclaiming Democracy: The Social Justice and Political Economy of Gregory Baum and Kari Polanyi Levitt*. Montreal: McGill-Queen's University Press.

Menger, Carl. 1883. *Untersuchungen über der Socialwissenschaften und der politischen Oekonomie insbesondere*. Leipzig: Duncker und Humblot.

Mercer, David, Jerry Ravetz, Stephen P. Turner, and Steve Fuller. 2005. "A Parting Shot at Misunderstanding: Fuller vs. Kuhn." *Metascience* 14:3–32.

Merton, Robert K. 1937. "The Sociology of Knowledge." *Isis* 27:493–503.

———. 1941. "Karl Mannheim and the Sociology of Knowledge." *Journal of Liberal Religion* 2:125–47.

———. 1942. "A Note on Science and Democracy." *Journal of Legal and Political Sociology* 1:115–26.

———. 1945. "The Sociology of Knowledge." In *Twentieth-Century Sociology*, edited by Georges Gurvitch and Wilbert E. Moore, 366–405. New York: Philosophical Library.

———. 1972. "Insiders and Outsiders: A Chapter in the Sociology of Knowledge." *American Journal of Sociology* 77:9–47.

———. 1973a. "Paradigm for the Sociology of Knowledge." In Merton 1973b, 7–40.

———. 1973b. *The Sociology of Science: Theoretical and Empirical Investigations*, edited by N. W. Storer. Chicago: University of Chicago Press.

Merton, Robert K., with Harriet Zuckerman. 1973. "Institutionalized Patterns of Evaluation in Science." In Merton 1973b, 460–96.

Merz, John. 1965. *A History of European Thought in the Nineteenth Century. Scientific Thought*. Vols. 1–2, *Scientific Thought*. New York: Dover.

Meyer, Kurt, and Hermann Mark. 1928. "Über den Bau des kristallisierten Anteils der Cellulose." *Berichte der deutschen chemische Gesellschaft* 61:593–614.

Mill, John Stuart. 1865. *Auguste Comte and Positivism*. London: Trübner.

Mirowski, Philip. 1997. "On Playing the Economics Trump Card in the Philosophy of Science: Why It Did Not Work for Michael Polanyi." *Philosophy of Science (Proceedings)* 64:S127–38.

———. 1998–1999. "Economics, Science, and Knowledge: Polanyi versus Hayek." *Tradition and Discovery* 25:29–42.

Mises, Ludwig von. 1969. *The Historical Setting of the Austrian School of Economics*. Rochelle, NY: Arlington House.

Mitchell, Mark T. 2006. *Michael Polanyi*. Wilmington, DE: ISI Books.

Moggridge, Donald E. 2004. "A Vested Intellectual Interest . . . Tinged with a Dose of Pique? D. H. Robertson on the National Debt Inquiry." *History of Political Economy* 36:187–93.

Moleski, Martin X, SJ. 2005–2006. "The Man Who Fell Among Theologians." *Tradition and Discovery* 32 (3): 35–39.

———. 2006–2007. "Polanyi vs. Kuhn: Worldviews Apart." *Tradition and Discovery* 33 (2): 7–24.

———. 2007–2008. "Provocative Questions, Abbreviated Answers." *Tradition and Discovery* 34 (3): 42–43.

Montagu, Ivor. 1999. "The Peacemonger." In Swann and Aprahamian 1999, 212–34.

Moore, Walter. 1989. *Schrödinger: Life and Thought*. Cambridge: Cambridge University Press.

Morawetz, Herbert. 1995. "Hermann Francis Mark." *National Academy of Sciences Biographical Memoirs* 68:195–209.

Morgan, Mary S. 2003. "Economics." In *The Cambridge History of Science*. Vol. 7, *The Modern Social Sciences*, edited by Theodore M. Porter and Dorothy Ross, 275–305. Cambridge: Cambridge University Press.

Morgan, Mary S., and Marcel Boumans. 2004. "Secrets Hidden by Two-Dimensionality: The Economy as a Hydraulic Machine." In *Models: The Third Dimension of Science*, edited by Soraya de Chadarevian and Nick Hopwood, 369–401. Stanford, CA: Stanford University Press.

Morgan, Mary S., and Malcolm Rutherford, eds. 1999. *From Interwar Pluralism to Post War Neo-Classicism*. Durham, NC: Duke University Press.

Morse, P. M. 1930. "Diatomic Molecules According to the Wave Mechanics. II. Vibrational Levels." *Physical Review* 34:57–64.

Morton, George A. 1980. "Tribute to Hartmut Kallmann." *IEEE Transactions on Nuclear Science* 27:13.

Mott, Nevill. 1986. *A Life in Science*. London: Taylor and Francis.

Mucsi, Ferenc. 1990. "The Start of Karl Polanyi's Career." In Polanyi-Levitt 1990a, 26–29.

Mullins, Phil. 1997. "Michael Polanyi and J. H. Oldham in Praise of Friendship." *Appraisal* 1 (4): 179–89.

———. 2002. "On Persons and Knowledge: Marjorie Grene and Michael Polanyi." In Auxier and Hahn 2002, 31–60.

———. 2009–2010. "In Memoriam: Marjorie Grene." *Tradition and Discovery* 36 (1): 55–69.

Musil, Robert. 1996. *The Man Without Qualities*. Vol. 1. Translated by Sophie Wilkins. New York: Vintage.

Nagy, Endre J. 1994. "After Brotherhood's Golden Age: Karl and Michael Polanyi." In *Humanity, Society and Commitment: On Karl Polanyi*, edited by Kenneth McRobbie, 81–112. Montreal: Black Rose Books.

Nall, Joshua F. K. 2007. "'The Struggle between Truth and Propaganda': Michael Polanyi and the Social Relations of Science." Master's diss., University of Cambridge.

Nash, Leonard K. 1944. "The Gases in Meteorites." PhD diss., Harvard University.

———. 1963. *The Nature of Natural Sciences*. Boston: Little, Brown.

Neale, Walter C. 1990. "Karl Polanyi and American Institutionalism: A Strange Case of Convergence." In Polanyi-Levitt 1990a, 145–51.

Needham, Joseph. 1936. *Order and Life*. Cambridge: Cambridge University Press.

Nemeth, Elisabeth. 2000. "Review." *British Journal for the Philosophy of Science* 54:515–20.

Nickles, Thomas. 1998. "Kuhn, Historical Philosophy of Science, and Case-Based Reasoning." *Configurations* 6:51–85.

Nye, Mary Jo, ed. 1984. *The Question of the Atom: From the Karlsruhe Congress to the First Solvay Conference. 1860–1911. A Selection of Primary Sources*. Los Angeles: Tomash and New York: American Institute of Physics.

———. 1986. *Science in the Provinces: Scientific Communities and Provincial Leadership in France, 1860–1930*. Berkeley: University of California Press.

———. 1993. *From Chemical Philosophy to Theoretical Chemistry: Dynamics of Matter and Dynamics of Disciplines 1800–1950*. Berkeley: University of California Press.

———. 1999. *Before Big Science: The Pursuit of Modern Chemistry and Physics 1800–1940*. Cambridge, MA: Harvard University Press.

———. 2000a. "Laboratory Practice and the Physical Chemistry of Michael Polanyi." In *Instruments and Experimentation in the History of Chemistry*, edited by F. L. Holmes and Trevor Levere, 367–400. Cambridge, MA: MIT Press.

———. 2000b. "Physical and Biological Modes of Thought in the Chemistry of Linus Pauling." *Studies in History and Philosophy of Modern Physics* 31:475–91.

———. 2001. "At the Boundaries: Michael Polanyi's Work on Surfaces and the Solid State." In *Chemical Sciences in the Twentieth Century*, edited by Carsten Reinhardt, 246–57. Berlin: Wiley-VCH.

———. 2002. "Michael Polanyi's Theory of Surface Adsorption: How Premature?" In *Prematurity and Scientific Discovery*, edited by Ernest B. Hook, 151–63. Berkeley: University of California Press.

————, ed. 2003. *The Modern Physical and Mathematical Sciences*. Vol. 5, *The Cambridge History of Science*. Cambridge: Cambridge University Press.

————. 2004. *Blackett: Physics, War and Politics in the Twentieth Century*. Cambridge, MA: Harvard University Press.

————. 2007a. "Working Tools for Theoretical Chemistry: Polanyi, Eyring and Debates over the 'Semi-Empirical Method.'" *Journal of Computational Chemistry* 28:98–108.

————. 2007b. "Historical Sources of Science-as-Social Practice: Michael Polanyi's Berlin." *Historical Studies in the Physical and Biological Sciences* 37 (2): 411–36.

————. 2008. "Re-reading Bernal: History of Science at the Crossroads in 20th-Century Britain." In *Aurora Torealis: Studies in the History of Science and Ideas in Honor of Tore Frängsmyr*, edited by Marco Beretta, Karl Grandin, and Svante Lindqvist, 237–60. Sagamore Beach, MA: Science History Publications.

————. 2010. "Science and Politics in the Philosophy of Science: Popper, Kuhn, and Polanyi." In *Science as Cultural Practice*. Vol. 1. *Cultures and Politics of Research from the Early Modern Period to the Age of Extremes*, edited by Moritz Epple and Claus Zittel, 201–16. Berlin: Akademie Verlag.

Oakeshott, Michael. 1958. "The Human Coefficient." *Encounter* 11:77–80.

Ogg, R. A. Jr., and M. Polanyi. 1935a. "Mechanism of Ionic Reactions." *TFS* 31:604–20.

————. 1935b. "Substitution by Free Atoms and Walden Inversion: the Decomposition and Racemisation of Optically Active sec-Butyl Iodide in the Gaseous State." *TFS* 31:482–95.

Olby, Robert. 1974. *The Path to the Double Helix: The Discovery of DNA*. New York: Dover.

Olby, Robert, G. N. Cantor, J. R. R. Christie, and M. J. S. Hodge, eds. 1992. *Companion to the History of Modern Science*. London: Routledge.

Olesko, Kathryn M. 1991. *Physics as a Calling: Discipline and Practice in the Königsberg Seminar for Physics*. Ithaca, NY: Cornell University Press.

Orowan, E. 1934. "Zur Kristallplastizität." *ZP* 89:605–59.

————. 1965. "Dislocations in Plasticity." In *The Sorby Centennial Symposium on the History of Metallurgy*, edited by C. S. Smith, 359–76. New York: Gordon and Breach.

Ortolano, Guy. 2009. *The Two Cultures Controversy: Science, Literature and Cultural Politics in Postwar Britain*. Cambridge: Cambridge University Press.

Paletschek, Sylvia. 2001a. "The Invention of Humboldt and the Impact of National Socialism: The German University Idea in the First Half of the Twentieth Century." In *Science in the Third Reich*, edited by Margit Szollösi-Janze, 37–58. Oxford: Berg.

————. 2001b. "Verebreitete sich ein Humboldtsches Modell an den deutschen Universitäten im 19. Jahrhundert?" In *Humboldt International: Der Export des deutschen Universitätsmodells im 19 und 20 Jahrhundert*, edited by Rainer C. Schwinges, 75–104. Basel: Schwabe.

Palló, Gábor. 1990. "Scientists' First Step of Emigration: From the Hungarian Periphery to the Centre." *Periodica Polytechnica* 34:319–23.

————. 1991. "Hungarians' Second Step of Emigration: Toward the New Centers." *Periodica Polytechnica* 35:78–86.

————. 1996a. "Polányi Mihály és a Kriptonlámpa" [Michael Polanyi and the krypton lamp]. *Fizikai Szemle* 9: 311–16.

————. 1996b. "Párhuzamok és metszéspontok: Lakatos Imre és Polányi Mihály" [Parallels and intersections: Imre Lakatos and Michael Polanyi]. *Replika* (23–24): 39–50.

————. 1998. "Michael Polányi's Early Years in Science." *Bulletin of the History of Chemistry* 21:39–43.

————. 2005. "Scientific Creativity in Hungarian Context." *Hungarian Studies* 19:215–31.

———. 2010. "The Advantage and Disadvantage of Peripheral Ignorance: The Gas Adsorption Controversy." *Ambix* 57:216–30.

Paneth, F. A. 1916. "Über den Element-und Atombegriff in Chemie und Radiologie." *ZpC* 91:171–98.

Paneth, F. A. and G. von Hevesy. 1914. "Zur Frage der isotopen Elemente." *Physikalische Zeitschrift* 15:797–805.

Park, Buhm Soon. 2009. "Between Accuracy and Manageability: Computational Imperatives in Quantum Chemistry." *Historical Studies in the Natural Sciences* 39:32–62.

Parsons, Keith, ed. 2003. *The Science Wars: Debating Scientific Knowledge and Technology.* Amherst, NY: Prometheus Books.

Pauling, Linus. 1939. *The Nature of the Chemical Bond and the Structure of Molecules and Crystals.* Ithaca, NY: Cornell University Press.

Pauling, Linus, and E. Bright Wilson. 1935. *Introduction to Quantum Mechanics with Applications to Chemistry.* New York: McGraw-Hill.

Pechukas, Philip. 1982. "Recent Developments in Transition State Theory." In Bunsen Gesellschaft 1982, 372–78.

Pelzer, H., and E. Wigner. 1932. "Über die Geschwindigkeitskonstante von Austauschreaktionem." *ZpC* B15:445–71.

Perutz, Max F. 1981. "A Sagacious Scientist." *New Scientist* 90:39–40.

———. 1990. "How Lawrence Bragg Invented X-Ray Analysis." *Proceedings of the Royal Institution* 62:183–98.

———. 1996. "The Cabinet of Dr. Haber." *New York Review of Books*, June 20, 31–36.

Pickstone, John V., ed. 2007. *The History of Science and Technology in the North West.* Vol. 18, *Manchester Region History Review.* Manchester, UK: Manchester Centre for Regional History.

Plesch, Peter H. 1946. "Researches into the Catalyzed Low-Temperature Polymerisation of Isobutene." PhD diss., University of Manchester.

———. 2007–2008. "On Working with Michael Polanyi." *Tradition and Discovery* 32 (2): 39–48.

Poincaré, Henri. 1913. *The Foundations of Science.* Translated by George B. Halsted. New York: Science Press.

Polanyi, John C. 2003. "Michael Polanyi, the Scientist." *Polanyiana* 2003 (1): 117–21.

Polanyi, Karl. 1922. "Sozialistische Rehnungslegung." *Archiv für Sozialwissenschaft und Sozialpolitik* 49:377–418.

———. 1925. "Die funktionelle Theorie der Gesellschaft und das problem der sozialistischen Rechnungslegung." *Archiv für Sozialwissenschaft und Sozialpolitik* 52: 218–28.

———. 1932–1933. "Wirtschaft und Demokratie." *Oesterreichische Volkswirt* 25:301–3.

———. 1937. "Europe Today." Preface by G. D. H. Cole. London: Workers' Educational Trade Union Committee.

———. 1944. *The Great Transformation: The Political and Economic Origins of Our Time.* New York: Rinehart and Company.

Polányi, Livia. 1995. "Cornucopions of History: A Memoir of Science and the Politics of Private Lives." In *Technoscientific Imaginaries: Conversations, Profiles, and Memoirs*, edited by George E. Marcus, 1–28. Chicago: University of Chicago Press.

Polanyi, Magda. 1956. *Dictionary of Textile Terms. Textil-Fachwörterbuch. German-English, English-German.* London: Pergamon.

———. 1967. *Dictionary of Textile Terms.* 2nd rev. enlarged ed. Oxford: Pergamon.

Polanyi, Michael. 1911a. "Beitrag zur Chemie der Hydrocephalusflüssigkeit." *Biochemische Zeitschrift* 34:205–10.

———. 1911b. "Untersuchungen über die Veränderung der physikalischen und chemischen Eigenschaften des Blutserums während des Hungerns." *Biochemische Zeitschrift* 34:192–204.

———. 1913a. "Eine neue thermodynamische Folgerung aus der Quantenhypothese." *VdpG* 15:156–61.

———. 1913b. "Neue thermodynamische Folgerungen aus der Quantenhypothese." *ZpC* 83:339–69.

———. 1914. "Über die Adsorption von Standpunkt des dritten Wärmesatzes." *VdpG* 16:1012–16.

———. 1916. "Adsorption von Gasen (Dämpfen) durch ein festes nichtpflüchtiges Adsorbens." *VdpG* 18:55–80.

———. 1917a. "To the Peacemakers." Translated by Endre J. Nagy in Polanyi 1997b, 15–28.

———. 1917b. "Gázok absorptiója szilárd, nem illanó adszorbensen" [Adsorption of gases by a solid non-volatile adsorbent]. PhD thesis, University of Budapest.

———. 1920a. "Zum Problem der Reaktionsgeschwindigkeit." *ZE* 26:223–331.

———. 1920b. "Über die nichtmechanische Natur der chemischen Vorgänge." *ZP* 1:337–44.

———. 1920c. "Zur Theorie der Reaktionsgeschwindigkeit." *ZP* 2:90–110.

———. 1920d. "Zum Ursprung der chemischen Energie." *ZP* 3:31–35.

———. 1921a. "Briefe." *Die Naturwissenschaften* 9:288.

———. 1921b. "Faserstruktur im Röntgenlichte." *Naturwissenschaften* 9:337–40.

———. 1921c. "Das Röntgen-Faserdiagramm." *ZP* 7:149–80.

———. 1928a. "[Reply to] 'Irrtümliche: Bestimmung des Zellulose-Raumgitters.'" *Naturwissenschaften* 15:263–64.

———. 1928b. "Geheimrat Fritz Haber Vollendet am 9 Dezember sein 60. Lebensjahr." *Metallwirtschaft* 7 (49): 1316–17.

———. 1929a. "Grundlagen der Potentialtheorie der Adsorption." *ZE* 35:431–32.

———. 1929b. "Betrachtungen über den Aktivierungsvorgang an Grenzflächen." *ZE* 35: 561–67.

———. 1930. "Förderung der Wissenschaft," *Deutsche Volkswirt* 5:1149–51.

———. 1931. "Über einfache Gasreaktionen." *ZpC* 12B:279–311.

———. 1932a. "Introductory Paper to Section III. Theories of the Adsorption of Gases. A General Survey and Some General Remarks." *TFS* 28:316–33.

———. 1932b. *Atomic Reactions.* London: Williams and Norgate.

———. 1934. "Über eine Art von Gitterstörung, die einen Kristall plastisch machen könnte." *ZP* 89:660–64.

———. 1935a. "Adsorption and Catalysis." *Journal of the Society of Chemical Industry* 54: 123–24.

———. 1935b. "USSR Economics—Fundamental Data, System, and Spirit." *Manchester School of Economic and Social Studies* 6:67–89.

———. 1936a. "The Value of the Inexact." *Philosophy of Science* 3:233–34.

———. 1936b. "Upon the Position of Jews: Address to the Jewish Medical Society of Liverpool, January 21, 1936." Typescript, MPP 25:8.

———. 1938. "An Outline of the Working of Money." Manchester: Norbury, Lockwood and Co.

——. 1939. "The Rights and Duties of Science." In Polanyi 1997b, 61–78.

——. 1940a. "Collectivist Planning." In Polanyi 1997b, 121–43.

——. 1940b. *The Contempt of Freedom: The Russian Experiment and After.* London: Watts, 1940.

——. 1941. "Cultural Significance of Science." *Nature* 147:119.

——. 1942. "Anti-Semitism." *New Statesman and Nation,* June 27. Newspaper clipping, MPP 47:2.

——. 1943a. "Jewish Problems." In Polanyi 1997b, 33–46.

——. 1943b. "The Autonomy of Science." *Memoirs and Proceedings of the Manchester Literary and Philosophical Society* 85:19–38.

——. 1944a. "Patent Reform." *Review of Economic Studies* 11:61–76.

——. 1944b. "Science and the Decline of Freedom." *Listener,* June 1, 599.

——. 1945a. "Rights and Duties of Science." *SFS Occasional Pamphlet* 2 (June).

——. 1945b. *Full Employment and Free Trade.* Cambridge: Cambridge University Press.

——. 1946a. "The Planning of Science," *SFS Occasional Pamphlet* 4 (February).

——. 1946b. "Authority and Conscience." In Polanyi 1964, 42–62.

——. 1946c. "Dedication or Servitude." In Polanyi 1964, 63–84.

——. 1946d. "Soviets and Capitalism: What Is the Difference?" *Time and Tide,* April 6, 317.

——. 1946e. "Profits and Polycentricity." In Polanyi 1951, 138–53.

——. 1947. "The Foundations of Academic Freedom." *SFS Occasional Pamphlet* 6 (September).

——. 1948. "Profits and Enterprise." In Polanyi 1997b, 145–56.

——. 1949. "Ought Science to Be Planned? The Case of Individualism." In Polanyi 1951, 111–37.

——. 1950. "Scientific Beliefs." *Ethics* 61:27–37.

——. 1951. *The Logic of Liberty.* London: Routledge and Kegan Paul, 1951.

——. 1951–1952. "Hypothesis of Cybernetics." *British Journal for the Philosophy of Science* 2: 312–15.

——. 1952a. "Some British Experiences. I. Michael Polanyi." *Bulletin of the Atomic Scientists* 8:223–29.

——. 1952b. "The Stability of Beliefs." *British Journal for the Philosophy of Science* 3: 217–32.

——. 1953. "Pure and Applied Science and Their Appropriate Forms of Organization." *SFS Occasional Pamphlet* 14 (December).

——. 1957. "Scientific Outlook: Its Sickness and Cure." *Science* 125:480–84.

——. 1958. *Personal Knowledge: Towards a Post-Critical Philosophy.* Chicago: University of Chicago Press.

——. 1959a. "The Two Cultures." In Polanyi 1969a, 40–46.

——. 1959b. *The Study of Man.* Chicago: University of Chicago Press.

——. 1961. "Commentary by Michael Polanyi." In Crombie 1963, 375–80.

——. 1962a. "The Republic of Science: Its Political and Economic Theory." *Minerva* 1 (1962): 54–74.

——. 1962b. "My Time with X-Rays and Crystals." In Polanyi 1969a, 97–104.

——. 1963. "The Potential Theory of Adsorption: Authority in Science Has Its Uses and Dangers." *Science* 141:1010–13.

——. 1964. *Science, Faith and Society.* The Riddell Memorial Lectures, University of Durham. Chicago: University of Chicago Press. Originally published by Oxford University Press, 1946.

———. 1969a. *Knowing and Being: Essays by Michael Polanyi*. Edited by Marjorie Grene. London: Routledge and Kegan Paul.

———. 1969b. "The Determinants of Social Action." In *Roads to Freedom: Festschrift for F. A. von Hayek*, edited by E. Stressler, 165–69. London: Routledge.

———. 1970. "Reflections on Viewing a Painting." Foreword to *Optics, Painting and Photography*, by Henri Pirenne. Cambridge: Cambridge University Press.

———. 1972. "Genius in Science." *Encounter* 38 (1): 43–50.

———. 1997a. "The Struggle between Truth and Propaganda." In Polanyi 1997b, 47–60.

———. 1997b. *Society, Economics and Philosophy: Selected Papers*. Edited by R. T. Allen. New Brunswick, NJ: Transaction Publishers.

———. 2009. *The Tacit Dimension. The Terry Lectures, Yale University, 1962*. Chicago: University of Chicago Press. Originally published London: Routledge and Kegan Paul, 1966.

Polanyi, Michael, and Harry Prosch. 1975. *Meaning*. Chicago: University of Chicago Press.

Polanyi, Michael, E. Schiebold, and K. Weissenberg. 1924. "Über die Entwicklung des Drehkristallverfahrens." *ZP* 23:337–40.

Polanyi, Michael, and K. Weissenberg. 1922. "Das Roentgen-Faserdiagramm." *ZP* 10:44–53.

Polanyi, Michael, and K. Welke. 1928. "Adsorption, Adsorptionswärme und Bindungscharakter von Schwefeldioxyd an Kohle bei geringen Belegungen." *ZpC* 132:371–83.

Polanyi-Levitt, Kari, ed. 1990a. *The Life and Work of Karl Polanyi: A Celebration*. Montreal: Black Rose Books.

———. 1990b. "The Origins and Significance of *The Great Transformation*." In Polanyi Levitt 1990a, 111–24.

———. 2007. "Preface: The English Experience in the Life and Work of Karl Polanyi." In *Karl Polanyi: New Perspectives on the Place of the Economy in Society*, edited by Mark Harvey et al., xi–xvi. Manchester, UK: Manchester University Press.

Popper, Karl. 1959. *The Logic of Scientific Discovery*. London: Hutchinson.

———. 1962. *Conjectures and Refutations: The Growth of Scientific Knowledge*. New York: Basic Books.

———. 1974. "Replies to My Critics." In *The Philosophy of Karl Popper*, edited by Paul Arthur Schilpp, 961–1197. Vol. 2. La Salle, IL: Open Court.

———. 1976. *Unended Quest: An Intellectual Biography*. La Salle and London: Open Court.

———. 2002. *The Logic of Scientific Discovery*. New York: Routledge Classics.

———. 2003. *The Open Society and Its Enemies*. Vol. 2., *Hegel and Marx*. New York: Routledge Classics.

———. 2004. *The Poverty of Historicism*. London and New York: Routledge.

Porter, George. 1992. "Chemistry in Microtime." In *The Chemical Bond: Structure and Dynamics*, edited by Ahmed Zewail, 113–48. San Diego, CA: Academic Press.

Porter, Roy. 1992. "The History of Science and the History of Society." In *Companion to the History of Modern Science*, edited by Robert Olby et al., 32–46. London: Routledge.

Price, Derek de Solla. 1963. *Little Science, Big Science*. New York: Columbia University Press.

Proctor, Robert N. 1991. *Value-Free Science? Purity and Power in Modern Knowledge*. Cambridge, MA: Harvard University Press.

Pugh, Martin. 1993. *The Making of Modern British Politics 1867–1939*. Oxford: Blackwell.

Pyenson, Lewis. 1983. *Neohumanism and the Persistence of Pure Mathematics in Wilhelmian*

*Germany.* Vol. 150, *American Philosophical Society Memoirs.* Philadelphia: American Philosophical Society.

Quine, W. V. O. 1951. "Two Dogmas of Empiricism." *The Philosophical Review* 60: 20–43.

Rabinbach, Anson. 1983. *The Crisis of Austrian Socialism: From Red Vienna to Civil War, 1927–1934.* Chicago: University of Chicago Press.

Radovic, Ljubisa R. 2002. "Polanyi's Adsorption Potential (Affinity): Concept Development From Pre- to Post-Maturity and Beyond." Manuscript copy, author's personal collection.

Ramsey, Jeffry L. 1997a. "Between the Fundamental and the Phenomenological: The Challenge of 'Semi-Empirical Methods.'" *Philosophy of Science* 64:627–53.

——. 1997b. "Molecular Shape, Reduction, Explanation, and Approximate Concepts." *Synthese* 111:233–51.

——. 1999. "Interpreting the 'Mona Lisa' of Chemical Reactions: Explanation, Mechanism, and Methodological Values." History of Science Society Annual Meeting, Pittsburgh, PA, October. Manuscript copy, author's personal collection.

——. 2006. "Philosophy of Chemistry." In *The Philosophy of Science: An Encyclopedia,* edited by Sahotra Sarkar and Jessica Pfeifer, 101–6. New York: Routledge.

Rauschning, Hermann. 1939. *Hitler Speaks: A Series of Political Conversations with Hitler on His Real Aims.* London: Butterworth.

Ravetz, Jerome R. 1971. *Scientific Knowledge and Its Social Problems.* Oxford: Oxford University Press.

Rawlins, F. I. G. 1958. "Detachment Relinquished." *Nature* 182 (4638): 757.

Reich, Leonard. 1985. *The Making of American Industrial Research: Science and Business at GE and Bell, 1876–1926.* Cambridge: Cambridge University Press.

Reichenbach, Hans. 1929. "Neue Wege der Wissenschaft: Philosophische Forschung." Reprinted as "New Approaches in Science: Philosophical Research." In *Hans Reichenbach: Selected Essays, 1909–1953,* edited and translated by Maria Reichenbach and Robert S. Cohen, 1:249–53. Dordrecht: Reidel.

——. 1938. *Experience and Prediction: An Analysis of the Foundations of the Structure of Knowledge.* Chicago: University of Chicago Press.

Reinisch, Jessica. 2000. "The Society of Freedom in Science, 1940–1963." Master's thesis, University of London.

Reis, Alfred. 1920. "Zur Kenntnis der Kristallgitter." *ZP* 1:204–20.

Reisch, George A. 2006. *How the Cold War Transformed Philosophy of Science: To the Icy Slopes of Logic.* Cambridge: Cambridge University Press.

Rhodes, Richard. 1986. *The Making of the Atomic Bomb.* New York: Touchstone Press.

Richardson, Alan. 2002. "Engineering Philosophy of Science: American Pragmatism and Logical Empiricism in the 1930s." *Philosophy of Science* 59:S36–S47.

——. 2006. "Remarks on George Reisch's *How the Cold War Transformed Philosophy of Science.*" Paper from the Philosophy of Science Association Meeting, Vancouver, BC, November.

——. 2008. "Scientific Philosophy as a Topic for History of Science." *Isis* 99:88–96.

——. 2009. "Philosophy and/as Science: Analytic Philosophy as Marginal Science." Vienna International Summer School Lecture, July 24.

Richardson, Alan, and Thomas Uebel, eds. 2008. *The Cambridge Companion to Logical Empiricism.* Cambridge: Cambridge University Press.

Rider, Robin E. 1984. "Alarm and Opportunity: Emigration of Mathematicians and Physicists to Britain and the United States," *HSPS* 15:107–76.

Ringer, Fritz K. 1969. *The Decline of the German Mandarins: The German Academic Community, 1890–1933*. Cambridge, MA: Harvard University Press.

———. 1978. *Education and Society in Modern Europe*. Bloomington: Indiana University Press.

———. 1990. "The Intellectual Field, Intellectual History, and the Sociology of Knowledge." *Theory and Society* 19:269–94.

———. 2004. *Max Weber: An Intellectual Biography*. Chicago: University of Chicago Press.

Roberts, Gerrylynn K. 1980. "The Liberally-Educated Chemist: Chemistry in the Cambridge Natural Sciences Tripos, 1851–1914." *HSPS* 11:157–83.

Roberts, Paul Craig. 2005. "Polanyi the Economist." In Jacobs and Allen 2005, 127–32.

Robinson, Robert. 1976. *Memoirs of a Minor Prophet: Seventy Years of Organic Chemistry*. London: Elsevier.

Rocke, Alan J. 2000. *Nationalizing Science: Adolphe Wurtz and the Battle for French Chemistry*. Cambridge, MA: MIT Press.

Roll-Hansen, Nils. 2005. *The Lysenko Effect: The Politics of Science*. New York: Humanities Books.

Rose, Hilary, and Steven Rose. 1970. *Science and Society*. Harmondsworth, UK: Penguin.

———. 1999. "Red Scientist: Two Strands from a Life in Three Colors."In Swann and Aprahamian 1999, 132–59.

Rosner, Peter. 1990. "Karl Polanyi on Socialist Accounting." In Polanyi-Levitt 1990a, 55–65.

Rowe, David E., and Robert Schulmann. 2007. *Einstein on Politics: His Private Thoughts and Public Stands on Nationalism, Zionism, War, Peace, and the Bomb*. Princeton, NJ: Princeton University Press.

Russell, Bertrand. 1952. *The Impact of Science on Society*. London: Allen and Unwin.

Russian Chemical Bulletin. 1993. "Mikhail Mikhailovich Dubinin." *Russian Chemical Bulletin* 42:1282.

Ruthenberg, Klaus. 1997. "Friedrich Adolf Paneth (1887–1958)." *HYLE* 3:103–6.

———. 2009. "Paneth, Kant, and the Philosophy of Chemistry." *Foundations of Chemistry* 11:79–91.

Rutherford, Malcolm. 2004. "Institutional Economics at Columbia University." *History of Political Economy* 36:31–78.

Sachse, Carola. 2009. "What Research, to What End? The Rockefeller Foundation and the Max Planck Gesellschaft in the Early Cold War." *Central European History* 42:97–141.

Sachse, Carola, and Mark Walker. 2005a. "Introduction: A Comparative Perspective." In *Politics and Science in Wartime: Comparative International Perspectives on the Kaiser Wilhelm Institute*, by Carola Sachse and Mark Walker, 1–20. *Osiris*, vol. 20. Chicago: University of Chicago Press.

———. 2005b. *Politics and Science in Wartime: Comparative International Perspectives on the Kaiser Wilhelm Institute*. *Osiris*, vol. 20. Chicago: University of Chicago Press.

Sanderson, Michael. 1972. *The Universities and British Industry, 1850–1970*. London: Routledge and Kegan Paul.

Sands, Donald E. 1975. *Introduction to Crystallography*. New York: Dover.

Sarton, George. 1931. *The History of Science and the New Humanism*. New York: H. Holt.

Saunders, Frances Stonor. 1999. *The Cultural Cold War: The CIA and the World of Arts and Letters*. New York: New Press.

Scerri, Eric. 1994. "Has Chemistry Been at Least Approximately Reduced to Quantum Mechanics?" *PSA* 1:160–74.

————. 2000. "Realism, Reduction, and the 'Intermediate Position.'" In *Of Minds and Molecules: New Philosophical Perspectives on Chemistry*, edited by Nalini Bhushan and Stuart Rosenfeld, 51–72. Oxford: Oxford University Press.

Schabas, Margaret. 2005. *The Natural Origins of Economics*. Chicago: University of Chicago Press.

Scherrer, Paul. 1962. "Reminiscences." In Ewald 1962, 642–46.

Schiebold, E. 1922. "Bemerkungen zur Arbeit: Das Roentgenfaserdiagramm von M. Polanyi." *ZP* 9:180–83.

Schrecker, Ellen. 1998. *Many Are the Crimes: McCarthyism in America*. Princeton, NJ: Princeton University Press.

Schroeder-Gudehus, Brigitte. 1966. *Deutsche Wissenschaft und internationale Zusammenarbeit, 1914–1918*. Geneva: Dumaret and Golay.

————. 1972. "The Argument for the Self-Government and Public Support of Science in Weimar Germany." *Minerva* 10:537–70.

————. 1978. *Les scientifiques et la paix: La communauté scientifique international au cours des années 20*. Montreal: Presses de l'Université de Montréal.

Schüring, Michael. 2006. "Expulsion, Compensation, and the Legacy of the Kaiser Wilhelm Society." *Minerva* 44 (3): 307–24.

Schwab, Georg-Maria. 1937. *Catalysis from the Standpoint of Chemical Catalysis*. Translated from the first German edition by Hugh S. Taylor and R. Spence. New York: Van Nostrand.

Schwarz, W. H. E., D. Andraea, S. R. Arnold, J. Heidberg, H. Hellmann, J. Hinze, A. Karachalios, M. A. Kovner, P. C. Schmidt, and L. Zülicke. 1999. "Hans G. A. Hellmann (1903–1938)." Translated by Mark Smith and W. H. E. Schwarz, with revisions by J. Hinze and A. Karachalios. *Bunsen-Magazin* 1:10–21, 2:60–70.

Schweber, S. S. 2000. *In the Shadow of the Bomb: Bethe, Oppenheimer, and the Moral Responsibility of the Scientist*. Princeton, NJ: Princeton University Press.

Schwinges, Rainer C., ed. 2001. *Humboldt International: Der Export des deutschen Universitätsmodells im 19 und 20 Jahrhundert*. Basel: Schwabe.

Scott, William T. 1982. "The Question of a Religious Reality: Commentary on the Polanyi Papers." *Zygon* 17:83–87.

————. 1983. "Michael Polanyi's Creativity in Chemistry." In *Springs of Scientific Creativity*, edited by Rutherford Aris et al., 279–307. Minneapolis: University of Minnesota Press.

————. 1998–1999. "At the Wheel of the World: The Life and Times of Michael Polanyi." *Tradition and Discovery: The Polanyi Society Periodical* 25 (3): 10–25.

Scott, William T., and Martin X. Moleski, S.J. 2005. *Michael Polanyi: Scientist and Philosopher*. Oxford: Oxford University Press.

Seguin, Eve. 2000. "Bloor, Latour, and the Field." *Studies in the History and Philosophy of Science* 31:503–8.

Seitz, Frederick, Erich Vogt, and Alvin M. Weinberg. 1995. "Eugene P. Wigner, November 17, 1902–January 1, 1995." *National Academy of Sciences Biographical Memoirs* 74:1–26.

Semenov, Nikolai N. 1935. *Chemical Kinetics and Chain Reactions*. Translated by Y. Shmidt-Chernuisheva and J. I. Frenkel. Oxford: Clarendon Press.

Service, R. F. 1996. "Materials Scientists View Hot Wires and Bends by the Bay." *Science* 272:484–85.

Shapin, Steven. 1982. "History of Science and its Sociological Reconstructions," *History of Science* 20:157–211.

———. 1992. "Discipline and Bounding: The History and Sociology of Science as Seen through the Externalism-Internalism Debate." *History of Science* 30:333–69.

———.1999. Foreword to the 1999 edition, *The Politics of Pure Science*, by Daniel S. Greenberg, xv–xxii. Chicago: University of Chicago Press.

———. 2008. *A Scientific Life: A Moral History of a Late Modern Vocation*. Chicago: University of Chicago Press.

———. 2010a. "The Darwin Show." *London Review of Books*, January 7, 3–9.

———. 2010b. *Never Pure: Historical Studies of Science as if It Was Produced by People with Bodies, Situated in Time, Space, Culture, and Society, and Struggling for Credibility and Authority*. Baltimore, MD: The Johns Hopkins University Press.

Shils, Edward. 1952. "Editorial: America's Paper Curtain." *Bulletin of the Atomic Scientists* 8:210–17.

———. 1954. "The Scientific Community: Thoughts after Hamburg." *Bulletin of the Atomic Scientists* 10:151–55.

———. 1973. "*Ideology and Utopia* by Karl Mannheim." *Daedalus* 103 (Special Issue: *Twentieth-Century Classics Revisited*): 83–89.

———. 1995–1996. "On the Tradition of Intellectuals: Authority and Antinomianism according to Michael Polanyi." *Tradition and Discovery* 22 (2): 10–26.

Shinn, Terry. 1979. "The French Science Faculty System 1808–1914: Institutional Change and Research Potential." *HSPS* 19:271–332.

Shinn, Terry, and Pascoul Ragouet. 2005. *Controverses sur la science: Pour une sociologie transversaliste de l'activité scientifique*. Paris: Raison d'agir Editions.

Siegmund-Schultze, Reinhard. 2009. *Mathematicians Fleeing from Nazi Germany*. Princeton, NJ: Princeton University Press.

Sime, Ruth Lewin. 1996. *Lise Meitner: A Life in Physics*. Berkeley: University of California Press.

Simões, Ana. 2003. "Chemical Physics and Quantum Chemistry in the Twentieth Century." In Nye 2003, 394–412.

———. 2004. "Textbooks, Popular Lectures and Sermons: The Quantum Chemist Charles Alfred Coulson and the Crafting of Science." *BJHS* 37:299–342.

Singer, Peter. 2003. *Pushing Time Away: My Grandfather and the Tragedy of Jewish Vienna*. New York: HarperCollins.

Sismondo, Sergio. 2004. *An Introduction to Science and Technology Studies*. Oxford: Blackwell.

———. 2008. "Science and Technology Studies and an Engaged Program." In Hackett et al. 2008, 13–32.

Smith, Alice Kimball. 1965. *A Peril and a Hope: The Scientists' Movement in America, 1945–1947*. Chicago: University of Chicago Press.

Snow, C. P. 1993. *The Two Cultures*. Canto ed. Cambridge: Cambridge University Press.

Snow, Joel A. 1968. "The Politics of Pure Science." *Bulletin of the Atomic Scientists* 24:34–36.

Society for Freedom in Science. 1951. "Speeches Made at the Dinner Held to Celebrate the Tenth Anniversary of the Foundation of the Society." *SFS Occasional Pamphlet* 11 (March): 4–11.

Söderbaum, H. G. 1934. "The Nobel Prize for Chemistry." In *Les Prix Nobel en 1932*, by Nobelstiftelsen [Nobel Foundation], 21–24. Stockholm: Imprimerie Royale.

Solomon, Miriam. 2008. "STS and Social Epistemology of Science." In Hackett et al. 2008, 141–258.

Somlyody, Laszlo, and Nora Somlyody, eds. 2003. *Hungarian Arts and Sciences 1848–2000*. New York: Columbia University Press.

Spiegel, Henry William. 1971. *The Growth of Economic Thought*. Rev. ed. Durham, NC: Duke University Press.

Sponsler, Olenus Lee, and Walter Harrington Dore. 1926. "The Structure of Ramie Cellulose as Derived from X-Ray Data." *Colloid Symposium Monographs* 4:174–202.

Stadler, Friedrich. 1997. *Studien zum Wiener Kreis: Ursprung, Entwicklung und Wirkung des Logischen Empirismus im Kontext*. Frankfurt: Suhrkamp.

———. 2001. *The Vienna Circle: Studies in the Origins, Development, and Influence of Logical Empiricism*. Translated by Camilla Nielsen et al. Vienna: Springer.

———, ed. 2003. *The Vienna Circle and Logical Empiricism: Re-evaluation and Future Perspectives*. Dordrecht: Kluwer.

Stent, Gunther S. 1972. "Prematurity and Uniqueness in Scientific Discovery." In *Advances in Biosciences* 8:433–449.

———. 2002. "Prematurity in Scientific Discovery." In *Prematurity in Scientific Discovery: On Resistance and Neglect*, edited by Ernest B. Hook, 22–33. Berkeley: University of California Press.

Stern, Fritz. 1999a. "Together and Apart: Fritz Haber and Albert Einstein." In *Einstein's German World*, 59–164. Princeton, NJ: Princeton University Press.

———. 1999b. *Einstein's German World*. Princeton, NJ: Princeton University Press.

———. 2006. *Five Germanys I Have Known*. New York: Farrar, Straus and Giroux.

Stolper, Toni. 1960. *Ein Leben in Brennpunkten Unserer Zeit: Wien, Berlin, New York, Gustav Stolper 1888–1947*. Tübingen: Rainer Wunderlich.

Stoltzenberg, Dietrich. 1994. *Fritz Haber. Chemiker, Nobelpreisträger, Deutscher, Jude*. Weinheim: VCH.

———. 2004. *Fritz Haber: Chemist, Nobel Laureate, German, Jew*. Philadelphia, PA: Chemical Heritage Press.

Stöltzner, Michael, and Thomas E. Uebel, eds. 2006. *Wiener Kreis: Texte zur Wissenschaftlicher Weltauffassung von Rudolf Carnap, Otto Neurath, Moritz Schlick, Phillip Frank, Hans Hahn, Karl Menger, Edgar Zilsel und Gustav Bergmann*. Hamburg: Meiner.

Storer, Norman W. 1973. "Introduction." In Merton 1973b, xi–xxxi.

Stressler, Erich, ed. 1969. *Roads to Freedom: Festschrift for F. A. von Hayek*. London: Routledge.

Striker, Barbara. 1999. "Re: 'Laura Polanyi 1882–1959: Narratives of a Life by Judit Szapor.'" *Polanyiana* 8 (1–2). Accessed July 3, 2010. http://www.kfki.hu/chemonet/polanyi/9912/striker.html.

Stuewer, Roger H. 1975. *The Compton Effect: Turning Point in Physics*. New York: Science History Publications.

Sugiura, Y. 1927. "Über die Eigenschaften der Wasserstoffmoleküls um Grundzustände." *ZP* 45:484–92.

Suits, C. Guy, and Miles J. Martin. 1974. "Irving Langmuir, January 31, 1881–August 16, 1957." *National Academy of Sciences Biographical Memoirs* 45:215–47.

Swann, Brenda, and Francis Aprahamian, eds. 1999. *J. D. Bernal: A Life in Science and Politics*. London: Verso.

Synge, Anne. 1999. "Early Years and Influence." In Swann and Aprahamian 1999, 1–15.

Szapor, Judit. 1997. "Laura Polanyi 1882–1957: Narratives of a Life." *Polanyiana* 6 (2). Accessed April 16, 2010. http://www.kfki.hu/chemonet/polanyi/9702/szapor.html.

Szilard, Leo. 1969. "Reminiscences." Edited by Gertrud Weiss Szilard and Kathleen R. Winser. In Fleming and Bailyn 1969, 94–151.

———. 1978. *Leo Szilard: His Version of the Facts: Special Recollections and Correspondence*, edited by Spencer R. Weart and Gertrud W. Szilard. Cambridge, MA: MIT Press.

Szöllösi-Janze, Margit. 1998. *Fritz Haber 1868–1934: Eine Biographie*. Munich: C. H. Beck.

Tansley, A. G. 1942. *The Values of Science to Humanity*. London: G. Allen and Unwin.

Taylor, F. Sherwood. 1945. "Is the Progress of Science Controlled by the Material Wants of Man?" *SFS Occasional Pamphlet* 1 (April).

Taylor, G. I. 1934. "The Mechanism of Plastic Deformation of Crystals." *Proceedings of the Royal Society* A145:362–415.

Taylor, Hugh S. 1959. "Personal Knowledge." *American Scientist* 47:52A–52B.

———. 1962. "Fifty Years of Chemical Kineticists." *Annual Review of Physical Chemistry* 13:1–18.

Taylor, Hugh S., ed. 1931. *A Treatise on Physical Chemistry: A Cooperative Effort by a Group of Physical Chemists* 2nd ed. New York: Van Nostrand.

Taylor, Hugh S., and Samuel Glasstone, ed. 1951. *A Treatise on Physical Chemistry*. 5 vols. Vol. 3., *States of Matter*. 3rd ed. New York: Van Nostrand.

Teller, Edward. 2001. *Memoirs: A Twentieth-Century Journey in Science and Politics*. With Judith L. Shoolery. Cambridge, MA: Perseus.

Thorpe, Charles. 2001. "Science against Modernism: The Relevance of the Social Theory of Michael Polanyi." *British Journal of Sociology* 52:19–35.

———. 2009. "Community and Market in Michael Polanyi's Philosophy of Science." *Modern Intellectual History* 6:59–89.

Todd, Alexander. 1983. *A Time to Remember: The Autobiography of a Chemist*. Cambridge: Cambridge University Press.

Tönnies, Ferdinand. 1887. *Gemeinschaft und Gesellschaft: Abhandlung des Communismus und des Socialismus als empirischer Culturformen*. Leipzig: Fues.

Torpey, John. 2000. *The Invention of the Passport: Surveillance, Citizenship and the State*. Cambridge: Cambridge University Press.

Toulmin, Stephen. 1958. *The Uses of Argument*. Cambridge: Cambridge University Press.

———. 1959a. "Concerning the Philosophy Which Holds That the Conclusions of Science are Never Final." *Scientific American* 200:189–96.

———. 1959b. "Review." *Universities Quarterly*, February, 212–16.

———. 1961. *Foresight and Understanding: An Enquiry into the Aims of Science*. New York: Harper and Row.

———. 1971. "New Directions in Philosophy of Science." *Encounter*, January, 53–64.

———. 1972. *Human Understanding: The Collective Use and Evolution of Concepts*. Princeton, NJ: Princeton University Press.

Treue, Wilhelm, and Gerhard Hildebrandt. 1987. *Berlinische Lebensbilder*. Vol. 1, *Berlinische Lebensbilder: Naturwissenschaftler*, edited by Wolfgang Ribbe. Berlin: Colloquium Verlag.

Turing, A. M. 1950. "Computing Machinery and Intelligence." *Mind* 59:433–60.

Turner, R. Steven. 1982. "Justus Liebig versus Prussian chemistry: Reflections on Early Institute-Building in Germany." *HSPS* 13:129–62.

Turner, Stephen P. 2005. "Polanyi's Theory of Science." In Jacobs and Allen 2005, 83–97.

———. 2007. "Merton's 'Norms' in Political and Intellectual Context." *Journal of Classical Sociology* 7:161–78.

———. 2008. "The Social Study of Science before Kuhn." In *The Handbook of Science and Technology Studies*, edited by Edward J. Hackett et al., 33–62. 3rd ed. Cambridge, MA: MIT Press.

Uebel, Thomas. 2008. "Writing a Revolution: On the Production and Early Reception of the Vienna Circle's Manifesto." *Perspectives on Science* 16:70–102.

Unguru, Sabetai. 1975. "On the Need to Rewrite the History of Greek Mathematics." *Archive for History of Exact Sciences* 15:67–110.

Van der Vet, Paul. 1977. "The Debate between F. A. Paneth, G. von Hevesy, and K Fajans on the Concept of Chemical Identity." *Janus: Revue Internationale de l'Histoire des Sciences de la Médecine et de la Technique* 66:285–303.

Van Vleck, John H., and Albert Sherman. 1935. "The Quantum Theory of Valence." *Reviews of Modern Physics* 7:167–227.

Van't Hoff, Jacobus H. 1884. *Etudes de dynamique chimique.* Amsterdam: Frederik Muller.

Veblen, Thorstein.1994. *The Theory of the Leisure Class.* New York: Dover.

Vezér, Erzséget. 1990. "The Polanyi Family." In Polanyi-Levitt 1990a, 18–25.

Vinti, Carlo. 2005. "Polanyi and the 'Austrian School.'" In Jacobs and Allen 2005, 133–48.

Vogt, Annette. 1999. *Wissenschaftlerinnen in Kaiser-Wilhelm-Instituten A-Z.* Vol. 12, *Veröffentlichungen aus dem Archiv zur Geschichte der Max-Planck-Gesellschaft.* Edited by Eckart Henning. Berlin: Max-Planck-Gesellschaft.

Vom Bruch, Rüdiger. 2001. "Berliner Universitätsgeschichte—Erreichts und Erstrebtes." *Dahlemer Archivsgespräche, Archiv zur Geschichte der Max-Planck-Gesellschaft* 7:31–45.

Von Leitner, Gerit. 1996. *Der Fall Clara Immerwahr: Leben für eine humane Wissenschaft.* Munich: C. H. Beck.

Waddington, C. H. 1941. *The Scientific Attitude.* Harmondsworth, UK: Penguin.

Walker, Mark. 1995. *Nazi Science: Myth, Truth and the German Atomic Bomb.* New York: Basic Books.

Wang, Jessica. 1999. *American Science in an Age of Anxiety: Scientists, Anticommunism, and the Cold War.* Chapel Hill: University of North Carolina Press.

Waters, C. Kenneth, and Albert Van Helden, eds. 1992. *Julian Huxley: Biologist and Statesman of Science.* Houston, TX: Rice University Press.

Watkins, John. 1997. "Karl Raimund Popper. 1902–1994." *Proceedings of the British Academy* 94:645–684. Accessed December 30, 2008. http://www.britac.ac.uk/pubs/src/popper/.

Watson, W. H. 1938. *On Understanding Physics.* Cambridge: Cambridge University Press.

Weaire, D. L., and C. G. Windsor, eds. 1987. *Solid State Science: Past, Present and Predicted.* Bristol, UK: Adam Hilger.

Weart, Spencer. 1992. "The Solid Community." In Hoddeson et al. 1992, 617–69.

Weaver, Warren. 1945. "Free Science." *SFS Occasional Pamphlet* 3 (November).

Webb, Sidney, and Beatrice Webb. 1935. *Soviet Communism: A New Civilisation?* London: Longmans, Greens, and Co.

———. 1941. *Soviet Communism: A New Civilisation.* London: Longmans, Green, and Co.

Weber, Max. 1923. *General Economic History: Sketch of a Universal History of the Economy and Society.* Translated by Frank H. Knight. London: George Allen and Unwin.

———. 1948a. "Science as a Vocation." In *From Max Weber: Essays in Sociology,* translated and edited by H. H. Gerth and C. Wright Mills, 129–56. London: Routledge and Kegan Paul.

———. 1948b. *From Max Weber: Essays in Sociology.* Translated and edited by H. H. Gerth and C. Wright Mills. London: Routledge and Kegan Paul.

Weibel, Peter, and Friedrich Stadler. 1995. *Vertreibung der Vernunft: The Cultural Exodus from Austria.* 2nd rev. enlarged ed. Vienna: Springer-Verlag.

Weinberg, Alvin. 1961. "Impact of Large-Scale Science on the United States." *Science* 134:161–64.

Weindling, Paul. 1996. "The Impact of German Medical Scientists on British Medicine: A Case Study of Oxford, 1933–1945." In Ash and Söllner 1996b, 86–114.

Weiner, Charles. 1969. "A New Site for the Seminar: The Refugees and American Physics in the Thirties." In Fleming and Bailyn 1969, 190–234.

Weissberg, Alex. 1952. *Conspiracy of Science*. Translated by Edward Fitzgerald. London: Hamish Hamilton.

Weissenberg, Karl. 1924. "Ein neues Röntgengoniometer." *ZP* 23:229–38.

Weitz, Eric D. 2007. *Weimar Germany: Promise and Tragedy*. Princeton, NJ: Princeton University Press.

Werskey, Gary. 1978. *The Visible College: A Collective Biography of British Scientists and Socialists of the 1930s*. London: Allen Lane.

———. 2007a. "The Marxist Critique of Capitalist Science: A History in Three Movements?" *Science as Culture* 16 (4): 397–461.

———. 2007b. "The Visible College Revisited: Second Opinions on the Red Scientists of the 1930s." *Minerva* 45:305–19.

Westfall, Catherine. 2008. "Wigner, Eugene Paul (Jenó Pál)." *New Dictionary of Scientific Biography*. Vol. 7, edited by Noretta Koertge, 293–97. New York: Charles Scribner's Sons.

Wheeler, John Archibald. 1998. *Geons, Black Holes, and Quantum Foam*. With Kenneth Ford. New York: W. W. Norton.

Whewell, William. 1847. *The Philosophy of the Inductive Sciences, Founded upon Their History*. 2nd ed. London: J. W. Parker.

White, Hayden. 1973. "The Politics of Contemporary Philosophy of History." *Clio* 3:35–53.

Wigner, E. P., and R. A. Hodgkin. 1977. "Michael Polanyi, 12 March 1891—22 February 1976." *Biographical Memoirs of Fellows of the Royal Society* 23:413–48.

Wigner, E. P., and Michael Polanyi. 1928. "Über die Interferenz von Eigenschwingungen als Ursache von Energieschwankungen und chemischer Umsetzungen." *ZpC* 139:439–52.

Wigner, Eugene. 1932. "Über das Überschreiten von Potentialschwellen bei chemischen Reaktionen." *ZpC* B19:203–16.

———. 1992. *The Recollections of Eugene P. Wigner As Told to Andrew Szanton*. New York: Plenum.

Wilkie, Tom. 1991. *British Science and Politics since 1945*. Oxford: Blackwell.

Willstätter, Richard. 1965. *From My Life: The Memoirs of Richard Willstätter*. Translated by Lilli S. Hornig. New York: W. A. Benjamin.

Wilson, David. 1983. *Rutherford: Simple Genius*. Cambridge, MA: MIT Press.

Wilson, Tuzo. 1968. "A Static or Mobile Earth: The Current Scientific Revolution." *American Philosophical Society Proceedings* 112:309–20.

Wittgenstein, Ludwig. 1953. *Philosophical Investigations*. Translated by G. E. M. Anscombe. New York: Macmillan.

Woodger, J. H. 1937. *The Axiomatic Method in Biology*. Cambridge: Cambridge University Press.

———. 1960. "Science and Persons." *British Journal for the Philosophy of Science* 11:65–71.

Woody, Andrea I. 2000. "Putting Quantum Mechanics to Work in Chemistry: The Power of Diagrammatic Representation." *Philosophy of Science Proceedings* 67:S612–S627.

Wulwick, Nancy J. 1989. "Phillips' Approximate Regression." *Oxford Economic Papers* 41:170–88.

Yearley, Steven. 2005. *Making Sense of Science: Understanding the Social Study of Science.* London: Sage.

Zammito, John H. 2004. *A Nice Derangement of Epistemes: Post-Positivism in the Study of Science from Quine to Latour.* Chicago: University of Chicago Press.

Zeise, H. 1929. "Die Adsorption von Gasen und Dämpfen und die Langmuirsche Theorie," *ZE* 35:426–31.

Zeisel, Hans. 1968. "Polanyi, Karl." *International Encyclopedia of the Social Sciences* 12:172. New York: Macmillan and Free Press.

Zilsel, Edgar. 1926. *Die Entstehung des Geniebegriffs: Ein Beitrag zur Ideengeschichte der Antike und des Frühkapitalismus.* Tübingen: Mohr.

———. 2000. *Edgar Zilsel: The Social Origins of Modern Science.* Edited by Diederick Raven, Wolfgang Krohn, and Robert S. Cohen. *Boston Studies in the Philosophy of Science.* Vol. 200. Dordrecht: Kluwer.

Ziman, John M. 1968. *Public Knowledge: An Essay Concerning the Social Dimension of Science.* Cambridge: Cambridge University Press.

———. 2000. "Commentary I." *Minerva* 38:21–25.

Zirkle, Conrad. 1959. *Evolution, Marxian Biology and the Social Scene.* Philadelphia: University of Pennsylvania Press.

# INDEX

*Page numbers in italics indicate figures.*

Aaserud, Finn, 21
Abelson, Philip. H., 289
Academic Assistance Council (Great Britain), 21–22, 196
Adams, Henry, 47
adsorption, 9–10, 38, 40, 86–99, 109–10; adsorption potential theory, 88, 91–92, 110; BET and other theories after 1940, 97–99; Bunsen Gellschaft papers, 41, 92; chemisorption and physisorption, 94, 95; Eucken-Polanyi theory, 96, 97; Faraday Society meeting, Oxford, 94–95, 109; Polanyi's differences with Langmuir, 92, 94–95, 322n46; Polanyi's doctoral thesis, 39–40, 88–89. *See also* Langmuir; London
AEG (Allgemeine Elektrizitätswerke Gesellschaft), 56, 61, 63, 150
Allison, Samuel, 28, 313n98
Allmand, Arthur John, 69, 72, 73
Anderson, John, 197
Andrássy, Count Gyula, 7–8
anomalous results: Kuhn, 251; Polanyi, 111, 234, 245, 264, 284; Popper, xiii. *See also* Popper, Karl: falsifiability
anti-Communism, 1, 3, 249; McCarthy era in USA, xvi, 26–30, 211
anti-Fascism, 196–97, 232
anti-Semitism, 1–3, 14–17, 45–46, 70–71, 75, 79; German laws and their effects,

20–21, 44–45, 74; in Great Britain, 23–24, 32; and Jewish identity, 33–36; Numerus Clausus laws, 16–17, 20; in USA, 23–25, 312n86
apprenticeship (and skills), xviii, xix, xx, 112, 203, 234, 245, 263, 303; craftsmanship, xviii, 42; Kuhn on, 243; Ziman on, 293
Armstrong, Henry J., 72, 132
Arnold, Matthew, 44
Aron, Raymond, 211, 304
Arrhenius, Svante, 46, 52, 68; Arrhenius activation equation, 114, 119, 120
Ash, Mitchell, 23
Ashby, Eric, 208
Asherson, Neal, 304
Auer Gas Light Company (DGA), 52, 54, 55, 80
Augustine, Saint, 265, 268, 271, 273
authority in science, 111–12, 221, 251. *See also* elites, scientific; Polanyi, Michael (Mihály Polányi): authority structure of science
autonomy of science, 67, 249, 250, 298–99, 347n136; Polanyi, xviii, 164, 179, 180, 189, 204, 207, 209, 261

Baeyer, Adolf von, 56
Baker, John, 184, 204–5, 206, 208, 210
Balázs, Béla, 10, 280